LABORATORY NOTES ON ELECTRICAL AND GALVANOMAGNETIC MEASUREMENTS

MATERIALS SCIENCE MONOGRAPHS

Advisory Editor: C. LAIRD

Vol. 1 Dynamic Mechanical Analysis of Polymeric Material (Murayama)

Vol. 2 Laboratory Notes on Electrical and Galvanomagnetic Measurements (Wieder)

MATERIALS SCIENCE MONOGRAPHS, 2

LABORATORY NOTES ON ELECTRICAL AND GALVANOMAGNETIC MEASUREMENTS

H. H. WIEDER

Department of Physics, Colorado State University, Fort Collins, Colorado 80523

and

Electronic Material Sciences Division, Naval Ocean Systems Center San Diego, Carlifornia 92152, U.S.A.

ELSEVIER SCIENTIFIC PUBLISHING COMPANY

Amsterdam – Oxford – New York 1979

ELSEVIER SCIENTIFIC PUBLISHING COMPANY
335 Jan van Galenstraat
P.O. Box 211, 1000 AE Amsterdam, The Netherlands

Distributors for the United States and Canada:

ELSEVIER/NORTH-HOLLAND INC.
52, Vanderbilt Avenue
New York, N.Y. 10017

Library of Congress Cataloging in Publication Data

Wieder, H H 1919-
Laboratory notes on electrical and galvanomagnetic measurements.

(Materials science monographs ; v. 2)
Includes bibliographical references.
1. Electric measurements--Laboratory manuals.
2. Galvanometer--Laboratory manuals. I. Title.
II. Series.
QC535.W63 530.4'1'028 78-23238
ISBN 0-444-41763-X

ISBN 0-444-41763-X (Vol. 2)
ISBN 0-444-41685-4 (Series)

Printed in The Netherlands

PREFACE

This monograph is concerned primarily with electrical and galvanomagnetic measurements and with their use for exploring the basic electronic properties of solids. It is intended for physicists engaged in applied research in solid-state physics and for electronic engineers concerned with the development of solid-state electronic devices and components. It is also intended to serve as a laboratory reference and guide for graduate students in physics and engineering with an adequate background in physics of the solid state.

Excellent textbooks and monographs which describe the charge carrier transport parameters of crystalline solids starting from first principles are currently available. However, most of these contain only cursory descriptions of experimental measurement techniques and the apparatus used to gather and process the data needed for deriving these parameters. The focus of this monograph is on electrical, galvanomagnetic, thermoelectric and thermomagnetic measurements which have influenced and, in turn, have been affected by the evolution of basic and applied research in solid-state physics. New instruments and techniques have led to changes in conventional experimental procedures, new device applications have imposed more stringent requirements on the precision and accuracy of the measurements. This book is not - and is not meant to be - a comprehensive survey of the field. It deals with measurements which require relatively simple tools and apparatus but also provides the reader with an introduction to more complex computer-assisted laboratory measurements. It grew, almost of its own accord, from notes originally addressed to graduate students at Colorado State University perplexed by artifacts encountered in experimental investigations on semiconductors. In the course of time and with some feedback from faculty and students there came the belated realization that a systematic extension and ordering of these notes might, perhaps, be of benefit to others.

The indispensable person in this undertaking has been my secretary, Mrs. E. B. Davis, who labored mightily to bring order out of chaos taking an essentially unwieldy manuscript through its various stages from a rough draft to its final typewritten form. I am grateful to my colleagues and associates, J. Sites, D. Lile, W. Y. Lum, A. R. Clawson and L. Meiners who read the original manuscript, pointed out various errors of omission and commission and made many useful and constructive comments. Any residual errors are, obviously, my own responsibility and I trust that readers will bring them to my attention. To quote Sir Francis Bacon "Truth comes out of error more easily than out of confusion."

San Diego
30 August 1978

H. H. Wieder

CONTENTS

Page

CHAPTER 1

RESISTIVITY AND CONDUCTIVITY

1.0 Introduction

The electrical resistivity ρ and its reciprocal, the conductivity $\sigma = 1/\rho$ of a material, are characteristic parameters which represent the proportionality between the current density vector $\overline{J}$ and the electric field $\overline{E}$ in terms of Ohm's law,

$$\overline{J} = \hat{\sigma}\overline{E} \qquad \overline{E} = \hat{\rho}\overline{J} \tag{1.1}$$

In general, $\hat{\sigma}$ and $\hat{\rho}$ are second order tensors; they can be reduced to a diagonal form by a suitable rotation of axes, leaving three principal conductivity or resistivity tensor components to be determined parallel to the principal crystallographic axes. The resistivity of a single crystal cut in the form of a parallelipiped whose edges make the angles α_1, α_2, and α_3 with the crystallographic axes is

$$\rho = \rho_1\cos^2\alpha_1 + \rho_2\cos^2\alpha_2 + \rho_3\cos^2\alpha_3 \tag{1.2}$$

Symmetry considerations require that for cubic crystals $\rho_1 = \rho_2 = \rho_3$; for tetragonal, hexagonal, and trigonal crystals which have a single axis of more than two-fold symmetry, $\rho_1 = \rho_2 \neq \rho_3$ and monoclinic, orthorhombic and triclinic crystals have 3 different tensor components of ρ.

Charge carrier transport involved in the measurement of ρ and σ is an irreversible thermodynamic process. Therefore, if no magnetic fields are acting on the material to be measured, then Onsager's principle (ref. 1) applies and the tensors are symmetric, $\rho_{k\ell} = \rho_{\ell k}$ and $\sigma_{\ell k} = \sigma_{k\ell}$.

The current density applied to a specimen cannot be made arbitrarily large without inducing Joule heat in it; if the specific resistivity is temperature-dependent for the heat absorbed by the specimen, then $\overline{E}$ is no longer a linear function of $\overline{J}$. The rate of generation of Joule heat is a scalar product of the current density and the electric field

$$J_kE_k = \rho_{k\ell}J_kJ_\ell = \rho J^2 \tag{1.3}$$

Therefore, the heat produced per unit volume per unit time is the product of the current density squared and of the resistivity in the direction of current flow.

Thermal gradients and fluxes may appear if Joule heat or heat absorbed from an external source is distributed inhomogenously and the interaction between them and electric charge transport may introduce appreciable errors in the measurement of ρ or σ if they cannot be eliminated or explicitly accounted for. Such interactions are represented by the Seebeck and Peltier thermoelectric phenomena. The Seebeck effect is the electric field established by a temperature gradient, $\overline{\nabla T}$, across a junction of dissimilar materials such as that between different metals or between a metal and a semiconductor. Its converse, the Peltier effect, is the heat flux density, $\overline{J_h}$, which accompanies an electric current density flowing across such a junction under isothermal conditions:

$$\begin{aligned} \overline{E} &= S\cdot\overline{\nabla T} \quad \text{for } \overline{J} = 0 \\ \overline{J_h} &= P\cdot\overline{J} \quad \text{for } \overline{\nabla T} = 0 \end{aligned} \tag{1.4}$$

Here, S and P are, respectively, the Seebeck and Peltier coefficients; they are also second rank tensors related to each other at the absolute temperature T by Kelvins's equation $P_{k\ell} = TS_{\ell k}$, a subset of Onsager's reciprocal relations.

Experimentally it is more convenient to measure S than P; $S = -\Delta V/\Delta T$ is much easier to measure in terms of a potential difference ΔV and temperature difference ΔT than heat flux density. Typical values of S in $\mu V/°K$ are in the range between $|1$ and $10|$ for metals, between $|10$ and $100|$ for semi-metals and between $|10^2$ and $\sim 5\times10^2|$ for semiconductors.

The errors introduced by electrothermal and thermoelectric effects in electrical measurements of the resistivity or conductivity of materials can be reduced or eliminated by performing such measurements under isothermal conditions, by commutating the polarity of the current applied to a specimen and averaging algebraically the potential difference measured on it, by making such measurements with alternating currents, or by using small currents in order to minimize Joule heat.

1.1 Two-probe resistivity measurements

The resistivity of a material can be calculated from measurements made of the resistance of a specimen such as the parallelipiped shown in Fig. 1.1. Electrical measurements are needed in order to determine its resistance in terms of a current applied to it from a constant current source and the potential difference generated between its point-contact probes. Mechanical measurements are also required in order to determine its dimensions and the distance between its potential probes.

Consider the uniform, homogenous and isotropic parallelipiped shown in Fig. 1.1. A current, i, is applied to the ohmic contacts which cover the area dw; therefore, the current density is $J = i/wd$. Two spring-mounted, hardened tool steel or tungsten carbide probes are placed near the center of one face, in electrical contact with it, oriented along the direction of current flow with a separation D between them. If the specimen boundaries are at least 10D in any

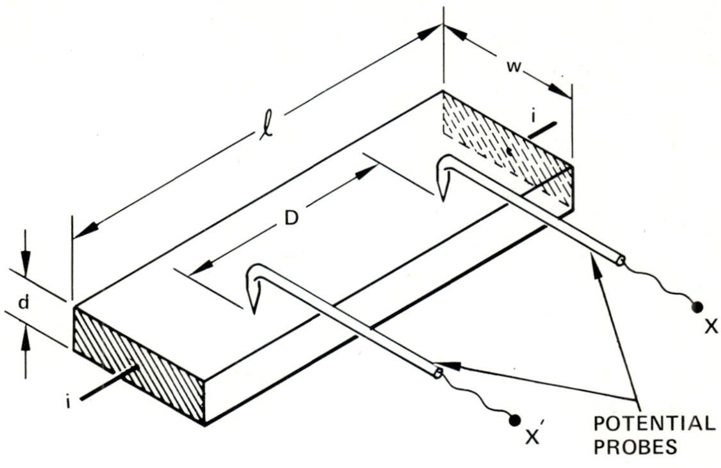

Fig. 1.1. Material specimen prepared for resistivity measurements shows schematically the ohmic contact electrodes covering area dw and approximate location of potential probes.

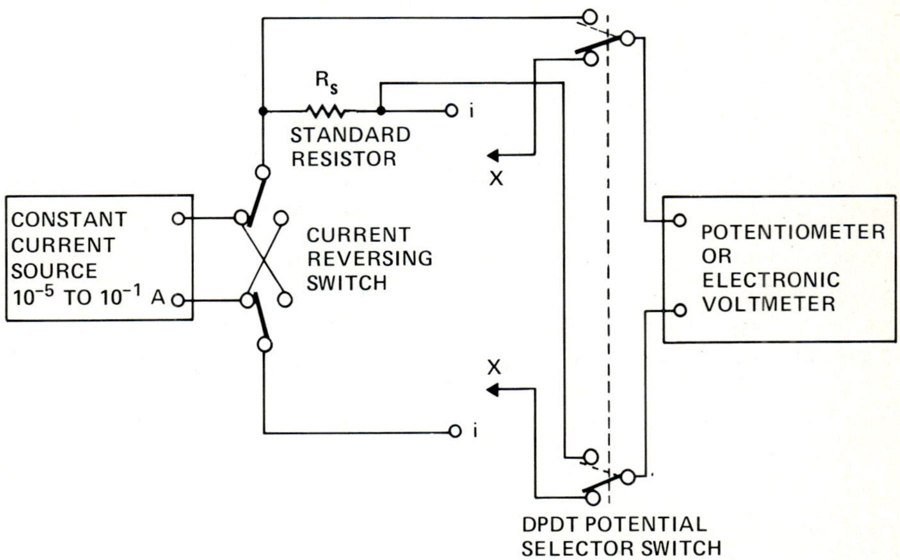

Fig. 1.2. Electrical circuit used for two-point probe resistivity measurements.

direction from the potential probes, then the electric field, in terms of the potential difference, V, between them is $E = V/D$. Therefore the resistivity of the specimen is simply

$$\rho = (V/i)\cdot(wd/D) \tag{1.5}$$

Figure 1.2 shows schematically the apparatus required to perform such measurements. A constant current source is used to supply the current, i, to a specimen such as shown in Fig. 1.1. A constant current source is a current generator which can supply (within preset limits) a constant current regardless of the resistance of the load connected to it. Regulation is a measure of the "constancy" of a current source and is defined as the change in output current which occurs when the load is changed from a short-circuit to the value that requires the current source to supply full output voltage. In general, the regulation should be better by one order of magnitude than the desired accuracy of measurement; a typical value is $\pm$ 0.5%. The current reversing switch in Fig. 1.2 is used to commutate the current polarity; the standard resistor should have a value within a factor of 10^2 of that of the specimen to be measured and the double-throw double-pole potential selector switch is used to select either the potential difference across the standard resistor or that across the potential probes, X-X. The potentiometer or electronic voltmeter should be capable of measuring potential differences between 10^{-4} V and 1 V of either polarity with an accuracy of $\pm$ 0.5% and should have an input impedance at least 10^5 times that of the resistance to be measured. A micrometer or vernier caliper with an accuracy of $\pm$ 0.5% should be used to measure the dimensions of the specimen. A calibrating microscope should be used to determine the probe spacing with a tolerance of $\pm$ 0.5%. The probes should have a nominal radius of the order of 25-50 μm and should be loaded with a force of 1.5 $\pm$ 0.5 N.

To perform the measurements, the current leads of the specimen are connected to terminals i - i in Fig. 1.2 and the current is adjusted until the electric field in the specimen is of the order of 1 V/cm. The potential, V_1, across the standard resistor, R_s, is measured first, then that across the potential probes, V_p, and then the potential across the standard resistor, V_2, is remeasured. The resistivity, ρ_1, is then calculated from

$$\rho_1 = (2R_s wd/D)[V_p/(V_1 + V_2)] \tag{1.6}$$

The polarity of the current is reversed and the resistivity, ρ_2, is then determined by the same procedure described above. The average resistivity is

$$\rho_{av} = (\rho_1 + \rho_2)/2 \tag{1.7}$$

From such measurements made at +23°C on Si wafers and bars at nine different laboratories the typical precision is estimated to be ±6%.

A number of perturbations, in addition to the electrothermal and thermoelectric effects described in the previous section, can introduce errors in the measurement of ρ. These include photoinjection of charge carriers from ambient radiation, induced potentials, particularly in the voltage probe circuit, from adjacent high-frequency power sources, and mechanical vibration of the specimen which may cause noise modulation of the probe contact resistance. Procedures for overcoming these problems will be described in a following chapter.

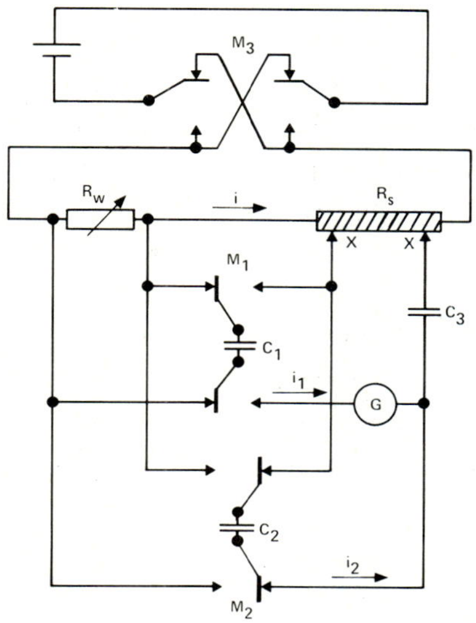

Fig. 1.3. Potentiometric apparatus for two-probe resistivity measurements; M_1, M_2 and M_3 are synchronized choppers, R_W is a variable precision resistor, C_1, C_2 and C_3 are $\sim$ 80 μF polystyrene capacitors, R_S is the specimen to be measured and G is a galvanometer or differential μVoltmeter. [after T. M. Dauphinee and E. Mooser, Rev. Sci. Instr., 26(1955)660]

A potentiometric apparatus for measuring the resistivity of specimens such as shown in Fig. 1.1 has been described by Dauphinee (ref. 3) and by Dauphinee and Mooser (ref. 4). Figure 1.3 shows schematically the main features of the circuit. A double-pole double-throw make-before-break chopper, M_3, converts the current, i, flowing through the reference resistor, R_W, and specimen, R_S, into an $\sim$ 35 Hz square wave alternating current. Two choppers, M_1 and M_2, (break-before-make) driven synchronously with M_3 so that if the current, i_f, is flowing in a forward direction, then capacitor C_1 is connected to R_W and C_2 to R_S, and if it is flowing in a reverse direction, i_r, then the capacitor connections are exchanged. The difference in potentials $V_f(R_W)$ and - $V_r(R_S)$ produces a pulsating direct current, i_1, which flows through the galvanometer G. In a similar manner, a current, i_2, is generated owing to the difference in potential between $V_r(R_w)$ and $-V_f(R_s)$. The chopper timing is adjusted so that the currents i_1 and i_2 are 180° out-of-phase forming an alternating current $(i_1 + i_2)$ which can pass through capacitor C_3.

The potential differences producing i_1 and i_2 and ΔV_1 and ΔV_2, respectively, are

$$\begin{aligned} \Delta V_1 &= V_f(R_w) + V_r(R_s) \\ \Delta V_2 &= V_r(R_w) + V_f(R_s) \end{aligned} \qquad (1.8)$$

However, $V_f(R_w) = -V_r(R_w)$ and $V_f(R_s) = -V_r(R_s)$; therefore R_w can be adjusted so that $\Delta V_1 = \Delta V_2$ and consequently, $i_1 = i_2$; this balance is indicated by a null reading on the galvanometer, G, and the resistance of the specimen, $R_s = R_w$. Its resistivity can now be calculated by means of Eq. (1.5). The measurements are not affected by essentially DC thermoelectric potentials which are blocked by C_3. The condition for balance and hence the value of R_s remains unaffected by unequal reverse and forward currents, $i_f \neq i_r$ and thus only the arithmetic average of the resistance, V/i, can be obtained with this circuit. Furthermore, since zero current flows between the potential probes at equilibrium, the contact resistances do not affect the measurement. Resistance measurements can be made within the range of 10^{-3} ohms and 10^5 ohms and even with simulated thermoelectric potentials of 1.5 V present the accuracy of the measurement was determined to be one part in 10^5 (ref. 4).

Modifications to the circuit shown in Fig. 1.3 which allow specimens resistances up to 10^8 ohms to be measured with an error not exceeding 1% while retaining a low resistance sensitivity of better than 10^{-4} ohms have been described by Edwards (ref. 5). However, large errors arise when the chopping cycle time constant of $\sim 2.8 \times 10^{-2}$ sec. approaches that of the RC time constant due to stray capacitance, the current source and sample resistance. This problem can be overcome by reducing the chopper frequency to $\sim$ 1.9 Hz; the resultant penalty is a reduction in the sensitivity of the potentiometer.

1.2 Collinear four-probe resistivity measurements

The measurement techniques described thus far have some inherent disadvantages. These include the need for specimens with at least two well-defined ohmic contacts. The metallurgical problems encountered in making such contacts, particularly to semiconductors whose bandgaps exceed 1 eV, can be severe. The restriction on specimen shape chosen to provide a simple distribution of equipotential planes and current streamlines is a further disadvantage of two-probe resistivity measurements.

The collinear four-probe array, shown schematically in Fig. 1.4, overcomes these disadvantages. It provides for the measurement of the resistivity of specimens having a wide variety of shapes, including those having irregular boundaries and also of the resistivity of smaller regions included in a matrix with different electrical properties. Figure 1.4 shows four pointed equispaced

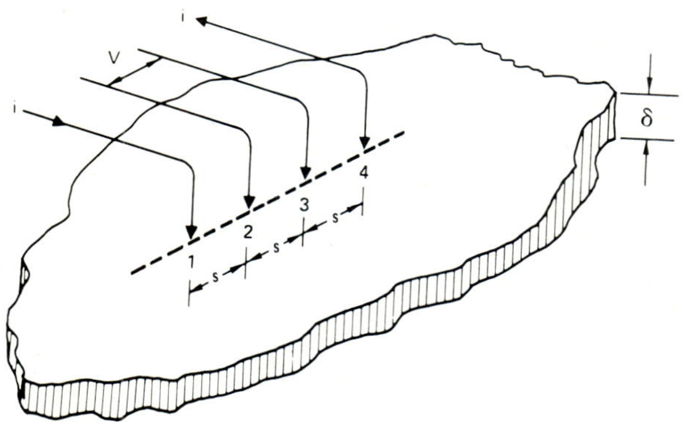

Fig. 1.4. Collinear four-probe array on a semi-infinite sheet of thickness, δ; outer probes 1 and 4 are the current input and output probes, inner probes 2 and 3 measure the potential difference, V, between them.

probes in contact with a plane surface of a uniform isotropic material to be measured. The probes are considered to be far from any of the other boundaries of the specimen so that it may be considered to be an infinite area of uniform resistivity. Further constraints are:

(a) the diameter of the contact between each probe and the specimen surface should be small with respect to the distance, s, between the probes,

(b) the surfaces should have a high recombination rate so that any electrons or holes injected by the probes recombine in their immediate vicinity and their effect on the conductivity is negligible, and

(c) the boundary between the current-carrying electrodes and the bulk material is small in diameter and hemispherical in shape.

Let a current, i, be impressed on the two outer probes (1 and 4) while the potential difference, V, is measured between the inner probes (2 and 3). The current streamlines are considered to have radial symmetry. The potential, V_f, at a distance, r, from a current-carrying electrode in a specimen whose thickness $\delta >> s$ is

$$V_f = \rho i/(2\pi r) \tag{1.9}$$

Since there are two current-carrying electrodes, $V_f = (\rho i/2\pi)(1/r_1 - 1/r_4)$ where r_1 is radial distance from probe 1 and r_4 is distance from probe 4. The floating potentials at probes 2 and 3 are

$$V_{2f} = (\rho i/2\pi)(1/s - 1/2s) \qquad V_{3f} = (\rho i/2\pi)(1/2s - 1/s) \tag{1.10}$$

The potential difference between the probes, $(V_{2f} - V_{3f}) = V$ and the current i are used to calculate the resistivity,

$$\rho_o = 2\pi s \cdot (V/i) \tag{1.11}$$

This relation and one developed for a collinear array having unequal spacing between its probes were derived by Valdes (ref. 6), who also calculated by the method of images the correction factors $F(\ell/s)$ applicable to the resistivity, $\rho = \rho_o \cdot F(\ell/s)$ under the additional constraint that one of the exterior probes of the array is at a distance, ℓ, from either a parallel or a perpendicular non-conducting or a conducting specimen boundary, and Uhlir (ref. 7) extended these calculations to include correction factors for finite specimen thickness. If Eq. (1.11) is used to calculate the resistivity of a specimen whose thickness, $\delta \geq 10s$ with the four-probe array oriented perpendicular to and at a distance, ℓ, from a nonconducting boundary, then the error in ρ is equal to or less than 2.5% provided that $\ell/s \geq 2.5$. If the array is parallel to a nonconducting boundary, then this error $\leq$ 2.5% if $\ell/s \geq 3$. For a conducting boundary perpendicular to the array the error in ρ is $\leq$ 2.5% if $\ell/s \geq 2.5$; it is the same for a parallel conducting boundary if $\ell/s \geq 1.5$.

Collinear four-probe measurements are used extensively in the silicon device and integrated circuit technology for the evaluation of Si wafers, epitaxial layers, ion implanted regions and impurity diffusion processes. The specimen or layer thickness is a significant parameter in the evaluation of the actual resistivity if the interprobe spacing, s, is comparable to the thickness, δ, shown in Fig. 1.4. In the limit, if the thickness $\delta \ll s$, the specimen may be considered as essentially two-dimensional, an infinite sheet having an infinitesimal thickness, δ. Since the current streamlines have radial symmetry in the homogenous isotropic sheet, Ohm's law is $(\partial V/\partial r) = -\rho \bar{J}$. The current density $\bar{J}$ and potential difference, $V = V_2 - V_3$, between probes 2 and 3 are, respectively,

$$\bar{J} = \frac{i\hat{r}}{2\pi\delta r} - \frac{i\hat{r}'}{2\pi\delta r'}, \qquad V = -\rho\int_{2s}^{s} J dr \tag{1.12}$$

where the unit vectors $\hat{r} = -\hat{r}'$ and $r' = 3s - r$; therefore,

$$V = \frac{i\rho}{2\pi\delta}\int_{s}^{2s}\left(\frac{1}{r} + \frac{1}{3s - r}\right)dr \tag{1.13}$$

The solution of this integral then leads to the resistivity,

$$\rho = \frac{V}{i} \cdot \frac{\pi\delta}{\ln 2} = 4.5324\delta R \qquad (1.14)$$

Equation (1.14) is adequate for determining the sheet resistivity, $\rho_s = \rho/\delta$, of an essentially infinite conducting sheet of infinitesimal thickness in comparison with the interprobe spacing, s. Smits (ref. 8) has employed the method of images to calculate a correction factor, C, for such sheet resistivities, measured by means of a centered four-probe array placed on a line of symmetry such as the diameter, d, on a circular specimen or on the bisectrix, d/2, of a rectangular ℓ x d specimen. The correction factor was found to be dependent on both the ratio d/s and ℓ/d; thus ρ_s = RC(d/s, ℓ/d). For small d/s the sheet resistivity $\rho_s \simeq$ Rd/s. Table 1 shows the calculated dependence of the correction factor C on d/s and ℓ/d.

TABLE 1.1

Correction factor C for the calculation of the sheet resistivity measured with collinear four-point probes placed on a symmetry axis. [after F. M. Smits, Bell Syst. Tech. J., 37(1958)711]

d/s	Circle C for dia. d	Square C for ℓ/d = 1	Rectangle C for ℓ/d = 2	Rectangle C for ℓ/d = 3	Rectangle C for ℓ/d $\geq$ 4
1.0				0.9988	0.9994
1.5			1.4788	1.4893	1.4893
2.0			1.9454	1.9475	1.9475
3.0	2.2662	2.4575	2.7000	2.7005	2.7005
4.0	2.9289	3.1137	3.2246	3.2248	3.2248
5.0	3.3625	3.5098	3.5749	3.5750	3.5750
10.0	4.1716	4.2209	4.2357	4.2357	4.2357
20.0	4.4364	4.4516	4.4553	4.4553	4.4553
∞	4.5324	4.5324	4.5324	4.5324	4.5324

For an essentially infinite conducting sheet of thickness, δ, Eq. (1.14) is adequate provided that $\delta/s \leq 0.4$. A thickness correction factor, $F(\delta/s)$, has been tabulated by Smits (ref. 8) and should be used to multiply Eq. (1.14) for $0.4 < (\delta/s) < 2.0$. It is a relatively weak function of the ratio of thickness to interprobe spacing; for $(\delta/s) = 1$, $F(\delta/s) = 0.9214$.

For finite size specimens of finite thickness a further correction is required; for circular or rectangular specimens it may be expressed, to first order, in terms of Table 1.1 and Eq. (1.14) as

$$\rho = R\delta \cdot C(d/s, \ell/d) \cdot F(\delta/s) \tag{1.15}$$

The resistivity of a disc of radius r measured by means of a collinear four-probe array, with $(\delta/s) \leq 0.4$, placed at the center of the disc, can be expressed (ref. 9) as $\rho = R\delta C_o$, with the geometrical correction factor $C_o = (\pi/\ell n2)[1 + \eta]^{-1}$, where

$$\eta = (1\ell n2) \cdot \ell n \left[\frac{1 + 3(s/2r)^2}{1 - 3(s/2r)^2} \right] \tag{1.16}$$

If the array is displaced a distance x from the center of the disc or, its equivalent, a distance k = r - x from the periphery of the disc, then the value of C_o decreases by the fractional amount

$$\Delta C = (1/C_o)(C_1 - C_o) \tag{1.17}$$

Using conformal mapping techniques Perloff (ref. 9) has calculated the dependence of C_1 on s, k, and r for an arbitrary orientation of the array at an angle θ with respect to the radius of the disc and displaced from its center. Subject to the requirement that

$$\left(1 - \frac{k}{r} + \frac{3s}{2r}\cos\theta\right)^2 + \left(\frac{3s}{2r}\sin\theta\right)^2 \leq 1 \tag{1.18}$$

the angular and displacement dependence of the geometrical correction factor, $C_1(\theta)$ is

$$C_1(\theta) \simeq \pi\left[\ell n2 + (1/4)\ell n\alpha\right]^{-1} \tag{1.19}$$

where the parameter α is a function of the effective probe separation $s^1 = s[1 - (k/2r)]^{-1}$ and

$$\alpha = \frac{[(k/s^1)^2 + \sin^2\theta - (1/4)\cos^2\theta]^2 + \cos^2\theta\sin^2\theta}{[(k/s^1)^2 + (1/4)\sin^2\theta - \cos^2\theta]^2 + \cos^2\theta\sin^2\theta} \tag{1.20}$$

The calculated $C_1(\theta)$ was found to be in good agreement with experimental measurements made on a 7.62 cm dia. boron implanted ($5 \times 10^{14}/cm^2$, 70 keV) 10 ohm-cm, annealed Si disc using s/r = 0.0417 and k/r = 0.0667.

If $(k/s^1)^2 \gg 1$, then C_1 is independent of θ and the position-dependence of C_1 is then expressed by

$$C_1 \simeq \pi\left\{\ell n2 + 1/2\ \ell n\left[1 + \frac{3}{4}\cdot\left(\frac{s^1}{k}\right)^2\right]\right\}^{-1} \tag{1.21}$$

The angular independence of C_1 was confirmed experimentally (ref. 9) by using a smaller probe spacing on the same wafer, s/r = 0.0167 and k/r = 0.0667. The geometric correction factor C_1 must be taken into account in the evaluation of the topographic distribution of the resistivity of circular wafers.

1.3 Square four-probe resistivity measurements

A square four-probe array such as shown in Fig. 1.5 provides some advantages over a colinear array for measuring the resistivity of small area specimens. For a thin semi-infinite conducting sheet, $\delta << 0.5s$, measured by means of a square four-probe array the resistivity (ref. 7) is

$$\rho \simeq 2\pi\delta R/\ln 2 \qquad (1.22)$$

Uhlir (ref. 7) has tabulated the thickness correction factors for $0.1 < (s/\delta) < 10$. Further corrections are required if the measurements are made

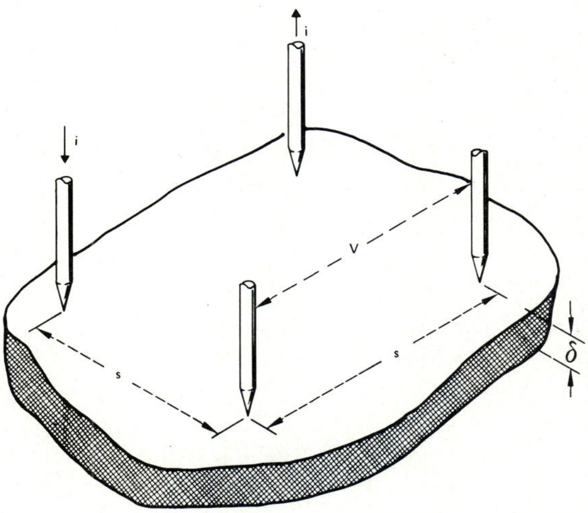

Fig. 1.5. Square four-probe array on a semi-infinite surface of thickness, δ; current input and output points constitute current dipole; interprobe spacing is s.

on specimens of finite size and specific shape. Consider a square four-probe array with interprobe spacing, s, located at the center of a circle of diameter, d, such as shown in Fig. 1.6. A current, I, is introduced at point A and removed at point D and the potential difference is measured between points B and C. Conformal mapping is used with the bilinear transformation

$$W = \frac{Z - i}{Z + i}, \qquad Z = \frac{i(1 + W)}{1 - W} \qquad (1.23)$$

to map a unit circle on the W-plane to the upper half of the Z-plane. Images on the Z-plane are located by inspection and transformed back to the W-plane. Thus, image currents A_i and D_i appear on the W-plane, shown in Fig. 1.6. In addition, if the circular specimen has a thickness δ then an infinite array of images appear perpendicular to the W-plane. The potential difference between points A and B measured for a current density, $J = I/2\pi r^2$ is

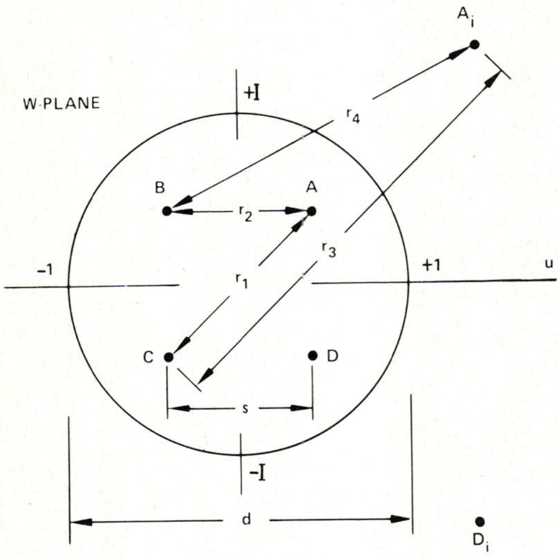

Fig. 1.6. Measurement of the resistivity of a disc of diameter d by means of a centered square four-probe array; A_i and D_i are images of the current introduced at points A and D in plane W of the model.

$$V = \frac{\rho I}{2\pi}\cdot\left(\frac{2}{Q} + \frac{M}{\delta}\right) \tag{1.24}$$

where the parameters Q and M are, respectively,

$$\begin{aligned} 1/Q &= -1/r_1 + 1/r_2 - 1/r_3 + 1/r_4 \\ M &= M(r_1/2\delta) - M(r_2/2\delta) + M(r_3/2\delta) - M(r_4/2\delta) \end{aligned} \tag{1.25}$$

The function M has been tabulated by Uhlir (ref. 7); for $\delta/s \ll 0.5$ it leads to

$$\rho = \pi\delta\cdot\left(V_{BC}/I_{AD}\right)\cdot\left[\ell n\left(\frac{r_1 r_3}{r_2 r_4}\right)\right]^{-1} \tag{1.26}$$

The resistivity can be expressed, therefore, as

$$\rho = \frac{\pi\delta R}{\ell n 2}\cdot F(d, s) \tag{1.27}$$

where $F(d, s) = \ell n2 \cdot [\ell n(r_1 r_3/r_2 r_4)]^{-1}$. This geometrical correction factor is shown as a function of d/s in Table 1.2.

TABLE 1.2

Geometrical correction factor for resistivity measurements made with a square four-probe centered array on a circular specimen.

d/s	$\sqrt{2}$	2.0	3.0	4.0	5.0	10.0	20.0	∞
F(d, s)	1.000	1.082	1.325	1.518	1.645	1.892	1.971	2.000

The geometrical correction factors applicable for the measurement of the resistivity of a circular specimen by means of a square four-probe array displaced from its center have been calculated by the method of images (ref. 10) and by conformal mapping techniques (ref. 11). For a disc of radius r and thickness $\delta < 0.5s$ Vaughan (ref. 10) has shown that in terms of the displacement of the center of the array from the center of the disc by a distance ($\alpha s/2$) in a direction parallel to the current dipole and its displacement by a distance ($\beta s/2$) in a direction perpendicular to the current dipole the resistivity is

$$\rho = \frac{2\pi R\delta}{\ell n 2} F(\alpha, \beta) \tag{1.28}$$

where the geometrical correction factor $F(\alpha, \beta)$ is

$$F(\alpha, \beta) = \frac{[ACP^4 - (A + C - 4)P^2 + 1]\cdot[BDP^4 - (B + D - 4)P^2 + 1]}{[ADP^4 - (A + D - 8)P^2 + 1]\cdot[BCP^4 - (B + C - 8)P^2 + 1]} \tag{1.29}$$

where $P = s/2r$

$$A = (\alpha - 1)^2 + (\beta + 1)^2, \qquad B = (\alpha + 1)^2 + (\beta + 1)^2$$
$$C = (\alpha - 1)^2 + (\beta - 1)^2, \qquad D = (\alpha + 1)^2 + (\beta - 1)^2 \tag{1.30}$$

Geometrical correction factors for the resistivity of a square specimen measured by means of a square array were calculated by Mircea (ref. 12) by a conformal transformation of the square into a circle. He found analytically and experimentally that the angular orientation of the array is relatively insensitive (to within 5%) to the coordinates of the specimen provided that the center of the array is located at its center.

A conformal transformation of a rectangle into a semi-infinite plane was used by Perloff (ref. 13) to calculate the geometrical correction factors used to determine the sheet resistivity, ρ_s = Rk of a square specimen of side a measured by means of a square array of side s centered or displaced by Δ_x, Δ_y from the center of the specimen. The correction factor was calculated in terms of the maximum displacement $\Delta^* = \Delta^*_x = \Delta^*_y$ with the current dipole oriented either parallel to or at 45° with respect to the side of the specimen. The latter is shown in Fig. 1.7; it represents the calculated dependence of $k(\Delta_x, \Delta_y)$ on $a\cdot(2s)^{-1/2}$ with the nature of the displacement as a running parameter. The k(0, 0) curve corresponds to that of the centered array, the $k(\pm\Delta^*, 0)$ curve corresponds to the displacement of the array to either of the extrema along the x-axis only, etc. If the electrodes are at the midpoints of the sides, $\alpha \cdot (2s)^{-1} = 1$ of the square sample, then $k(0, 0) = \pi/\ell n2$ and if $a >> s$ then the correction factor in Fig. 1.7, $k(0, 0) \to 2\pi/\ell n2$, the limiting case of a square array on an infinite conducting sheet. It is evident from Fig. 1.7 that the correction factor $k(\Delta_x, \Delta_y)$ can differ substantially from the value $k(0, 0)$ for a given probe spacing and sample size due to the influence of the nonconducting

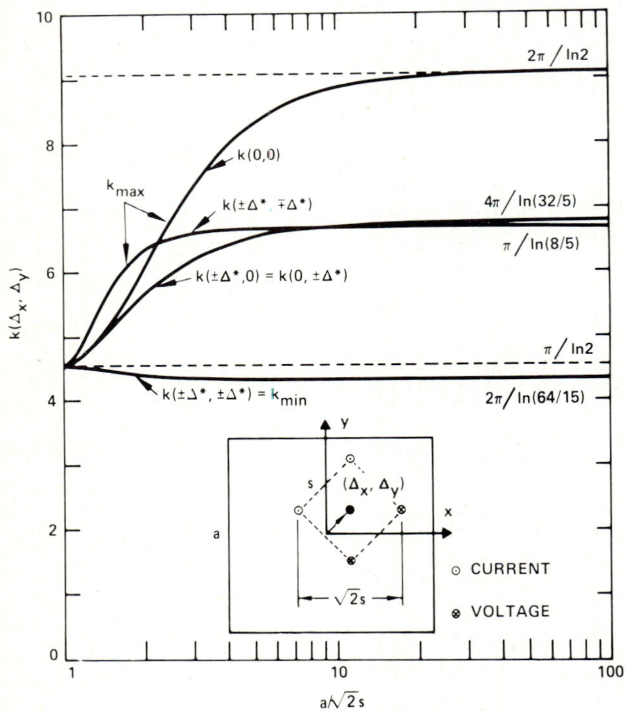

Fig. 1.7. Correction factors $k(\Delta_x, \Delta_y)$ for various locations of a square four-probe array oriented at $\theta = 45°$ with respect to the side of a square specimen of side a. The quantity $\Delta^* = (a-\sqrt{2}s)/2$ is the maximum displacement of the center of the probe. [after D. S. Perloff, Solid-State Electron., 20(1977)681]

boundaries on the potential distribution within the square sample. Similar calculations were made by Perloff for a parallel probe and specimen orientation and for a collinear four-probe array displaced from the center of a square specimen. Figure 1.8 shows that the ratio (k_{max}/k_{min}), considered to represent the positioning error of the square array is a function of the ratio $a \cdot [s(\cos\theta + \sin\theta)]^{-1}$, where θ is the angle between the current dipole and the side of the specimen;

it shows that the sensitivity of the geometrical correction factor to positional errors may be minimized by orienting the array at $\theta = \pi/4$, as shown in Fig. 1.7, rather than in the parallel configuration, $\theta = 0$, particularly for $a[s(\cos\theta + \sin\theta)]^{-1} \geq 1.2$. Furthermore, if the displacement error of a collinear and square four-probe array are compared by differentiating the relevant expressions for their geometrical correction factors, then it appears (ref. 10) that the probable displacement error of a collinear array is approximately 4 times larger than that of the square array.

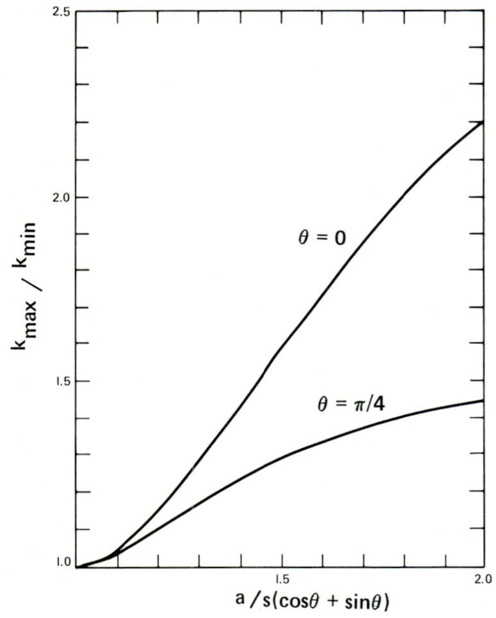

Fig. 1.8. Positioning error of geometrical correction factor, k_{max}/k_{min}, as a function of the angular orientation, θ, of the square array on a specimen of side a. [after D. S. Perloff, Solid-State Electron., 20(1977)681]

Geometrical correction factors for an arbitrary placement of current probes on one side of a rectangular bar have been calculated by Reber (ref. 14) by solving Laplace's equation for the potential distribution in the bar, including dimensional factors as well as the presence of two metallic equipotential end-surfaces, of utility in the evaluation of two-probe resistivity measurements. Mircea (ref. 15) has calculated the geometric correction factors by the application of Green's function to a parallelipiped and Stephens *et al.* (ref. 16) have extended these calculations to include: (a) non-equidistant point probes located symmetrically on the top face, and (b) point-probe potential contacts located symmetrically on the top face and point-probe current contacts centered on the end faces of the parallelipiped. Their results lead to the following generalizations for case a:

(1) The apparent, measured resistivity is always greater than or equal to the correct value.

(2) Longer samples and thinner samples give apparent resistivities that are more nearly correct.

(3) Smaller potential contact separations give apparent resistivities that are more nearly correct.

(4) The longer and thinner the sample, the closer the potential contacts can be placed to the current contacts while still measuring a resistivity equal to the correct resistivity.

Specifically, if the current contacts are located one-tenth the length of the sample from the ends, then the actual resistivity can be calculated from the relation, $\rho = 2\pi Rs$, provided the sample is at least 4 times as long as it is wide, the sample is at least as thin as half its width, and the voltage-contact separation is not greater than two-thirds the current-contact separation. For case (b) the applicable generalizations are:

(1) Longer samples give apparent resistivities that are more nearly correct.

(2) Except for very short samples, smaller potential contact separations give apparent resistivities that are more nearly correct.

(3) The longer the sample, the closer the voltage contacts can be placed to the ends of the sample while still obtaining apparent resistivity equal to correct resistivity.

(4) For large potential contact separation, samples thinner than half their width give apparent resistivities greater than the correct resistivity, while samples thicker than their width give apparent resistivities less than the correct value.

The actual resistivity of a specimen using this probe configuration can be calculated by means of $\rho = 2\pi Rs$ provided that the sample is at least 3 times as long as it is wide, is not thicker than it is wide, is not thinner than half its width, and provided the potential contact separation is not greater than half the length of the sample.

1.4 Probes and instruments

Four-point probe instruments are available commercially in the price range of several hundred to several thousand dollars. Such instruments are made to close mechanical tolerances in order to minimize errors caused by probe displacement and contact pressure variations. A simple collinear four-point probe design has been described by Kennedy (ref. 17). Figure 1.9 shows a sectional and an assembled view of its probe head. Tungsten wire $\sim$ 127 μm in diameter is

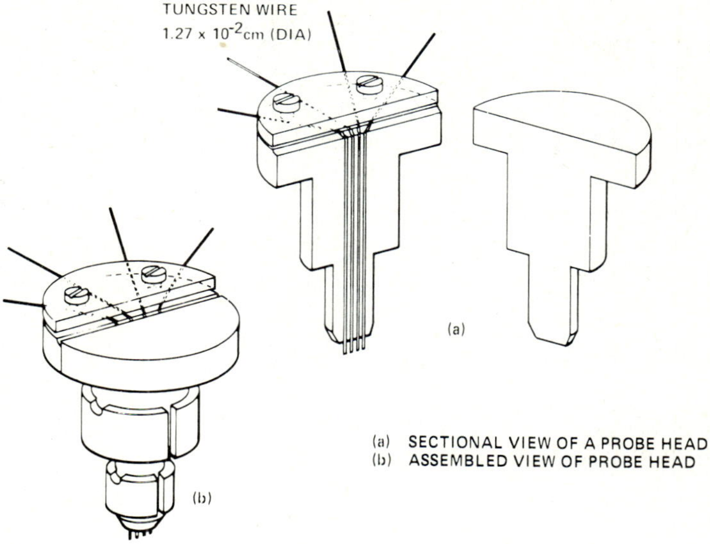

Fig. 1.9. Sectional and assembled collinear four-point probe head.
[after J. K. Kennedy, Rev. Sci. Instr., 33(1962)773]

used for the probes; their terminals are sharpened electrolytically. The body of the probe head is made of teflon because this material provides only slight friction for the longitudinal motion of each probe in its slot. The slots were milled to a depth and width of $\sim$ 152 μm, equispaced by $\sim$ 1.59 mm from each other. The pointed terminals of each probe extend $\sim$ 0.8 mm out of the probe head. The other end emerging from its slot is bent at 90° to the probe body and is secured by a plexiglass cap which prevents its lateral motion. Lowering the probe head onto a specimen provides the spring tension required for positive uniform contact with a specimen surface. A more complex and also a more versatile collinear four-point probe has been described by Paulnack and Chaplin (ref. 18). It is made of two independently suspended plastic moldings each with a pair of probes and an axle which is mutually perpendicular to a line joining the probes and the direction of the force exerted by the probes on the specimen. The independent suspension together with the rotation of each point-pair about its axis equalizes the force between each probe point and the specimen even when the specimen is not planar. Pressure is applied to the probe points by loading the axle through its supporting mechanism with appropriate weights. Removal of the weights from the two current probes enables the other probe pair to be used as a two-point fixed-spacing voltage probe. The construction of a simple collinear four-point probe, suitable for measuring resistivities in the range between 10^{-3} and 3×10^{5} ohm-cm, was described by Klimenko (ref. 19) and Zrudsky *et al.* (ref. 20) have described the design construction and performance of a square four-point probe array system.

A direct reading four-probe instrument used for measuring the resistivity of silicon and germanium wafers with a moderate precision of ∿ 5% was developed by Swartzendruber *et al.* (ref. 21). The unique feature of this analog instrument is the inclusion of a variable resistance network which, when adjusted for the specimen thickness and mean diameter, provides, for a given interprobe spacing, a direct readout of the resistivity in ohm-cm.

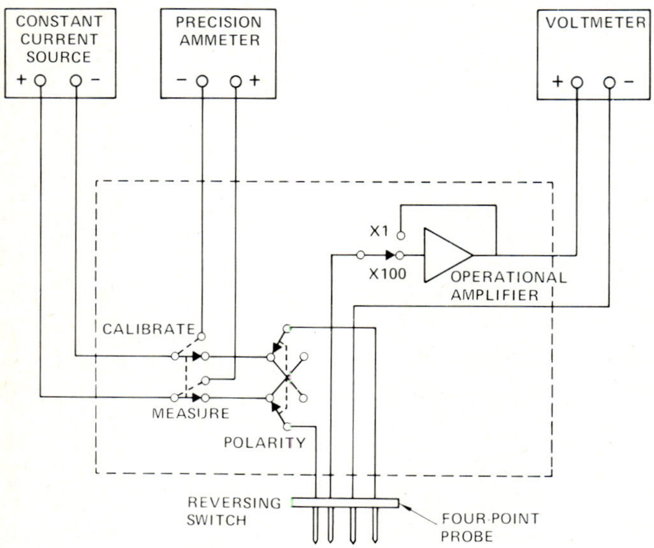

Fig. 1.10. Apparatus used for precise four-point probe measurements includes a precision current monitor used for calibration, current polarity reversing switch and operational amplifier for measuring low resistivities.

Figure 1.10 is a block diagram of auxiliary apparatus used for four-point probe measurements. It includes a constant current source equipped with a voltage limit control rheostat. The latter is used to protect thin film or epitaxial layer specimens from voltage breakdown which might otherwise occur following initial contact between probe and specimen. A "calibrate-measure" switch is used to connect an external precision ammeter for precise setting of the input current. The current polarity switch is used for identifying unipolar conduction and the operational amplifier provides for the amplification of the output of the voltage probes when measuring low resistivity specimens with small input currents.

An automatic instrument used for measuring the resistivity of silicon boules by the two-probe method (ref. 22) is shown in Fig. 1.11. A precise measurement is made of the boule diameter at the location where the resistivity is to be measured; the instrument also measures the potential difference between the probes and divides this by the current density. Translation of the probes along

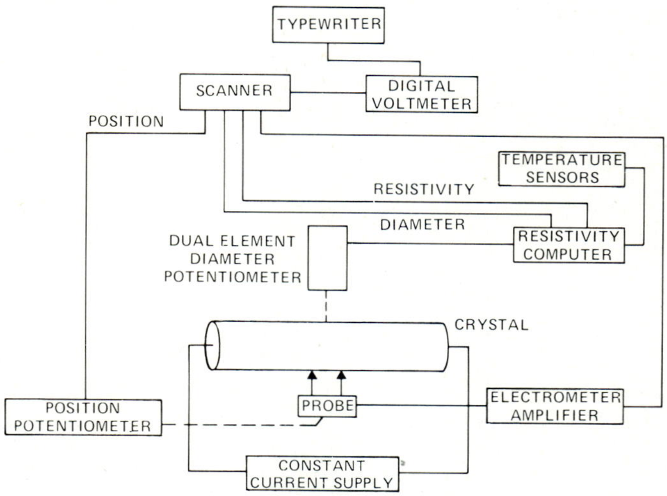

Fig. 1.11. Automatic two-probe instrument for measuring the axial variation of the resistivity along single crystal boules. [after Earleywine *et al.*, Semicond. Prod. Solid-State Techn., 8(1965)17]

the boule provides for the characterization of the axial resistivity of the boules, their radial resistivity assumed, for these purposes, to be homogenous. Such an instrument has been used to determine with a precision of $\sim 1.5\%$ the resistivities of 1 to 4 cm diameter silicon boules in the range between 0.1 and 10^3 ohm-cm.

The accuracy of two-probe and four-probe resistivity measurements is limited by systematic and random errors. The variation in interprobe spacing is a random error; when a probe makes contact with a wafer then any displacement directed along the line of the probe has a pronounced effect on the error of measurement. Let the sheet resistivity of a wafer considered to be semi-infinite, $\rho_s = 2\pi sR$, be measured by means of a collinear four-point probe array with interprobe spacing s. If the displacements of the four probes on contacting the surface of the wafer are, respectively, Δx_1, Δx_2, Δx_3, and Δx_4 then (ref. 23) the error in the measured resistivity is

$$\frac{d\rho}{\rho} = \frac{1}{4s}\left(3\Delta x_1 - 5\Delta x_2 + 5\Delta x_3 - 3\Delta x_4\right) \tag{1.31}$$

If the probe displacements are random with a standard deviation α then the error in the measurement of the resistivity,

$$\frac{d\rho}{\rho} = \frac{\sqrt{17}\alpha}{2s} = 2.06\alpha/s \qquad (1.32)$$

For random probe displacement of a two-point probe the error of measurement is

$$\frac{d\rho}{\rho} = \sqrt{2}\alpha/s = 1.41\alpha/s \qquad (1.33)$$

Consequently, the random displacement errors of four-point probes is $\sim$ 1.46 times larger than that of two-point probes.

For essentially two-dimensional thin layer specimens Eq. (1.14) applies for measurements made by four-point probes, $\rho_s = \pi R/\ell n2$; the error in the resistivity due to a random displacement α of the equispaced probes (ref. 24) is

$$\frac{d\rho}{\rho} = \frac{\sqrt{5}\alpha}{2s\ell n2} = 1.613\alpha/s \qquad (1.34)$$

The precision can be improved by averaging the result of several independent measurements with the probe head lifted and replaced on the specimen between successive measurements. A detailed procedure for measuring the resistivity of silicon wafers by means of a collinear four-point probe is given by the American Society for Testing Materials (ref. 25). It is intended for use on single crystals of silicon in the form of circular wafers with a diameter greater than 16 mm and thickness less than 1.6 mm. The spacing between the four probes is determined from measurements of indentations made by the probe tips on a polished Si slice. In order to minimize positioning errors the probe assembly is centered to within 0.25 mm of the center of the wafer. A calibrated mechanical or electronic thickness gauge is used for measuring the wafer thickness to ± 1% at nine different positions on the wafer; a micrometer or vernier caliper is used for measuring the wafer diameter. Probes made of tungsten carbide are recommended with a nominal tip radius of 25 to 50 μm and the force applied to each probe is to be 1.75 ± 0.25 N. The electrical isolation of each probe with respect to the probe assembly should be at least 10^8 ohms. The interprobe spacing should have a nominal value of 1.59 mm. The specimen should be supported by a copper block $\sim$ 10 cm in diameter and insulated from it by a 12 to 25 μm thick layer of mica. Mineral oil or silicon heat sink compound is used to provide good thermal conduction between the specimen and the copper block used as heat sink. The auxiliary apparatus for supplying the current to the probe and for monitoring the probe output voltage is presumed to be similar to that shown in Fig. 1.2. On homogenous silicon wafers with room temperature resistivities in the range between 8×10^{-4} and 120 ohm-cm a precision of ± 2% is quoted for measurements made at 6 different laboratories on 13 different wavers. On wafers with resistivities in the range between 120 and 500 ohm-cm the interlaboratory precision is quoted as ± 5% and for wafers whose resistivities are in the range between 5×10^2 and 2×10^3 ohm-cm, the quoted precision is equal to or better than ± 15%.

1.5 Resistivity of lamellae of arbitrary shape

For measurements made on large bandgap semiconductors and for low temperature measurements, point contacts are less than adequate because the contact resistances may be of the order of megohms and because random fluctuations in contact resistance may reduce the accuracy of measurement. For such applications bridge-shaped specimens such as shown in Fig. 1.12 are preferable. They

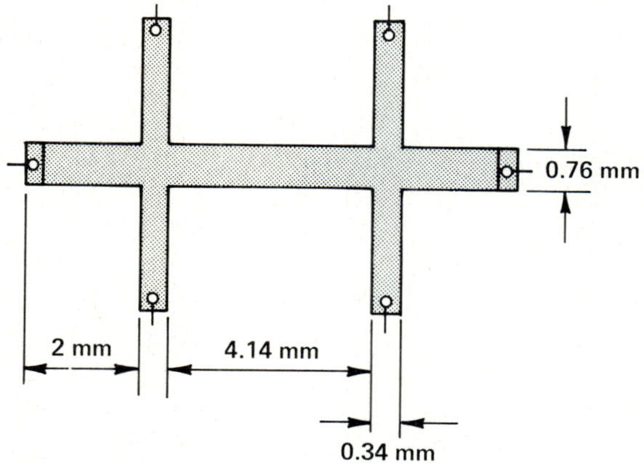

Fig. 1.12. Bridge-shaped specimen used for the measurement of the resistivity and Hall coefficient. Representative dimensions shown are those of an epitaxial layer on a semi-insulating substrate processed by photolithography; 0.76 mm-wide current contacts and two sets of 0.34 mm-wide voltage probes are used. Bulk specimens are cut out ultrasonically in a similar configuration.

provide voltage and current contacts covering relatively large areas. These can be made to yield stable low resistance contacts; the specific shape of the main body provides a simple pattern of equipotentials and current streamlines. It is not always convenient to process a material in the configuration shown in Fig. 1.12. Often, measurements need to be made on specimens having an irregular contour or on lamellae of arbitrary shape whose contacts are on their periphery. The theoretical foundation of such measurements is based on the conformal representation developed by van der Pauw (refs. 26, 27) for mapping a specimen of arbitrary shape on a semi-infinite half-plane.

Consider a continuous single-valued function of the form $z = f(t) = x(u, v) + jy(u, v)$ which defines a complex variable, $z = x + jy$ as a function of another variable $t = u + jv$. Let a particular value, t', represent a point in the complex plane of t. Then in the complex plane of z some particular value (or values), z', corresponds to t'. There is a similar correspondence for other points on the respective complex planes of t and z so that a curve on the t-plane is transformed by "conformal mapping" from the t-plane to the z-plane. Conformal mapping is an analytical technique used by van der Pauw (ref. 26) to demonstrate a procedure for measuring the resistivity of a specimen having an arbitrary shape and thickness with four, not necessarily equispaced, line-electrodes on its periphery. Consider the semi-infinite plane of thickness, δ, shown in Fig. 1.13. Line-electrodes 1, 2, 3, and 4 on the plane boundary are separated from each other by the dimensions a, b, and c. Let a current, i, be

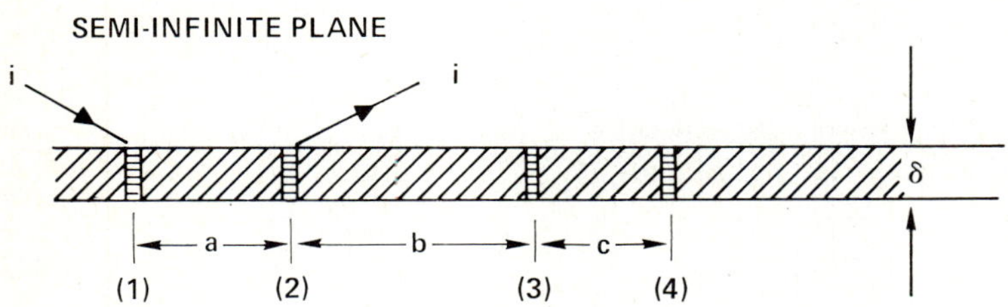

Fig. 1.13. Representation of a semi-infinite half-plane of thickness, δ, with line-electrodes 1 and 2 for current terminals and 3 and 4 used for voltage probes.

introduced at electrode 1 and removed at electrode 2. The appropriate current density, J, electric field, E, and the potential difference developed between electrodes 3 and 4, $V_{3,4}$, for the current, i, flowing <u>into</u> the half-plane are

$$V_4 - V_3 = \left(-\frac{\rho i}{\pi\delta}\right)\int_{a+b}^{a+b+c}\frac{dr}{r} \tag{1.35}$$

Solving the integral in Eq. (1.35) leads to

$$V_{3,4} = -\frac{\rho i}{\pi\delta}\,\ell n\left(\frac{a+b+c}{a+b}\right). \tag{1.36}$$

The potential $V_{3,4}$ developed by the current, i, flowing <u>out</u> of the half plane, between electrodes 3 and 4 is

$$V_{3,4} = \frac{\rho i}{\pi\delta}\int_{b}^{b+c}\frac{dr}{r} = \frac{\rho i}{\pi\delta}\,\ell n\left(\frac{b+c}{b}\right). \tag{1.37}$$

The potential difference between Eqs. (1.36) and (1.37) is, therefore,

$$V_4 - V_3 = V_{3,4} = \frac{\rho i}{\pi\delta} \cdot \ell n \frac{(a+b)(b+c)}{b(a+b+c)} \tag{1.38}$$

and the resistance determined from Eq. (1.38) is

$$R_{12,34} = \frac{V_{3,4}}{i} = \frac{\rho}{\pi\delta} \cdot \ell n \frac{(a+b)(b+c)}{b(a+b+c)} \tag{1.39}$$

If the currents and voltage electrodes are commutated so that 2 and 3 are current electrodes and 1 and 4 are potential probes, then, solving the analog equations for the resistance leads to

$$R_{23,41} = \left(\frac{\rho}{\pi\delta}\right) \cdot \ell n \left[\frac{(b+c)(a+b)}{ac}\right] \tag{1.40}$$

From Eqs. (1.39) and (1.40),

$$\exp\left[- \frac{\pi\delta}{\rho} R_{12,34}\right] + \exp\left[- \frac{\pi\delta}{\rho} R_{23,41}\right] = 1 \tag{1.41}$$

Let $\pi\delta\ R_{12,34} = x_1$ and $\pi\delta\ R_{23,41} = x_2$; then expressing x_1 and x_2 in symmetric and anti-symmetric terms,

$$\frac{1}{2} = e^{-\left(\frac{x_1+x_2}{2\rho}\right) \cdot \cosh\left(\frac{x_1-x_2}{2\rho}\right)} \tag{1.42}$$

Now choose a function, f, so defined that

$$f = \left(\frac{2\rho}{x_1+x_2}\right) \cdot \ell n2 \tag{1.43}$$

Then the resistivity, $\rho = (f/\ell n2)\ (x_1+x_2)/2$, is

$$\rho = \left(\frac{\pi\delta}{\ell n2}\right) \cdot \left(\frac{R_{12,34} + R_{23,41}}{2}\right) \cdot f \tag{1.44}$$

The function f can be evaluated by introducing Eq. (1.43) into (1.42)

$$\frac{R_{12,34} - R_{23,41}}{R_{12,34} + R_{23,41}} = \left(\frac{f}{\ell n2}\right) \cdot \text{arc cosh}\left[\frac{\exp(\ell n2/f)}{2}\right] \tag{1.45}$$

The function, f, on the right hand-side of this equation depends only on the ratio $R_{12,34}/R_{23,41}$ (i.e., on the commutated current and measured voltages). Figure 1.14 presents, in graphical form, the dependence of f (ref. 26) on the ratio $R_{12,34}/R_{23,41}$. This curve has a maximum error of about 2%. A useful analytical relation for calculating f is that given by Wasscher (ref. 28); provided that $-(1/2) < \alpha < (1/2)$, with α determined from

$$\frac{R_{12,34}}{R_{23,41}} = \frac{\ell n(1/2 - \alpha)}{\ell n(1/2 + \alpha)} \qquad f = \frac{\ell n(1/4)}{\ell n(1/2 + \alpha) + \ell n(1/2 - \alpha)} \tag{1.46}$$

A more accurate procedure suggested by Price (ref. 29) is the use of Newton's iterative method employed for solving nonlinear equations. Writing Eq. (1.41) in the form

$$f(\rho) = \exp(-\pi\delta R_{12,34}/\rho) + \exp(-\pi\delta R_{23,41}/\rho) = 0 \tag{1.47}$$

let ρ_0 be a trial approximation to a solution of Eq. (1.47) with successive approximations ρ_1, ρ_2, ρ_3, etc., generated by the iteration

$$\begin{aligned} \rho_1 &= \rho_0 - f(\rho_0)/f'(\rho_0) \\ \rho_2 &= \rho_1 - f(\rho_1)/f'(\rho_1) \\ \rho_{n+1} &= \rho_n - f(\rho_n)/f'(\rho_n) \end{aligned} \tag{1.48}$$

where $f'(\rho)$ denotes a derivative with respect to ρ. Writing α for $\exp(-\pi\delta R_{12,34}/\rho_n)$ and β for $\exp(-\pi\delta R_{23,41}/\rho_n)$ then

$$\rho_{n+1} = \rho_n - \left[\frac{\rho_n{}^2(1 + \alpha + \beta)}{\pi\delta(R_{12,34} + R_{23,41})}\right] \tag{1.49}$$

Van der Pauw (ref. 27) has shown that a lamina of arbitrary shape can be mapped by a conformal transformation onto the semi-infinite plane of Fig. 1.13 provided that: $\nabla\cdot\bar{J}=0$ and $\nabla X\bar{J}=0$, that the lamina is simply-connected, that it is homogenous and isotropic, that it is uniform in thickness, its four "ohmic" line-electrodes are on its periphery, and the line-electrode projections on the surface of the lamina are point-contacts. The resistivity of a lamina or wafer of arbitrary shape can be determined by applying a current, i, to line-electrodes 1 and 2 along its thickness dimension and measuring the potential difference between electrodes 3 and 4; this leads to the resistance $R_{12,34}$. In a similar manner, after commutating the current and voltage electrodes, $R_{23,41}$ is obtained. Figure 1.14 or Eqs. (1.46) to (1.48) are then used for the calculation of the appropriate $f(R_{12,34}/R_{23,41})$ and thereafter of the resistivity. In a practical example the resistivity accurate to four significant figures was obtained (ref. 29), after only four iterations, starting from a value of ρ_0 greater or smaller from the final value by about a factor of 3.

Note that if the specimen used were a perfect circle with the electrodes along orthogonal diameters then $R_{12,34} = R_{23,41}$. Therefore a solution of Eq. (1.45) leads to f = 1; the sheet resistivity $\rho_s = (\pi/\ell n2)\, R_{12,34}$ is a factor of two smaller than that determined by means of Eq. (1.22), that of a square four-point probe touching a semi-infinite sheet. Rymaszewsky (ref. 30) has shown that for a collinear equispaced four-point probe used on a thin film, the sheet resistivity can be determined by means of a correction factor based on conformal representation similar to that used by van der Pauw. For two commutated measurements such that

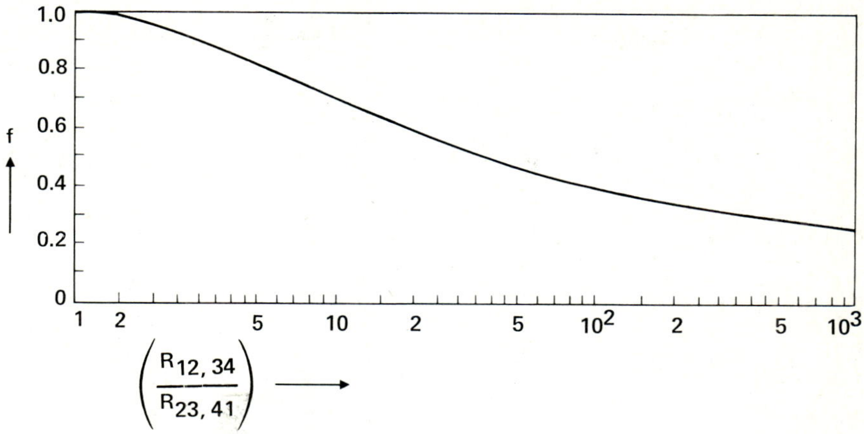

Fig. 1.14. The function f used to determine the specific resistivity of a specimen with an arbitrary contour as a function of the resistance ratio $R_{12,34}/R_{23,41}$. [after L. J. van der Pauw, Philips Res. Repts., 13(1958)1]

$$R_1 = V_{32}/i_{41}, \qquad R_2 = V_{34}/i_{21} \tag{1.50}$$

related by

$$\exp(-2\pi R_1/\rho_s) + \exp(-2\pi R_2/\rho_s) = 1 \tag{1.51}$$

ρ_s can be expressed in the form

$$\rho_s = \frac{\pi R_1}{\ell n 2}\left(1 + \frac{R_2}{R_1}\right)\cdot f\left(\frac{R_1}{R_2}\right) \tag{1.52}$$

Thus Eq. (1.14) is multiplied by a correction factor which includes van der Pauw's f-function and the function of the commutated resistances, R_2/R_1. He suggested that this method can be applied to a broad class of point-contact probes disposed either in line or along the circumference of a specimen.

Jandl *et al.* (ref. 31) have analyzed theoretically and experimentally the properties of a bridge-shaped specimen such as shown in Fig. 1.12, and found that the presence of lateral arms instead of true point contacts has a negligible effect on the distribution of the equipotential planes in it. The bridge-shaped specimen is, in fact, a special case subject to the same theoretical analysis developed by van der Pauw. Typically, the potential measured between the ends of the lateral arms with half the cross-section of the main sample body and separated by about eight sample widths is only 1% less than the voltage between two point contacts, separated by the same distance placed on the main body of the specimen.

1.6 Error introduced by contact size, position and geometry

The finite size of the electrodes applied to the periphery of arbitrary contour specimens affects the measurement of their resistivity made by means of van der Pauw's method (refs. 26, 27). The errors introduced by finite electrodes or by the displacement of an electrode from the specimen periphery were evaluated by van der Pauw (ref. 27) for a circular specimen of diameter D, whose contacts are placed symmetrically at 90° intervals; only one of the contacts is assumed to deviate from the ideal, as shown in Fig. 1.15; the relative

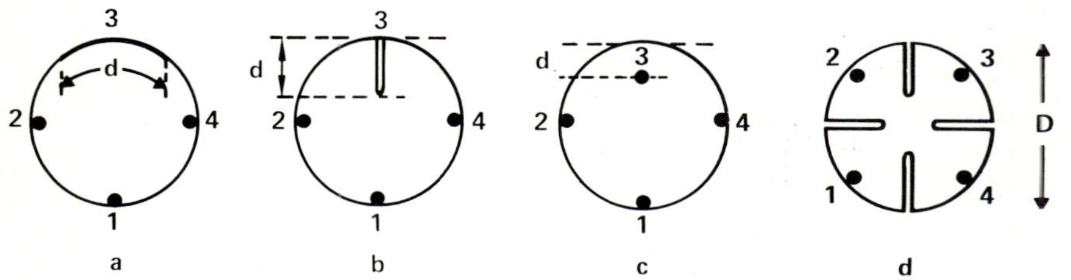

Fig. 1.15. Error in contact size and position: (a) one of the contacts has a length, d, along the periphery; (b) one of the contacts has a length, d, perpendicular to the periphery; (c) one of the point contacts is at a distance, d, from the periphery; (d) clover-leaf shaped specimen.

fractional errors in resistivity for the three cases illustrated here are:

$$\begin{aligned} &(a) \quad \Delta\rho/\rho \simeq - d^2 (16D^2 \ell n2)^{-1} \\ &(b) \quad \Delta\rho/\rho \simeq - d^2 (4D^2 \ell n2)^{-1} \\ &(c) \quad \Delta\rho/\rho \simeq - d^2 (2D^2 \ell n2)^{-1} \\ &(d) \quad \Delta\rho/\rho \simeq 0 \end{aligned} \qquad (1.53)$$

If none of the contacts are ideal then, to a first approximation, the total error is the sum of all the errors. The influence of the contact and the errors associated with then can be eliminated by the use of a clover-leaf shaped specimen such as that shown in Fig. 1.15d. Such a structure is easier to make than the bridge-shaped samples shown in Fig. 1.12 and less susceptible to breakage when made of brittle materials. Van Daal (ref. 32) investigated experimentally the magnitude of the error introduced by the size of the contacts relative to the diameter of p-type Ge discs before and after fashioning the discs into a clover-leaf shaped configuration. The results are shown in Fig. 1.16 where the relative error, $\Delta\rho/\rho$, is presented as a function of $(d/D)^2$ with d and D, respectively, the diameters of the contact and sample and, ρ, the resistivity determined for negligibly small contacts, $(d/D < 0.01)$. It is apparent from the figure that the relative error decreases with increasing sample thickness and that the clover-leaf shape reduces the error of measurement considerably. A clover-leaf configuration, applied to thin films or epitaxial layers, requires complex photolithographic processing which may alter

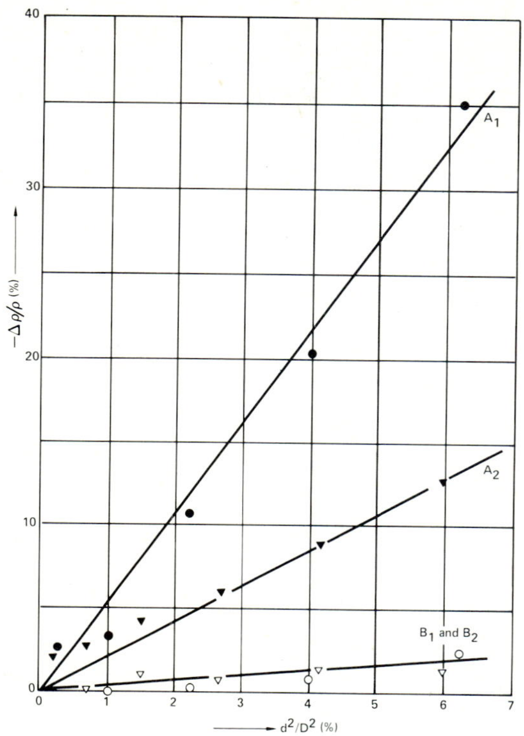

Fig. 1.16. Relative systematic error in the resistivity calculated by van der Pauw's equations as a function of the ratio, $(d/D)^2$, of contact size to specimen diameter with the thickness, δ, as a running parameter. For curve A_1, δ = 0.2 mm and D = 20mm; for curve A_2, δ = 1.3 mm and D = 24.5 mm. B_1 and B_2 were measured on the corresponding specimens after they were converted into clover-leaf configurations. [after H. J. van Daal, Phillips Res. Repts., (Suppl. 3)(1965)8]

their surfaces; such a structure usually occupies too much of the useful area of a specimen removing that portion from consideration for other applications. For this reason, simple circular or square structures with contacts which can be applied directly to the corners or edges of an epitaxial layer or thin film are considered desirable.

Chwang *et al.* (ref. 33) have investigated, analytically and experimentally, the errors introduced in the resistivity calculated by van der Pauw's method for triangular or square surface electrodes applied to the corners of square specimens such as shown in Fig. 1.17. They reached the conclusion that electrodes of finite size produce a larger deviation from the ideal van der Pauw case when they are located in the interior rather than the periphery of a specimen. Their calculations of the electric field distribution are based on the method of finite differences used to solve Laplace's equation in a specimen.

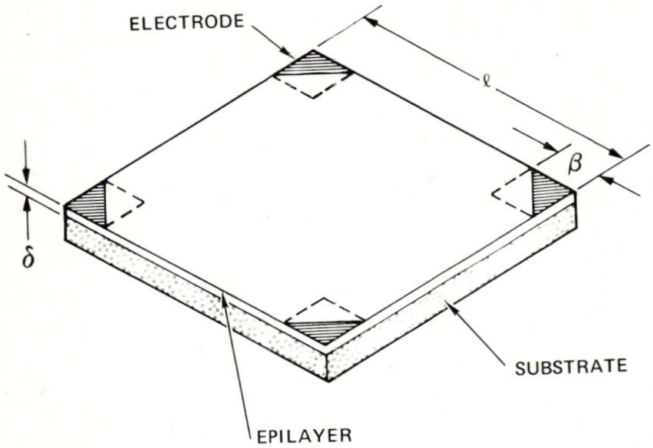

Fig. 1.17. Corner contacts made to an epitaxial layer of thickness, δ, on a square semi-insulating substrate; triangular contacts or square contacts with side-dimension, β, are shown on a square specimen of side-dimension, ℓ.

Experimental verification was obtained by means of an electrolytic tank equipped with suitable electrodes used to determine the analog potential distributions. Fig. 1.18 summarizes their results; the experimental data is in good agreement with theory; the potential difference measured between finite-size potential probes is always smaller than that to be expected from theory between idealized point-contact electrodes; therefore, the measured resistivity, ρ_m, is always smaller than the actual resistivity, ρ. For triangular electrodes, the resistivity correction factor, ρ/ρ_m, appears to increase monotonically with the ratio, (β/ℓ). For conditions usually encountered in practice ℓ may be greater by a factor of 5 to 10 in comparison with β; therefore, the errors introduced by the size of the contact electrodes are relatively small. For example: for $(\beta/\ell) \simeq 0.125$, $\rho/\rho_m = 1.0025$ for triangular electrodes while for square contacts and the same (β/ℓ) ratio $\rho/\rho_m \simeq 1.004$. Perloff (ref. 13) has investigated, analytically and experimentally, the correction factors relevant to the placement of peripherally contacted square and rectangular specimens. For corner-contacted square specimens or for rectangular specimens, with electrodes at the midpoints of their sides, the sheet resistivity determined by the commutation of the current electrodes and van der Pauw's equations is reduced to Eq. (1.14), $\rho_s = \pi R/\ell n2$. However, for corner-contacted rectangular specimens, large errors are introduced in ρ_s by the strong dependence of the geometrical correction factors on the length to width ratio of the specimens.

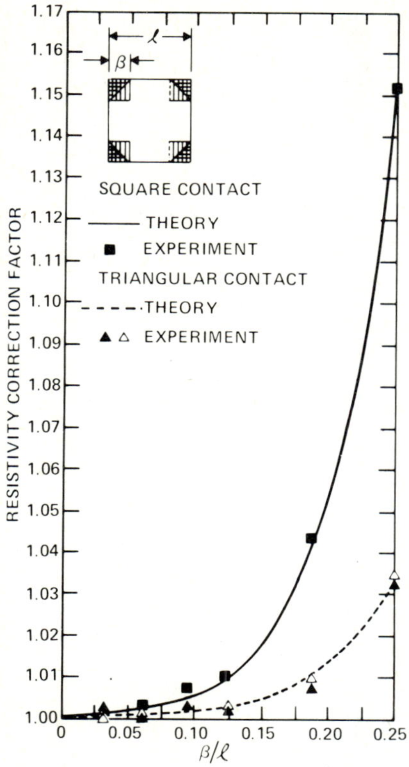

Fig. 1.18. Correction factors as a function of β/ℓ. Solid and dashed curves were calculated for square and triangular contacts, respectively. [after R. Chwang, B. J. Smith and C. R. Crowell, Solid State Electron., 17(1974)1217]

In the silicon integrated circuit technology there is a need for measurements to be made of the sheet resistivity of microstructures placed on an integrated circuit chip; the contact dimension of such a structure can be an appreciable fraction of the device perimeter. Consider the square configuration shown in the insert of Fig. 1.19 with side-centered contacts. Let the idealized van der Pauw sheet resistivity for this structure be $\rho_s = \pi R/\ell n2$ and let the measured sheet resistivity be ρ_{sm}; then the error of measurement is considered to be $E = 1-\rho_s/\rho_{sm}$. Figure 1.19 shows the dependence of E on the ratio of contact length to the side dimension of the square specimen (ref. 34); the error of measurement fits the empirical relation

$$E \simeq (0.595\pm0.015)\cdot(D/S)^{1.82\pm0.03} \qquad (1.54)$$

If $D/S < 0.1$ then the geometrical error associated with this structure is less than 1%. A different configuration also subject to a van der Pauw-type of

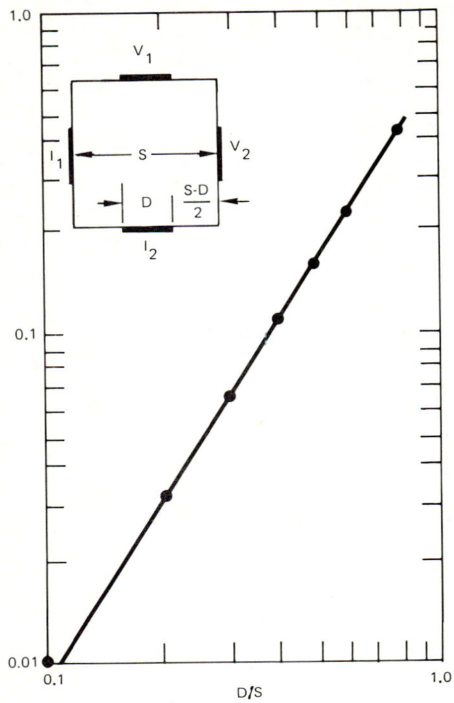

Fig. 1.19. Normalized sheet resistance error, E, with contacts centered on each side of a square specimen with dimensions as shown in the insert as a function the ratio of contact length to side of the square. [after J. M. David and M. G. Buehler, Solid-State Electron., 20(1977)539]

analysis is the Greek cross shown in the insert of Fig. 1.20. Except for very short cross-arm lengths such that $A/S < 0.1$ the error of measurement fits the empirical relation

$$E = (0.59\pm0.006)\ \exp[-(6.23\pm0.02)\ A/S] \tag{1.55}$$

If $A/S > 0.6$ then the geometrical error is less than 1% and if $A/S > 1.02$, a condition achieved easily in practice, then the corresponding error is less than 0.1%. The technological implication of the test structure, shown in Fig. 1.20 for planar integrated circuit process control, is its use for measuring accurately the sheet resistivity of a wafer; and yet it can be made to very small sizes because the width of the arms of the Greek cross is limited only by the minimum linewidths which can be obtained by conventional photolithography.

A different test pattern suitable for measuring the sheet resistivity of thin films or epitaxial layers by van der Pauw's method is illustrated in Fig. 1.21(a). This "pin-wheel" structure has four-fold rotational symmetry and its normalized error dependence on geometrical parameters has been calculated by David (ref. 35) for the "worst case" equivalent model shown in Fig. 1.21(b). The fabrication of planar four-probe test structures for measuring the bulk or collector resistivity

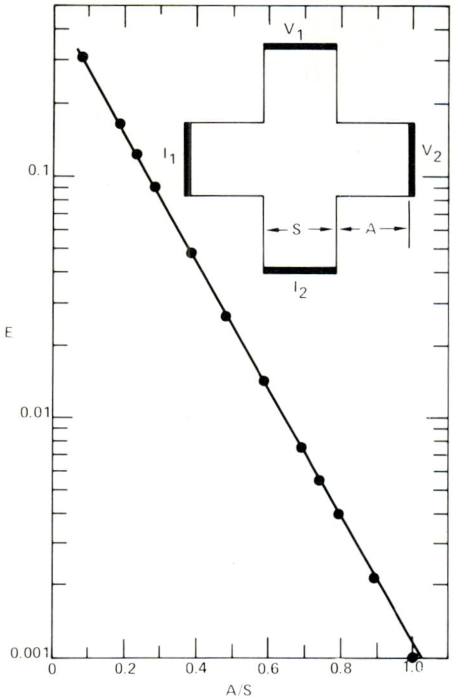

Fig. 1.20. Normalized sheet resistance error resulting from contacts applied to the cross-arm terminals of a Greek cross as a function of the length to width ratio of a symmetrical cross-arm as shown in insert. [after J.M. David and M. G. Buehler, Solid-State Electron., 20(1977)539]

of silicon wafers has been described by Buehler and Thurber (ref. 36). It is entirely compatible with the fabrication processes employed for bipolar transistors and occupies a square ∿ 318 μm per side. It can be placed either in an unused corner of a component chip or it can be included in a specific test pattern. When included on each component chip, automatic wafer probing techniques can be used to provide a map of the iso-resistivity lines on the wafer such as discussed in Chapter 7. On n-type wafers, whose resistivities were in the range between 0.013 and 12 ohm-cm, and on p-type wafers, whose resistivities were in the range between 0.07 and 30 ohm-cm, they found agreement to within ± 3% between collinear four-point probe measurements made on the wafers before fabrication and measurements made subsequently on the planar arrays fabricated on the same wafers.

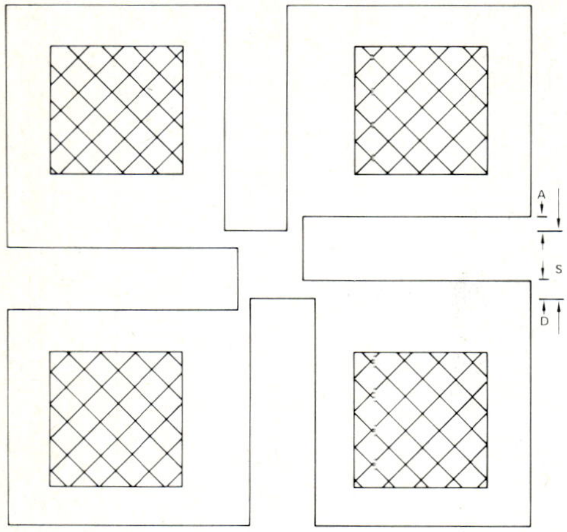

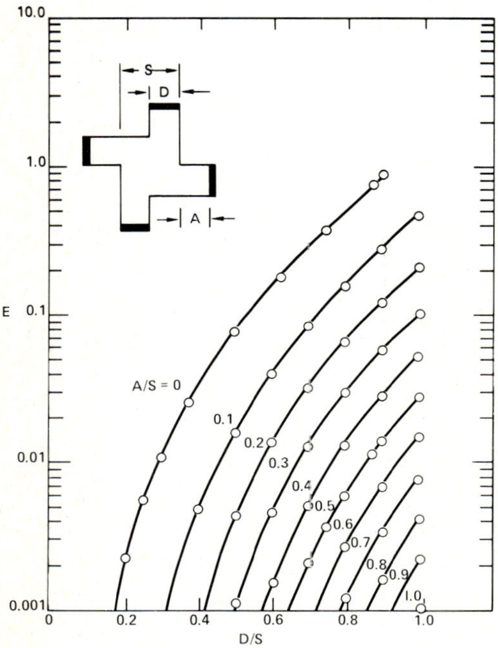

Fig. 1.21. (a) Orthogonal pin-wheel structure adaptable for microcircuit fabrication for measuring the sheet resistivity by van der Pauw's method; (b) calculated measurement error of equivalent model as a function of geometrical parameters of structure shown in insert. [after J. M. David and M. G. Buehler, Solid-State Electron., 20(1977)539]

1.7 Electrodeless measurement of resistivity

Electrodeless methods used for measuring the conductivity of liquids or of chemically active substances have many advantages over conventional methods which require contacts for introducing currents and potential probes on solid specimens such as described in the preceding sections of this chapter. Contacts between electrolytes and electrodes lead to problems with contact barrier potentials, electrode corrosion and chemical contamination of the substance whose conductivity is to be determined. Such specimens can be investigated by electromagnetic induction processes; their resistivity can be determined by means of the eddy currents induced in them while they are contained in a suitably shaped, chemically inert and electrically non-interactive container. A review of the many interesting eddy current methods developed in recent years has been published by Delaney and Pippard (ref. 37). Two relatively simple techniques have received the major share of attention: (a) the current decay method in which the resistivity is calculated from the decay of a current transient induced in a specimen by switching a small DC magnetic field impressed on it on or off (ref. 38); (b) an AC method in which the resistivity is calculated from the apparent susceptibility of a specimen placed in an alternating magnetic field of constant frequency (ref. 39). The theoretical basis for measurement of the conductivity of cylindrical and spherical specimens placed in the core of a sinusoidally-excited solenoid and then relating the changes in impedance reflected into the solenoid to the conductivity of the specimens were developed by Nyberg and Burgess (ref. 40). However this technique has a relatively limited utility and a low precision. The method of measurement described by Bauhofer (ref. 41) provides some advantages in this respect; it depends on the change in mutual impedance induced by a conductive specimen introduced between two coupled solenoids such as shown in Fig. 1.22. The mutual impedance, Z, between them can be represented by $Z = R+j\omega M$ where M is a mutual inductance and the angular frequency, $\omega = 2\pi\nu$, with ν a sinusoidal driving frequency. For a skin depth $\delta \geq 2r$ where r is the radius of a cylindrical specimen, $\Delta R \simeq \gamma\omega^2 M_0 r^2 \sigma/8$ where ΔR, the change in the resistive component of the mutual inductance, due to the insertion of the specimen between the coils, is related to the specimen conductivity, σ, (to an accuracy of better than 1% for nonmagnetic samples), to the mutual inductance of the empty coils, M_0, and to a filling factor, γ. ΔR is determined by measuring in the secondary coil, the change of induced voltage, ΔV_r, which is in phase with the current, i_p, in the primary coil, $\Delta V_r = \Delta R i_p$. The temperature dependent resistivity is determined by a substitution and calibration process; $\rho(T) = C(T)\cdot i_p \omega^2/\Delta V_r(T)$ where C(T) is a temperature-dependent calibration factor determined on a specimen of known resistivity with the same diameter as that of the specimen whose resistivity is to be measured. The apparatus shown in Fig. 1.22 used between 2 Hz and 20 kHz consists of a "lock-in" amplifier which is used as an oscillator driving a constant current 60 mA AC source; it is also used as a phase-sensitive detector. The coils are made of 10^3 turns of 10^2 μm copper wire wound coaxially on thin-walled teflon cylinders 3 to 12 mm in diameter. The mutual inductance between the coils is in the range between 5 and

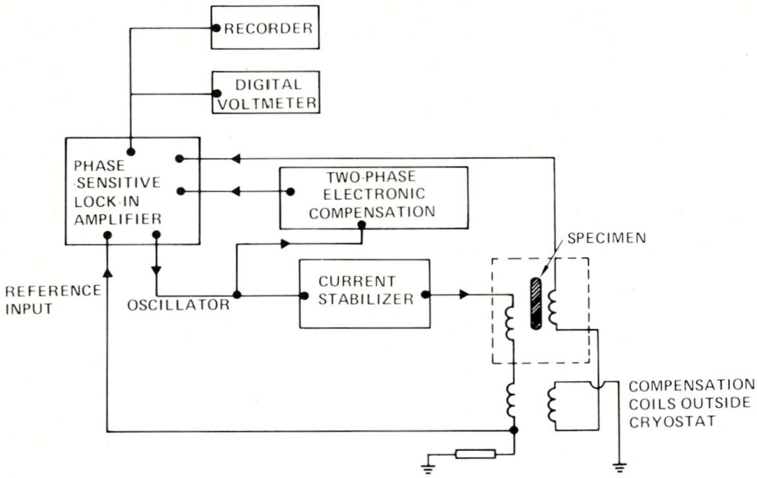

Fig. 1.22. Block diagram of apparatus for electrodeless eddy current measurement of resistivity of cylindrical conductive specimens. [after W. Bauhofer, J. Phys. E., Sci. Instr., 10(1977)1212]

100 mH. The apparatus was applied to quasi-continuous measurements of resistivity changes which accompany phase transitions as a function of temperature for resistivities in the range between 10^{-5} and 10^{-11} ohm-cm and specimen diameters of 1 to 10 mm. The reproducibility of the measurements for a specific sample is reported to be better than 2% and the absolute error of measurement of the resistivity smaller than 4%.

An analysis and comparison of the relative merits of the eddy current decay and steady state AC measurement methods have been discussed by Wejgaard and Tomar (ref. 42). They found that although theoretically the methods are equivalent, experimental consideration lead to a preference for the AC steady-state method for all except extremely low resistivity values. An analytical and experimental investigation of an electrodeless method of measuring the resistivity of electrolytes and aqueous solutions by a difference field technique was developed by Minnot *et al.* (ref. 43). If the fields of two similar solenoids with a common axis and a constant phase displacement of 180° are balanced then, if a conducting specimen is inserted between them, the induced eddy currents in the specimen produce an unbalance which can be detected by a separate sensing coil. The apparatus was used for conductivity measurements in the 10^{-3} to 1 ohm-cm range with an error of less than 2% in the frequency range between 20 to 500 kHz.

Apparatus and techniques for measuring the position-dependent electrical resistivity can be made by the induced eddy current method (ref. 44). The method is useful for investigating surface phenomena such as the anomalous skin effect. It provides a relative measurement of bulk resistivity on short thick specimens with an accuracy of at least 0.05% over frequency range of 10^2 to 10^5 Hz; it can be used to determine surface-related imperfections or mechanical damage in the surface layers of cold-worked metal specimens.

A radio frequency spreading resistance method for determining the resistivity of a semiconductor in the range between 0.1 and 10^2 ohm-cm was developed by Bryant and Gunn (ref. 45). Figure 1.23 shows the salient features of the apparatus. Radio frequency current flowing from the center electrode spreads the impedance, Z = R+jX which appears between the electrodes. The electrical length of the line is made exactly equal to an equal number of half-wavelengths at the measuring frequency of $\sim$ 0.45 GHz so that the RF bridge measures directly the value of $Z \simeq R_s(1+\omega^2\tau^2)^{-1}\cdot[1-j(1+\omega^2\tau^2\eta)/\omega\tau\eta]$ where $\eta = C_1/C_s$ and C_1 is a coupling capacitance due to the finite air gap between the specimen and the inner electrode, R_s and C_s are, respectively, the spreading resistance and spreading capacitance and $\tau = R_sC_s$. For air gaps of the order of 1 μm, $\eta \simeq 30$; accurate measurements are feasible in the range between 0.5 and 100 ohm-cm, the lower limit being reached when the phase angle is too large and the upper limit when the correction for the spreading capacitance becomes excessive.

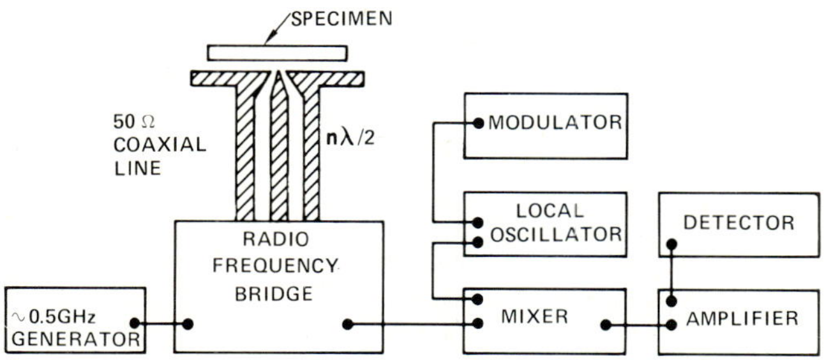

Fig. 1.23. Block diagram of rf probe for spreading resistance measurements made on semiconductor specimens. [after C. A. Bryant and J. B. Gunn, Rev. Sci. Instr., 36(1965)1614]

A contactless apparatus for the measurement of the conductivity of semiconductors using their radio frequency free-carrier power absorption was developed by Miller *et al.* (ref. 46). It is based on the properties of marginal oscillators similar to those used to investigate various nuclear resonance phenomena in which the Q of their resonant circuits is reduced due to power absorbed from them by a coupled nuclear spin system. However, the method is not limited to power absorption from resonant circuits and, therefore, can be used for a direct measurement of the conductivity of a semiconducting wafer by a procedure such as shown in Fig. 1.24. A high permeability ferrite core of circular cross-section is coupled to a radio frequency resonant circuit and a semiconducting wafer is placed between its pole-pieces. The oscillating magnetic field produces eddy currents in the semiconducting wafer. Assuming no flux leakage and negligible skin effect phenomena in the semiconductor, the power absorbed is

$$P_S = (E_T{}^2/8\pi N^2)\cdot\sigma\delta \tag{1.56}$$

where E_T is the rms radio frequency voltage, N represents the number of turns of the solenoid, L, on the magnetic core, σ is the conductivity, and δ the thickness

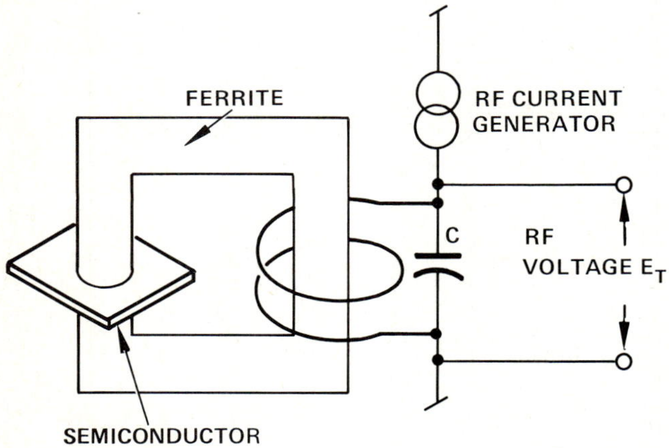

Fig. 1.24. Schematic circuit for measuring conductivity of a semiconductor wafer coupled to a radio frequency magnetic field.

of the semiconductor wafer. The power absorbed is, therefore, directly proportional to $(\sigma\delta)$ provided that $\delta \ll \delta_S$, where δ_S is the skin depth. In the unity coupling limit, $P_S = E_T I_T$ where I_T is the in-phase radio frequency drive current; therefore, the power flowing into the circuit can be determined by monitoring I_T. In practice the coupling between the resonant circuit and the specimen is not unity; this reduces the dependence of I_T on $(\sigma\delta)$ but the inherent linearity between I_T and $(\sigma\delta)$ is retained; thus

$$I_T = K(E_T/N^2)\cdot\sigma\delta \tag{1.57}$$

The magnitude of the power absorption can be used to determine the conductivity by coupling the semiconductor to an amplitude-stabilized marginal oscillator and noting the power needed to maintain the required level of oscillation.

Miller *et al.* (ref. 46) have described, in detail, the design, construction, and performance of such a circuit operating at a frequency of ∿ 10 MHz. One of the advantages of this type of measurement is that semiconducting wafers may be stacked upon each other. Their conductivities add linearly, irrespective of the nature of the charge carrier distribution in each one and, therefore, the apparatus can be used to determine specific changes in a wafer or wafers following impurity diffusion, ion implantation, or thermal annealing procedures. However, the measurements are only relative and need to be calibrated in terms of a standard specimen. Such instruments exhibit ∿ 1% linearity over a 100:1 range of conductivities, a resolution of 1 part in 10^4 and a limiting sensitivity of ∿ 10^{-5} mho/cm^2; they are suitable for the measurement of any conducting materials, in the above range, from semiconductors to metals.

The microwave conductivity of semiconductors has frequently been determined from transmission or reflection measurements of a TE_{01} mode rectangular waveguide section completely filled with the semiconductor to be measured (ref's. 47-51) or by measuring the attenuation suffered by a guided electromagnetic wave upon reflection from a specimen placed at the end of the waveguide. The attenuation and phase of the reflected wave are a function of the specimen resistivity and thickness. Champlin *et al.* (ref. 52) have shown that these techniques are, in fact, not electrodeless. They depend critically on the quality of the contact to the broad waveguide wall where the $\bar{E}$ field meets the specimen surfact at right angles. A series contact impedance arises from gaps between specimen and waveguide, the oxide layer on the semiconductor surface and surface charges all produces significant errors in the calculated resistivity. These errors increase with increasing conductivity of the specimens. Champlin *et al.* (ref. 53) have described a technique which eliminates this source of error by the use of the TE_{01} mode in an oversize circular waveguide reflection coefficient bridge circuit. From measurements made on intrinsic germanium specimens, they found good agreement with conductivity measurements made at DC in the conductivity range of 2 to 2 x 10^3 mho/m. In contrast with this, a conventional rectangular waveguide transmission bridge was found to saturate at ∿ 50 mho/m due to the above described series contact impedance.

REFERENCES

1 J. F. Nye, Physical Properties of Crystals, Oxford, Clarendon Press (1960) p.207.
2 Standard Method of Test for Resistivity of Semiconducting Materials, American Society for Testing Materials, 1971,, p.F43-71.
3 T. M. Dauphinee, Can. J. Phys., 31(1953)577.
4 T. M. Dauphinee and E. Mooser, Rev. Sci. Instr., 26(1955)660.
5 W. D. Edwards, J. Sci. Instr., 42(1965)432.
6 L. B. Valdes, Proc. IRE, 42(1954)420.
7 A. Uhlir, Jr., Bell Syst. Tech. J., 34(1955)105.
8 F. M. Smits, Bell Syst. Tech. J., 37(1958)711.

9 D. S. Perloff, J. Electrochem. Soc., 123(1976)1745.
10 D. E. Vaughan, Brit. J. Appl. Phys., 12(1961)414.
11 L. J. Swartzendruber, Solid-State Electron., 7(1964)413.
12 A. Mircea, J. Sci. Instr., 41(1964)679.
13 D. S. Perloff, Solid-State Electron., 20(1977)681.
14 J. M. Reber, Solid-State Electron., 7(1964) 525.
15 A. Mircea, Solid-State Electron., 6(1963)459.
16 A. E. Stephens, H. J. Mackey and J. R. Sybert, J. Appl. Phys., 42(1971)2592.
17 J. K. Kennedy, Rev. Sci. Instr., 33(1962)773.
18 C. L. Paulnack and N. J. Chaplin, Rev. Sci. Instr., 33(1962)873.
19 A. G. Klimenko, Instr. and Exper. Tech., 5(1964)991.
20 D. R. Zrudsky, H. D. Bush and J. R. Fassett, Rev. Sci. Instr., 37(1966)885.
21 L. J. Swartzendruber, F. H. Ulmer and J. A. Coleman, Rev. Sci. Instr., 39(1968)1858.
22 E. Earleywine, L. P. Hilton and D. Townley, Semicond. Prod. Solid-State Tech., 8(1965)17.
23 J. K. Hargreaves and D. Millard, Brit. J. Appl. Phys., 13(1962)231.
24 R. Hall, J. Sci. Instr., 44(1967)53.
25 Standard Method for Measuring Resistivity of Silicon Slices with a Collinear Four-Probe Array, American National Standards Institute F84-73 (corrected June 1975).
26 L. J. van der Pauw, Philips Res. Repts., 13(1958)1.
27 L. J. van der Pauw, Philips Tech. Rev., 20(1958)220.
28 J. D. Wasscher, Philips Res. Repts., 8(1969)16.
29 W. L. V. Price, Solid-State Electron., 16(1973)753.
30 R. Rymaszewsky, Electron. Lett., 3(1967)57.
31 S. Jandl, K. D. Usadel and G. Fischer, Rev. Sci. Instr., 45(1974)1479.
32 H. J. van Daal, Philips Res. Repts., (Suppl. 3),(1965)8.
33 R. Chwang, B. J. Smith and C. R. Crowell, Solid-State Electron., 17(1974) 1217.
34 M. G. Buehler and J. M. David, Nat'l Bureau of Stds Special Publ. 400-29 (1976)64.
35 J. M. David, Nat'l Bureau of Stds Special Publ. 400-19(1975)44.
36 M. G. Buehler and W. R. Thurber, IEEE Trans. Electron Dev., ED-23(1976)968.
37 J. A. Delaney and A. B. Pippard, Rep. Prog. Phys., 35(1972)677.
38 R. C. Callarotti, J. Appl. Phys., 43(1972)3949.
39 A. J. Hamdani, J. Appl. Phys., 44(1973)3486.
40 D. W. Nyberg and R. E. Burgess, Can. J. Phys., 40(1962)1174.
41 W. Bauhofer, J. Phys. E. Scient. Instr., 10(1977)1212.
42 W. Wejgaard and V. S. Tomar, J. Phys. E, Sci. Instr., 7(1974)395.
43 D. A. Minnot, S. L. Sandiford and E. T. Agard, J. Phys. E, Scient. Instr., 6(1973)229.
44 E. W. Johnson and H. H. Johnson, Rev. Sci. Instr., 35(1964)1510.
45 C. A. Bryant and J. B. Gunn, Rev. Sci. Instr., 36(1965)1614.
46 G. L. Miller, D. A. H. Robinson and J. D. Wiley, Rev. Sci. Instr., 47(1976)799.
47 T. S. Benedict and W. Shockley, Phys. Rev., 91(1953)1565.
48 H. Jacobs, F. A. Brand, J. D. Meindl, M. Benanti and R. Benjamin, Proc. IRE, 49(1961)928.
49 B. R. Nag and S. K. Roy, Proc. IEEE, 50(1962)2515.
50 K. S. Champlin, D. B. Armstrong and P. D. Gunderson, Proc. IEEE, 52(1964)677.
51 M. R. E. Bichara and J. P. R. Poitevin, IEEE Trans. Instrum. Meas., IM-13 (1964)323.
52 K. S. Champlin, G. H. Glover, J. Appl. Phys., 37(1966)2355.
53 K. S. Champlin, J. D. Holm and G. H. Glover, J. Appl. Phys., 38(1967)96.

CHAPTER 2

THE HALL EFFECT

2.0 Introduction

Electrons, considered as charged particles, drift in the direction of an electric field $\overline{E}$ with a velocity $\overline{v}$ driven by the force $\overline{F}_e = - e\overline{E}$. In the presence of a transverse magnetic field $\overline{B}$ the drifting electrons are subjected to the Lorentz force $\overline{F}_m = - e(\overline{v} \times \overline{B})$ and some of the electrons are deflected in a direction orthogonal to both the electric and magnetic field vectors producing a Hall field E_h which compensates the Lorenz force. The combination of electric and magnetically-induced forces

$$\overline{F} = - e(\overline{E} + \overline{v} \times \overline{B}) \tag{2.1}$$

act on the electrons and produce, under isothermal conditions, the two complementary galvanomagnetic phenomena, the Hall effect and the transverse magnetoresistance. In a rectangular plate such as that shown in Fig. 2.1 the Hall field

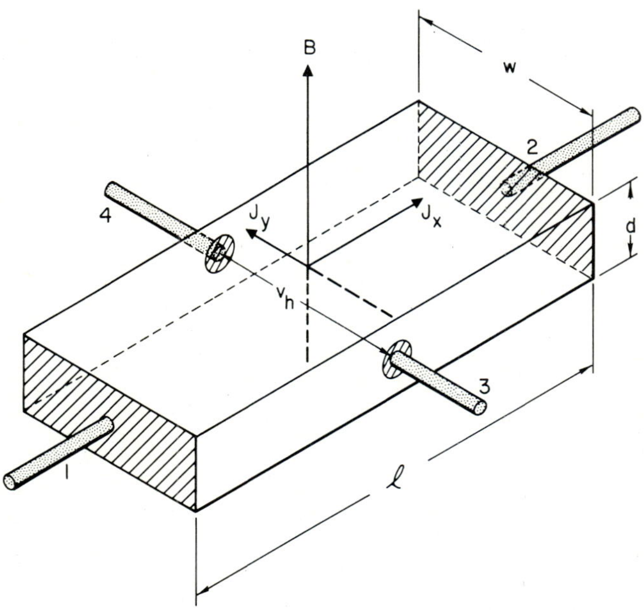

Fig. 2.1. Schematic representation of wafer in which electrons are the dominant charge carriers. Broad area faces, wd are current contacts designated by numerals 1 and 2. Hall voltage contacts transverse to magnetic field B and current density J_x are designated 3 and 4.

$E_h = E_y$ compensates the y-component of the force $\overline{F}$. Defining the electron mobility $\mu = - e\overline{v}/\overline{F}$ the components of the velocity vector $\overline{v}$ of the electrons are

$$v_x = \mu(E_x - v_y B)$$
$$v_y = \mu(E_y + v_x B) \qquad (2.2)$$
$$v_z = \mu E_z = 0$$

Ohm's law $\overline{J} = \sigma\overline{E}$ defines the linear relation between $\overline{E}$ and the current density vector $\overline{J} = ne\overline{v}$ in terms of the density of electrons n crossing a unit surface area which is perpendicular to their velocity vector in terms of Eq. (2.2). The components of $\overline{J}$ are

$$J_x = nev_x = \sigma\left(\frac{E_x - \mu B E_y}{1 + \mu^2 B^2}\right) \qquad J_y = nev_y = \sigma\left(\frac{E_y + \mu B E_x}{1 + \mu^2 B^2}\right) \qquad (2.3)$$

In B = 0 the conductivity σ or the resistivity ρ of a homogeneous isotropic solid are scalar quantities; for $B \neq 0$ these parameters become second order tensors, $\overline{J} = \hat{\sigma}\overline{E}$ and $\overline{E} = \hat{\rho}\overline{J}$, with the cap symbol representing a tensor. Neglecting higher order terms the Ohm-Hall equations can be expressed as

$$\overline{E} = \rho\overline{J} - R_h(\overline{J} \times \overline{B}) \qquad \overline{J} = \sigma\overline{E} - \sigma R_h(\overline{J} \times \overline{B}) \qquad (2.4)$$

where R_h is the coefficient of proportionality between the Hall field and the vector cross-product $(\overline{J} \times \overline{B})$. The conductivity tensor is

$$\hat{\sigma} = \begin{vmatrix} \sigma_{11} & \sigma_{12} \\ -\sigma_{12} & \sigma_{11} \end{vmatrix} = \begin{vmatrix} \dfrac{\rho}{\rho^2+R_h^2B^2} & \dfrac{R_hB}{\rho^2+R_h^2B^2} \\ \dfrac{-R_hB}{\rho^2+R_h^2B^2} & \dfrac{\rho}{\rho^2+R_h^2B^2} \end{vmatrix} \qquad (2.5)$$

It is apparent from Eq. (2.5) that all transverse "conductivities" are proportional to B and exhibit oblique symmetry $\sigma_{ij} = - \sigma_{ji}$; in accordance with the Onsager relations, $\rho_{ij}(\overline{B}) = \rho_{ji}(-\overline{B})$ (ref. 1).

The conductivity tensor is the reciprocal of the resistivity tensor; however, the elements of the resistivity tensor are not reciprocals of the conductivity tensor. If the specimen illustrated in Fig. 2.1 is characterized by the scalar conductivity $\sigma = ne\mu$ then, from Eq. (2.3) for $J_y = 0$

$$- E_y = \mu B E_x \qquad v_x = \mu E_x \qquad (2.6)$$

Introducing Ohm's law into Eq. (2.6) leads to

$$E_y = - R_h J_x B \tag{2.7}$$

In zero magnetic field the current streamlines are perpendicular to the equipotential planes. In a transverse magnetic field the current streamlines are rotated with respect to the equipotential planes by an angle θ defined as the Hall angle whose tangent, for $J_y = 0$, is $\tan\theta = - E_y/E_x$ hence $\tan\theta = - \sigma R_h B$.

To first order, in low or moderate magnetic fields, a Hall voltage V_h is generated between terminals 3 and 4 of Fig. 2.1 for an input current i applied between electrodes 1 and 2 and a transverse magnetic field B along the thickness d of the plate

$$V_h = \int_0^w E_y dy = - R_h(iB/d) \tag{2.8}$$

For electrons the Hall coefficient is inversely proportional to the electron density, $R_h = - 1/ne$. Therefore μ and n can be determined quantitatively from Hall and conductivity measurements.

The magnetoresistance effect in a homogenous isotropic material in which electrons are the charge carriers may be attributed to either a complete or a partial short circuit of the Hall field E_y. From Eq. (2.3) if $E_y = 0$ then the magnetic field-dependent resistivity $\rho(B) = E_x/J_x$ and the Hall current J_y are

$$\rho(B) = \rho(0)(1+\mu^2B^2), \qquad J_y/J_x = \mu B \tag{2.9}$$

where the scalar resistivity $\rho(0) = 1/\sigma$, and the magnetoresistance $\Delta\rho/\rho_0$ has a quadratic dependence on B,

$$\Delta\rho/\rho_0 = \mu^2B^2 \tag{2.10}$$

A partial short-circuit of E_y which allows a fractional Hall current to flow in a rectangular wafer produces in it a magnetoresistance which alters the linear relation expressed in Eq. (2.8) between V_h and B. Conversely, an incomplete short circuit of the Hall field, E_y reduces $\Delta\rho/\rho_0$ effective for a given B and may alter its quadratic dependence on B expressed in Eq. (2.10). Intuitively, it is to be expected that the geometrical aspect of a wafer used for Hall effect or for magnetoresistance measurements will affect the distribution of the equipotential planes in it. For rectangular plates such as shown in Fig. 2.1 the length to width ratio should be equal to or greater than 3 in order to minimize the electrostatic shorting effect of its electrodes.

If charge carriers are present in more than one conduction band then the total current density $\overline{J}$ in a material is the vector sum of the current densities of each band and

$$\hat{\sigma} = \sum_{i=1}^{n} \hat{\sigma}_i \tag{2.11}$$

where $\hat{\sigma}_i$ is the conductivity tensor of the i<u>th</u> conduction band. For two conduction bands, i = 1, 2, the resistivity tensor $\hat{\rho} = 1/\hat{\sigma} = (\hat{\sigma}_1+\hat{\sigma}_2)^{-1}$. The tensor components of a two-band system are, therefore,

$$\hat{\rho} = \begin{vmatrix} \dfrac{\alpha}{\alpha^2+\beta^2} & -\dfrac{\beta}{\alpha^2+\beta^2} \\ \dfrac{\beta}{\alpha^2+\beta^2} & \dfrac{\alpha}{\alpha^2+\beta^2} \end{vmatrix} \tag{2.12}$$

where $\alpha = \sigma_1^{11}+\sigma_2^{11}$ and $\beta = \sigma_1^{12}+\sigma_2^{12}$. Equation (2.2) is applicable to the two-band tensor as well; however the tensor components are then $\rho = \alpha\cdot[\alpha^2+\beta^2]^{-1}$ and $R_hB = \beta\cdot[\alpha^2+\beta^2]^{-1}$. The elements of the conductivity tensor can be expressed in terms of Eq. (2.3) for each band, i.e., σ_1, σ_2, R_{h_1} and R_{h_2}. The conductivity due to both conduction bands (refs. 2,3) is

$$\sigma = \frac{(\sigma_1+\sigma_2)^2 + B^2\sigma_1^2\sigma_2^2(R_{h_1}+R_{h_2})^2}{(\sigma_1+\sigma_2) + B^2\sigma_1\sigma_2(\sigma_1R_{h_1}^2+\sigma_2R_{h_2}^2)} \tag{2.13}$$

and the Hall coefficient of the two-band system is

$$R_h = \frac{\sigma_1^2R_{h_1} + \sigma_2^2R_{h_2} + B^2\sigma_1^2\sigma_2^2R_{h_1}R_{h_2}(R_{h_1}+R_{h_2})}{(\sigma_1+\sigma_2)^2 + B^2\sigma_1^2\sigma_2^2(R_{h_1}+R_{h_2})^2} \tag{2.14}$$

In consequence of two-band conduction Eq. (2.14) indicates that both the Hall coefficient and the conductivity become dependent on the magnetic field even if these parameters in their separate conduction bands are independent of B.

In a weak magnetic field, such that the terms multiplied by B^2 in Eqs. (2.13) and (2.14) are negligible, the conductivity and Hall coefficient approach the low field limit

$$\sigma_0 = \sigma_1 + \sigma_2 \qquad R_{h_0} = \frac{\sigma_1^2R_{h_1} + \sigma_2^2R_{h_2}}{(\sigma_1+\sigma_2)^2} \tag{2.15}$$

For two-band conduction in which electrons and holes are the charge carriers, Eq. (2.15) can be expressed as

$$\sigma_0 = e(n\mu_n+p\mu_p) \qquad R_{h_0} = -\frac{1}{e}\cdot\frac{n\mu_n^2 - p\mu_p^2}{(n\mu_n + p\mu_p)^2} \tag{2.16}$$

where n and p are the electron and hole densities per unit volume and μ_n and μ_p are, respectively, the electron and hole mobilities. By defining the ratio of their mobilities as $b = \mu_n/\mu_p$ and that of the carrier concentrations as $c = n/p$ (ref. 4) the magnetic field-dependent Hall coefficient is

$$R_h = R_{h_0} \cdot \frac{1 - X\mu_n^2B^2}{1 + Y\mu_n^2B^2}$$

$$X = \frac{1 - c}{b^2c-1}, \qquad Y = \frac{1 - c}{1+bc} \tag{2.17}$$

The Hall angles for each band and the corresponding vector diagram of the current densities are represented in Fig. 2.2. A magnetic field dependence

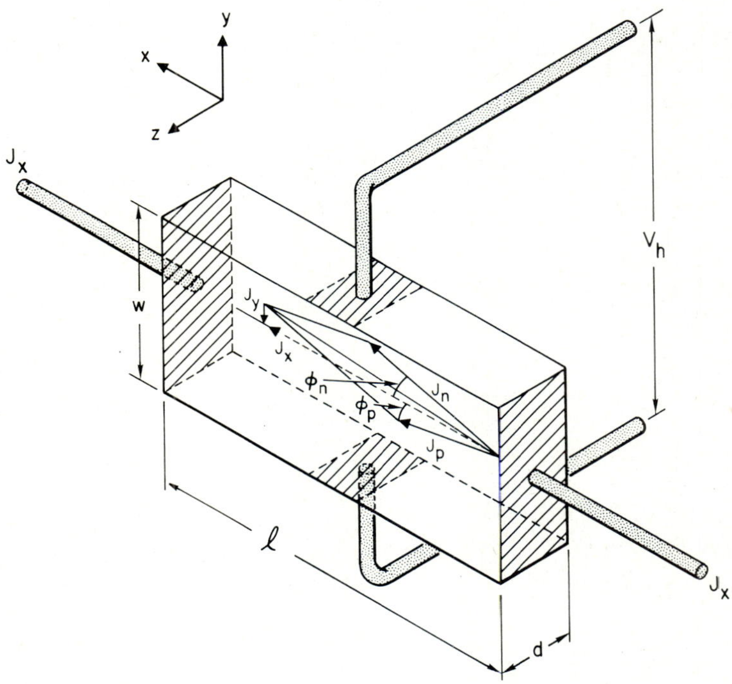

Fig. 2.2. The Hall effect in a rectangular plate in which electrons and holes are charge carriers. The magnetic field is oriented along z-axis. The vector diagram shows longitudinal electron and hole current densities J_n and J_p and the Hall current density J_y. The corresponding Hall angles are Φ_n and Φ_p.

of R_h or σ measured experimentally should not be considered as definitive proof of multi-band conduction. Inhomogeneities in a specimen can also give rise to $R_h(B)$ and $\sigma(B)$ and can lead to erroneous conclusions unless corroborative evidence is available that the specimen is isotropic and homogenous.

2.1 Measurement procedures

Hall effect measurements made on rectangular wafers such as shown in Fig. 2.1 require that the Hall voltage v_h be determined as a function of the magnetic field and DC input current in accordance with Eq. (2.8). The amplitude and polarity of v_h depends on the magnitude and polarity of i and on the direction of B relative to the coordinate axis. If terminal 1 is the positive polarity for a DC current i_+ and the direction of the steady-state magnetic field shown in Fig. 2.1 is defined as B_+ then the Hall voltage, measured between terminals 3 and 4 is such that terminal 3 is negative for an n-type material (dominant electronic conduction) and positive for a p-type material (dominant hole conduction). In an ideal situation these Hall electrodes are point contacts placed on an equipotential plane so that $v_h = 0$ for $B = 0$; in practice such conditions can be only approximated; in $B = 0$ there is usually a misalignment potential V_o between them. The potential measured between the Hall electrodes for $B \neq 0$ is, therefore, the algebraic sum of the misalignment potential and of the Hall voltage, $V = V_h + V_o$. The misalignment potential can be eliminated by measuring V for two opposite orientation equal amplitude magnetic fields while keeping all other parameters constant.

If the current polarity and magnetic f[illegible] orientations shown in Fig. 2.1 are indexed as i_+ and B_+, respectively, and [illegible]ntial difference measured between the Hall electrodes is $v(B_+i_+)$, [illegible] permutation of polarities, the Hall coefficient can be described i[illegible] Eq. (2.8) as

$$R_h(C/cm^3) = \frac{2.5x10^7 d}{|B|}\left[\frac{v(B_+,i_+)}{i_+} + \frac{v(B_+,i_-)}{i_-} \text{[illegible]} - \frac{v(B_-,i_+)}{i_+}\right] \tag{2.18}$$

The units of measurement usually encountered in practice and used in Eq. (2.18) are not a consistent set of units. They [illegible]pared in Table 2.1 against SI units.

The choice of a rectangular wafer with needle-like point-contacts is, in most cases, not a practical test structure for Hall effect measurements because of contact resistance fluctuations with time, temperature and other variables. The bridge-shaped test structures shown in Fig. 2.3 have some specific advantages for both Hall effect and resistivity measurements. Such bridge-shaped specimens can be made by photolithographic processing of epitaxial layers deposited on insulating or semi-insulating substrates or by cutting a similar contour ultrasonically from an appropriately oriented wafer and then making ohmic contacts to all of the terminals. A 6-arm bridge-shaped structure, such as that shown in Fig. 2.3, has been recommended as a standard (ref. 5). The companion asymmetric Hall bridge shown in Fig. 2.3 provides some advantages where space is at a premium. It can be used as a test structure on a portion of

TABLE 2.1.
Units of measurement.

Quantity	SI units (MKS System)	Units for Hall coefficient measurements
resistivity, ρ	ohm-m	ohm-cm
charge carrier density, n, p	m^{-3}	cm^{-3}
charge, e, q	C	C
mobility, μ	m^2/V-sec	cm^2/V-sec
Hall coefficient, R_h	m^3/C	cm^3/C
electric field	V/m	V/cm
magnetic flux density, B	T	T or gauss
current density, J	A/m^2	A/cm^2
potential difference, V	V	V
physical dimensions, ρ, w, d	m	cm

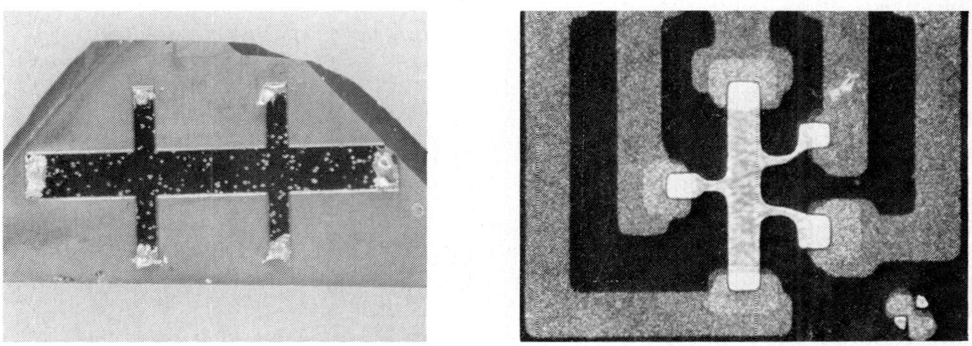

Fig. 2.3. Photolithographically-processed epitaxial GaAs layers made into bridge-shaped Hall effect structures. The device on the left is magnified 10X, that on the right, by 72X. Contacts shown are Au:Ge eutectic alloy covering edges of Hall structures and extending over semi-insulating substrates.

a wafer where it does not interfere with other devices and components. Furthermore, its misalignment voltage can be compensated by means of the circuit shown in Fig. 2.4b.

Compensation of the misalignment voltage of Hall test structures may be desirable for high-resolution Hall measurements in particular for the case where

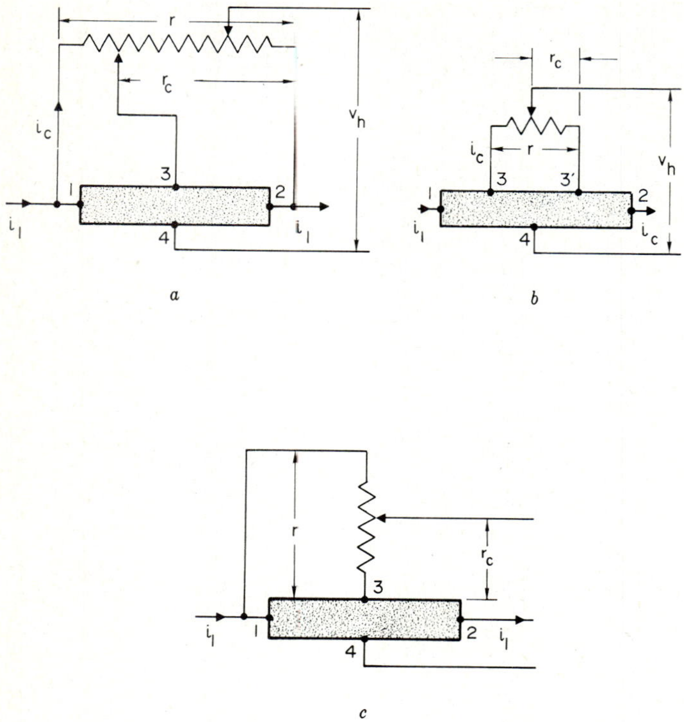

Fig. 2.4. External compensation circuits for eliminating the misalignment potential between Hall electrodes.

the Hall voltage is an order of magnitude smaller than the uncompensated misalignment voltage. The compensation circuits shown in Fig. 2.4 provide a temperature-independent compensation (ref. 6) provided that the input current i_1 is constant, that the branch current $i_c \ll i_1$, and that $i_c = i_1(r_c/r)R_i$. Here the resistances r and r_c are indicated in Fig. 2.4 and the internal resistance R_i is that present between contacts 1 and 2 for (a); between 3 and 3' for (b); and between 1 and 3 for (c). Only the compensation circuit shown in Fig. 2.4b is useful for large variations in magnetic field. The reason for this is that any magnetoresistance effects which influence R_i are the same as those which influence the test structure between contacts 1 and 2 at least to first order. For the other compensation schemes the misalignment voltage is itself a function of the magnetic field and, therefore, a potential source of magnetic field-dependent errors.

A variety of procedures have been used to make electrical contacts to specimens used for Hall effect measurements. In most cases the making of reliable ohmic contacts, particularly to large bandgap semiconductors, is an empirical art. The problem of contact resistance encountered with pressure-type contacts can lead to large random errors of measurement. Pressure contacts are common for routine room temperature measurements on metals and on narrow bandgap semiconductors. Pressure contacts are also used for high temperature Hall measurements, however, at low temperatures they tend to develop large contact resistances. Metallic points of tungsten or tungsten alloys are often used as pressure contacts; for high temperature applications or reactive materials graphite is used for pressure contacts.

Soldered contacts using indium or low temperature indium eutectic alloys can be used for making reliable contacts to low temperature bandgap semiconductors such as InSb or InAs. The soldering should preferably be done in an oxygen-free environment in order to prevent the formation of a "cold-solder joint." Indium-soldered contacts can also be made to n-type InP or GaAs with donor impurities of the order of $>10^{17}/cm^3$. Painted contacts made of a metallic powder such as silver or gold in a colloidal organic binder are used for rudimentary contacts on small bandgap semiconductors. For some of them air drying is sufficient; other types must be heated to temperatures of a few hundred degrees for volatilization of many of the binder constituents. Ohmic contact to p-type silicon can be made with aluminum deposited by vacuum evaporation followed by an alloying process; the specimen is heated just above 576°C the silicon-aluminum eutectic temperature. Since aluminum is a p-type dopant no surface barrier is formed with p-type silicon (ref. 7). For n-type silicon the doping level must exceed the solid solubility limit of aluminum ($\sim 2 \times 10^{19}/cm^3$), in order to over-ride the doping effect of the aluminum and thus prevent the formation of a p-n junction (ref. 8). Electroplated or electroless-plated contacts of copper, gold, or rhodium should be used with caution. At low temperatures they tend to have high and randomly variable contact resistances (refs. 9, 10). The most reliable ohmic contact made to n-type GaAs is the Au-Ge eutectic alloy developed by Braslau *et al.* (ref. 11). They used a source material in the form of 115 mg spheres of 88% Au and 12% Ge by weight placed together with a 0.5 dia. nickel wire $\sim$ 3.175 mm long on a tungsten evaporant source located $\sim$ 3 cm from the specimen to be coated in a vacuum chamber. The contents are evaporated completely producing an $\sim$ 0.3 μm thick eutectic alloy with the Ni primarily on its outer surface. The specimen is then alloyed in forming gas for $\sim$ 45 sec. in an oven held between 450 and 480°C. Nuese and Gannon (ref. 12) have described a process for making ohmic contacts to p-type GaAs by the vacuum deposition of an Ag-Mn alloy made of 96% by weight Ag and 4% Mn. The alloy is evaporated easily because Ag and Mn have roughly the same vapor pressures. The alloy is evaporated in vacuum to a thickness of $\sim$ 0.5 μm onto a GaAs substrate heated to $\sim$ 300°C. Sintering in hydrogen for 2 min. at a temperature of $\sim$ 550°C is desirable but it is not essential. Ohmic contacts can be made to GaAs, with acceptor concentrations greater than $10^{16}/cm^3$, on {100} and {111} surfaces. Ohmic contacts have also been made to high resistivity, ∿ 650 ohm-cm p-doped specimens and to p-doped $GaAs_xP_{1-x}$ ternary compounds.

A review of the basic theory of ohmic contacts and of their specific technology relevant to III-V compound semiconductors has been given by Rideout (ref. 13). He concludes that for the larger band semiconducting compounds no metal exists with a low enough work function which might yield a low contact barrier. Alloy regrowth, in-diffusion of a dopant contained in the specimen, epitaxial growth of a more heavily doped n- or p-layer, or ion implantation are promising techniques for the formation of ohmic contacts. Table 2.2 presents a selection of his collected data with particular emphasis on Hall effect measurements.

TABLE 2.2.

Ohmic contact technology for III-V compound semiconductors. [after V. L. Rideout, Solid-State Electron., 18(1975)541]

Compound	Conductivity type	Contact material	Deposition technique	Alloying temp., °C
AℓP	n	Ga-Ag	preform	150
AℓAs	n, p	In-Te	preform	160
	n, p	Au	preform	700
	n, p	Au-Ge	preform	
	n	Au-Sn	preform	
GaAs	n	Au-Ge(88:12)	evap.	350-450
	n	Au-Te(98:2)	evap.	500
	n	Au-Si(94:6)	evap.	300
	p	Au-Zn(99:1)	evap.	
	p	In-Au(80:20)	preform	
GaSb	p	In	preform	
	n	In	preform	
InP	n	Ag-Sn	preform, evap.	600
InAs	n	In	preform	
	n	Sn-Te(99:1)	preform	
InSb	n	In	preform	
	n	Sn-Te(99:1)	preform	

Six-arm bridge specimens can be made of various materials by means of an ultrasonically driven die, by sandblasting or spark erosion of the chosen contour by means of a template. For epitaxially-grown or diffused p-n junctions photolithographic techniques can be used to isolate the epitaxial layers. Such procedures are useful for making Hall test structures on silicon-integrated circuit wafers (ref. 14).

The essential components of apparatus for Hall effect measurements are: (a) a DC constant current source with provision for reversing its polarity; (b) an electrometer, galvanometer or high-impedance high-resolution digital or analog voltmeter with provision for reversal of its input polarity without affecting its accuracy or precision; (c) an electromagnet capable of providing an adjustable flux density from $\sim$ 0.05 T to $\sim$ 1.00 T with a homogeneity over the specimen volume of $\pm$ 1% and a precision and long-term stability of $\pm$ 1%. A nuclear resonance magnetic field probe or a precision-calibrated Hall generator should be available to monitor the magnetic field between the magnet poles in the course of the measurements. Adequate provision should be made for reversing the direction of the magnetic field. If a permanent magnet is used to provide the magnetic field then provision should be made for rotating the specimen 180° about its axis parallel to the direction of current flow. In other respects the provisions and specifications for resistivity measurements described in Section 1.1 apply here as well.

Figure 2.5 shows the disposition of apparatus (ref. 5) intended for Hall effect measurements made on 6-arm bridge specimens. The double-pole double-throw switch CR is used to reverse the polarity of the current provided by the constant current source C. The potentiometer, voltmeter, or electrometer D is used to measure the Hall voltage across the terminals 3 and 4 or across terminals 5 and 6. The choice of either terminal pair is made by means of a low resistance ganged switch S; the switch can also be used to shunt the voltmeter across the standard resistor R_{std} in order to measure the potential difference v_i; this provides a quantitative measurement of the input current to the test structure $i = v_i/R_{std}$. The calculation of the Hall coefficient can then proceed in the sequence and in accordance with Eq. (2.18). If the Hall coefficient R_{h_1} determined on the first set of Hall electrodes differs by more than 10% from R_{h_2}, that determined on the second set of Hall electrodes, then the measurements should be discarded. They indicate an unacceptably large inhomogeneity. If the results are consistent to within 10% then the average Hall coefficient may be calculated from $\langle R_h \rangle = (1/2)\cdot(R_{h_1}+R_{h_2})$. Hall effect measurements on metals are made difficult by their small Hall coefficients; Hall voltages are usually in the μV range. A high-sensitivity electrometer can be used as a microvolt detector and a differential amplifier interpolated between the Hall contacts and

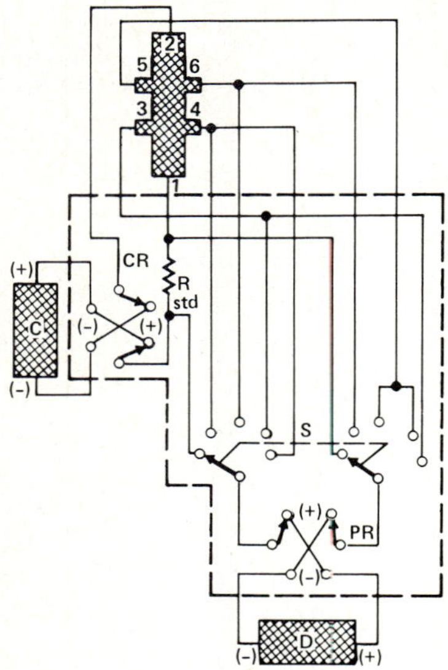

Fig. 2.5. Apparatus used for Hall measurements on 6-arm bridge specimens. C is a constant current source, CR a current polarity reversing switch, D is an electrometer, galvanometer or voltmeter, PR is a switch used to reverse polarity of voltage input to D, R_{std} is a standard resistor used to monitor current input to specimen, S is the switch used to select a pair of Hall output terminals for measuring Hall voltage or potential drop across R_{std}. [after American Soc. Testing Mat., F-76-78(1968)1785]

the electrometer allows both the specimen and the electrometer to be grounded. This reduces the residual common-mode error voltages across the electrometer.

2.2 Systematic errors

The calibration of the apparatus and fluctuations in experimentally measured parameters as well inherent or external noise sources limit the accuracy of Hall measurements. There are some specific errors of a systematic nature related to temperature gradients present in specimens subjected to Hall measurements. These have not been described in the earlier sections of this chapter in which the Hall effect is treated as an isothermal phenomenon without considering thermal or thermomagnetic effects. These are described in detail in Chapter 9. If a longitudinal temperature gradient ∇T_x is present in a specimen then a transverse voltage v_N appears in it by virtue of the Ettingshausen-Nernst effect and also a transverse temperature gradient by virtue of the Righi-Leduc effect. A transverse temperature gradient ∇T_y can generate a thermoelectric error voltage between the Hall electrodes if the specimen and its Hall contacts have a significant thermoelectric power between them and if the thermoelectric power is itself a significant function of the magnetic field. In most except ferromagnetic materials the latter mechanism is usually of negligible significance.

A detailed theoretical treatment of thermomagnetic phenomena has been given by Jaggi (ref. 15); adiabatic charge carrier transport has been treated in further detail in the monographs of Jan (ref. 16) and Beer (ref. 17). Figure 2.6 shows some of the principal thermomagnetic phenomena and their electrical,

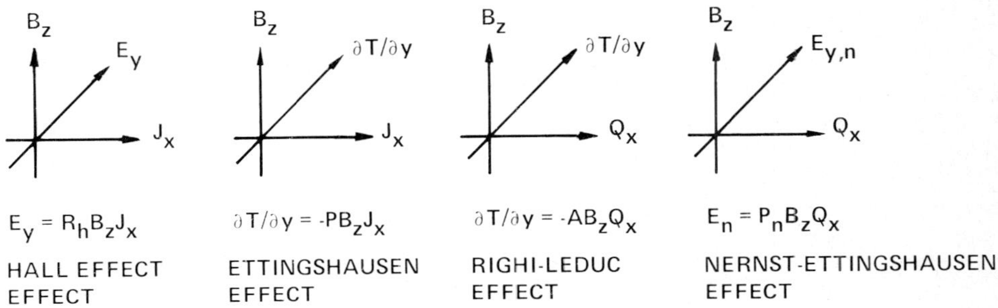

Fig. 2.6. Hall effect and principal thermomagnetic effects shown with their their orientation in cartesian coordinates. Q is the heat flux and $\partial T/\partial y$ is thermal gradient in the direction of Hall field.

magnetic and thermal components. The Ettingshausen-Nernst and Righi-Leduc effects can be eliminated by preventing longitudinal heat flow. However, the Ettingshausen effect is a function of electron flow in a magnetic field; consequently, the transverse temperature gradient $\partial T/\partial y$ is always present under usual experimental conditions and the Hall voltage is then adiabatic.

To avoid errors introduced by the Nernst-Ettingshausen and Righi-Leduc effects it is necessary to reduce or eliminate longitudinal heat flow such as may arise from localized Joule heating in an inhomogenous specimen from different amounts of heat transported along each of its current leads from the external environment and from the Peltier effect produced by differential thermoelectric heating of the junction between the current leads and the specimen. These effects can be reduced by providing a uniform thermal heat transfer, by thermal conduction to a heat sink, and by enclosing the specimen in a dewar thermostat in which a heat exchange gas such as helium is present. Such techniques have been described by Hurd (ref. 18) and by Alderson *et al.* (ref. 19). It is not feasible, as a rule, to eliminate all of the thermomagnetic effects by choosing an appropriate specimen symmetry or by the control of its environment. The errors introduced by such thermomagnetic effects on Hall measurements can be eliminated by a permutation of the current polarity and magnetic field orientation as described in Section 2.1 and Eq. (2.18). Let v_h be the true Hall voltage, v_o the misalignment voltage, v_e the Ettingshausen voltage, v_r the Righi-Leduc voltage, v_N the Nernst-Ettingshausen voltage and, v_T a thermoelectric error voltage generated by an externally imposed thermal gradient. Then Table 2.3 provides a list of the actual potentials, v_m, measured between the Hall contacts for the given permutations. Combining the measurements from Table 2.3 leads to

$$\frac{v_{m1} + v_{m4} - v_{m2} - v_{m3}}{4} = v_h + v_e = v_h + P\Pi \tag{2.19}$$

where P is the Ettingshausen coefficient and Π is the thermoelectric power of the specimen.

TABLE 2.3.

Permutations of current and magnetic field for Hall measurements.

Mode	Potential measured between Hall terminals
B_+, i_+	$v_{m1} = v_h + v_o + v_e + v_T + v_N + v_r$
B_-, i_+	$v_{m2} = - v_h + v_o - v_e - v_r + v_T - v_N$
B_+, i_-	$v_{m3} = - v_h - v_o - v_e + v_r + v_T + v_N$
B_-, i_-	$v_{m4} = v_h - v_o + v_e - v_r + v_T - v_N$

In order to determine the Hall coefficient from such Hall measurements the thickness d of a wafer must be determined by means of mechanical probe measurements, by the use of a calibrated microscope equipped with a filar eyepiece, or a precision vernier caliper. An optimum specimen thickness d_{opt} might be chosen in order to reduce the root mean square error in v_h and the rms error in d (ref. 18),

$$d_{opt} = (\Delta d/\Delta v_h)^{1/2} (R_h B i) \tag{2.20}$$

For a given R_h an optimum thickness exists because a decrease in d increases v_h thus increasing the accuracy with which v_h can be determined; however, by decreasing d the accuracy of the thickness measurement is reduced and the uniformity of d may also be affected. A minor source of systematic error is the inhomogeneity of the magnet as shown by de Mey (ref. 20) using an integral equation technique. As a rough approximation he found that if the magnetic field has a homogeneity better than x% then the error in the Hall voltage is less than 0.5x%. Koppe and Bryan (ref. 21) have considered an extreme case in which the magnetic field B, oriented as shown in Fig. 2.1 had a distribution, B(x), which decayed to half its peak value at a distance along the x-direction equal to half the width of the specimen; for such a case, the ratio of the Hall voltage v_{hi} to that measured in a homogenous magnetic field is $v_{hi}/v_h \simeq 0.8$. The fractional error in the value of the Hall coefficient can be expressed in terms of Eq. (2.8) in the form

$$\frac{\Delta R_h}{R_h} = \frac{\Delta v_h}{v_h} + \frac{\Delta d}{d} + \frac{\Delta i}{i} + \frac{\Delta B}{B} \tag{2.21}$$

For a specimen chosen to have the optimum thickness by means of Eq. (2.20) the fractional error in R_h is

$$\frac{\Delta R_h}{R_h} = \frac{\Delta i}{i} + \frac{\Delta B}{B} + 2\left(\frac{\Delta d \cdot \Delta v}{R_h i B}\right)^{1/2} \tag{2.22}$$

If the length of a rectangular wafer is not much greater than its width then a correction factor must be included in Eq. (2.8) to account for the reduction in Hall voltage produced by the electrostatic shorting of the Hall field by the current electrodes. Various analytical techniques (refs. 23-28) have been used to calculate the perturbation of the equipotentials and current streamlines in a rectangular plate produced as a function of its length to width, ℓ/w ratio, the extent of its contacts, and the Hall angle θ,

$$v_h = R_h \cdot (Bi/d) \cdot F(\theta, \ell/w) \tag{2.23}$$

The geometrical correction factor, $F(\theta,\ell/w)$ shown in Fig. 2.7, is normalized in terms of v_h for $\ell/w \rightarrow \infty$. Results obtained experimentally and justified theoretically by Drabble *et al.* (ref. 29) indicate that it is possible to obtain Hall voltages comparable (within 5%) to rectangular specimens with $\ell/w >> 1$ provided that the metal contacts of specimens with $\ell/w < 1$ are replaced by a material whose resistivity is much larger than that of the specimens and that these contacts have lengths that are at least equal to their widths.

2.3 Hall effect of arbitrary contour specimens

The Hall coefficients of a lamella with an arbitrary contour and a thickness δ can be determined by the method of van der Pauw (refs. 30, 31) provided that: (a) the lamella is simply connected and its thickness is uniform; (b) the contacts are sufficiently small to be considered as point contacts; (c) the magnetic field is perpendicular to the surface of the lamella. Figure 2.8 illustrates schematically such a lamella and suggests the distribution of the current streamlines considered to be invariant in a magnetic field. A current i is introduced at point contact M and extracted at contact O. The potential difference v is measured in B = 0 between point contacts N and P and the resistance is then calculated, $R_{MO,NP} = v/i$. If a transverse magnetic field is now applied

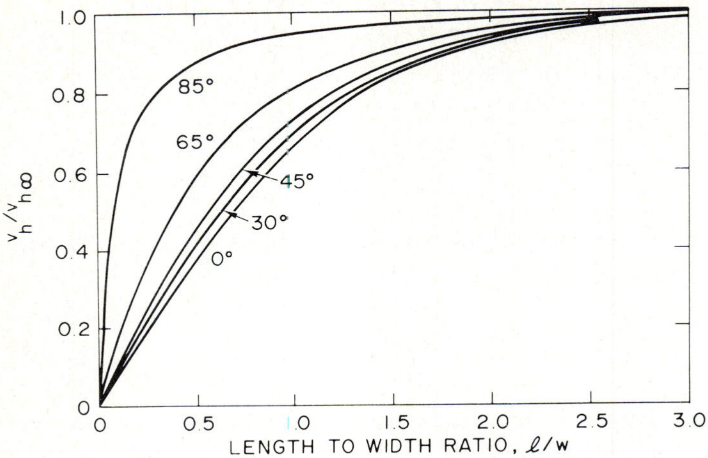

Fig. 2.7. Hall voltage v_h normalized in terms of $v_{h\infty}$, that of an infinite length rectangular wafer as a function of its length to width ratio with the angle θ as a running parameter.

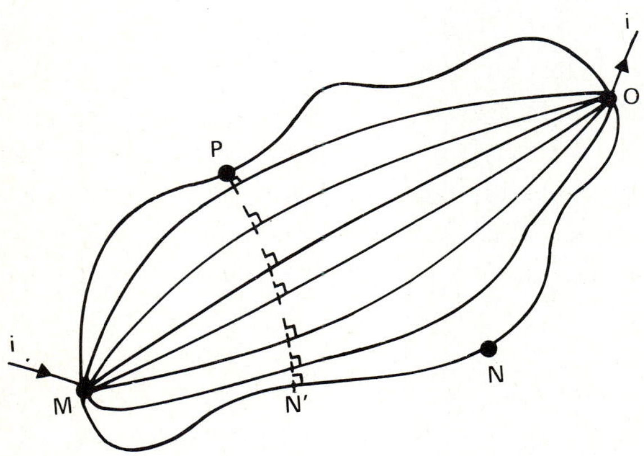

Fig. 2.8. A lamella of arbitrary contour with peripheral point contacts. A current i introduced and extracted at points M and O, respectively. The current streamlines are shown schematically and dashed line between P and N' is an equipotential.

to the lamina then $R_{MO,NP}$ changes by an amount $\Delta R_{MO,NP}$ and the Hall coefficient is

$$R_h = (\delta/B)\cdot\Delta R_{MO,NP} \tag{2.24}$$

The magnetic field produces a rotation of the equipotential planes in the lamella. The electric field and current density vectors are no longer collinear and the change in potential difference between P and N in Fig. 2.8, $\Delta(V_P-V_N)$ can be obtained by integrating the Hall field E_h between P and N', across the lamella, along a path s orthogonal to the current streamlines and then along a streamline between N' and N. The latter makes no contribution to the integral; therefore

$$\Delta(V_P-V_N) = \int_P^{N'} E_h ds = R_h B \int_P^{N'} J ds = R_h B i_{MO}/\delta \qquad (2.25)$$

which leads directly to Eq. (2.24).

The distribution of current streamlines and equipotentials in a disc of thickness δ with peripheral point contacts placed on orthogonal diameters is shown schematically in Fig. 2.9. For such a structure the Hall coefficient R_{h_1} can be determined by one measurement in which the current and voltage contacts

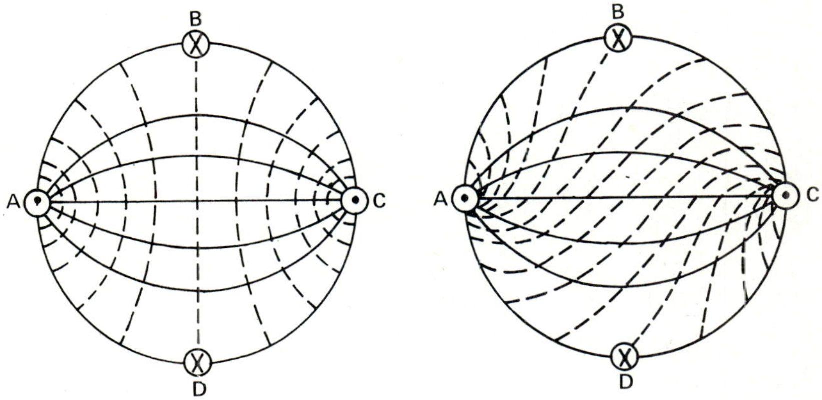

Fig. 2.9. Current streamlines (continuous lines) and equipotentials (dashed lines) in a disc; current electrodes are identified by dots, voltage probes by crosses. On the left B=0, on the right B is normal to plane of page.

are opposite each other. Thereafter a Hall coefficient R_{h_2} can be determined with the Hall contacts and current contacts interchanged. The magnetic field must be small enough to keep any magnetoresistance-induced nonlinearities between v_h and B to a negligible value. If R_{h_1} is ±10% of R_{h_2} then the average Hall coefficient is $\langle R_h \rangle = (1/2)(R_{h_1}+R_{h_2})$; if the discrepancy between R_{h_1} and R_{h_2} is greater than 10% then the specimen may be too inhomogenous for further investigation.

If the contacts are of finite size or if one or more of the point contacts are not located on the circumference of the disc then errors are introduced in the Hall measurements. These have been evaluated by van der Pauw (ref. 30) in terms of Fig. 1.15; the fractional error in R_h for (a) $\Delta R_h/R_h \simeq 2d\ (\pi^2 D)^{-1}$; for (b) $\Delta R_h/R_h \simeq -4d\ (\pi D)^{-1}$, and for (c) $\Delta R_h/R_h \simeq -2d\ (\pi D)^{-1}$. If none of the contacts are ideal then, to a first approximation, the total error is the sum of the errors of each contact. The use of the clover leaf-shaped specimen, shown in Fig. 1.15(d), reduces these errors. It provides a relatively larger Hall voltage than that of a bridge-shaped specimen for the same amount of heat dissipation. It has greater mechanical strength and is easier to make than bridge-shaped specimens. Jacobs and de Mey (ref. 32) have investigated the dependence of the Hall voltage measured on a clover leaf-shaped structure, such as shown in Fig. 2.10, as a function of contact size and geometry by solving Laplace's equation by an expansion technique. With the potential ψ expressed as a power series in the Hall tangent μB,

$$\psi = \psi_0 + \mu B \psi_1 + (\mu B)^2\ \psi_2 + \ldots \qquad (2.26)$$

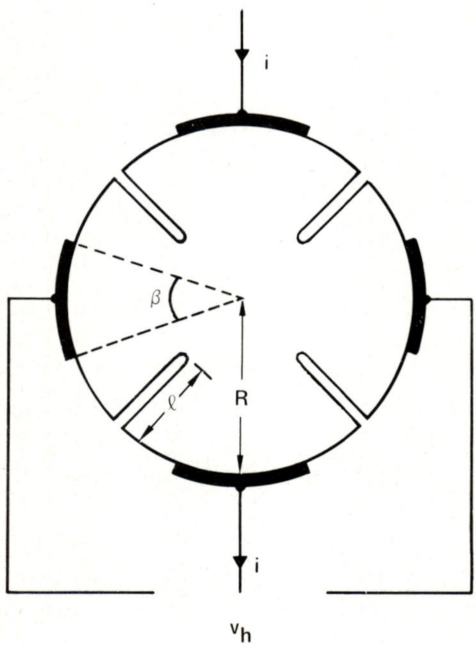

Fig. 2.10. Clover leaf-shaped structure for Hall effect measurements shows slots of length ℓ cut into disc of radius R. Contacts subtend angle β. [after E. Jacobs and G. de Mey, Electron. Lett., 12(1976)4]

The Hall voltage contains only odd powers of μB because of geometrical symmetry and, therefore,

$$v_h = v_{h_1} \cdot \mu B + v_{h_3} (\mu B)^3 + \ldots \qquad (2.27)$$

The first term of Eq. (2.27) is the conventional Hall effect and the higher order terms represent the nonlinear dependence of v_h on B. The calculations of Jacobs and de Mey (ref. 32) indicate the size of the contacts has a strong effect on the geometrical correction factor of the linear Hall effect and the nonlinearity of the measured Hall voltage of a clover leaf-shaped specimen is also a function of the size of the contacts and of the angle β. For small contacts the slots provide an improvement only if $\beta \to R$; however they do provide a decrease of the nonlinearity of v_h vs. B inherent in such measurements.

An experimental investigation made by van Daal (ref. 33) evaluated the errors introduced in the measurement of the Hall coefficient, $\Delta R_h/R_h$ as a function of the ratio of contact size to the diameter of disc-shaped specimens. He also determined the change in $\Delta R_h/R_h$ after radial slots were cut into the same specimens effectively converting them into clover-leaf configurations. His results, reproduced in Fig. 2.11 show a substantial reduction in the $\Delta R_h/R_h$ of a clover-leaf configuration compared to that of the same disc with non-ideal contacts on its periphery.

A symmetrical peripherally-contacted cross such as shown in Fig. 1.20 can be considered to be a special case of a clover-leaf structure. The analytical and experimental investigations of Haeusler and Lippmann (ref. 34) have shown that for a ratio of S/A=1 (refer to Fig. 1.20) the error in the linearity of v_h with respect to B is $\sim$ 0.3% for B=1T and $\sim$ 0.8% for B=2T for large Hall angles. Further investigations made by de Mey (ref. 35) indicate that R_h is essentially independent of B for Hall tangents < 0.10 and that the error in R_h is primarily of geometrical origin except in very high magnetic fields where magnetoresistance contributions can no longer be neglected (ref. 36). The symmetrical cross provides several advantages for low systematic error resistivity (Section 1.6 and Fig. 1.20) and Hall measurements. These include the simplification of lead attachment to the larger contacts of this structure, smaller current densities compared against the same current applied to rectangular or circular configurations, a reduction in Joule heating due to the better thermal conduction through the contacts of the cross compared to bridge-shaped specimens, smaller thermally-induced error voltages due to a reduction in thermal gradients and a reduction in contact-related noise.

Chwang *et al.* (ref. 37) have investigated the errors inherent in Hall measurements made on square specimens having triangular or square contacts. They indicated that the errors due to finite-size contacts can be interpreted as the sum of two contributions: (a) one due to the electrostatic shorting of the Hall field by the current electrodes, and (b) that due to the current shorting by the Hall electrodes; these cause a current loss around the edge of

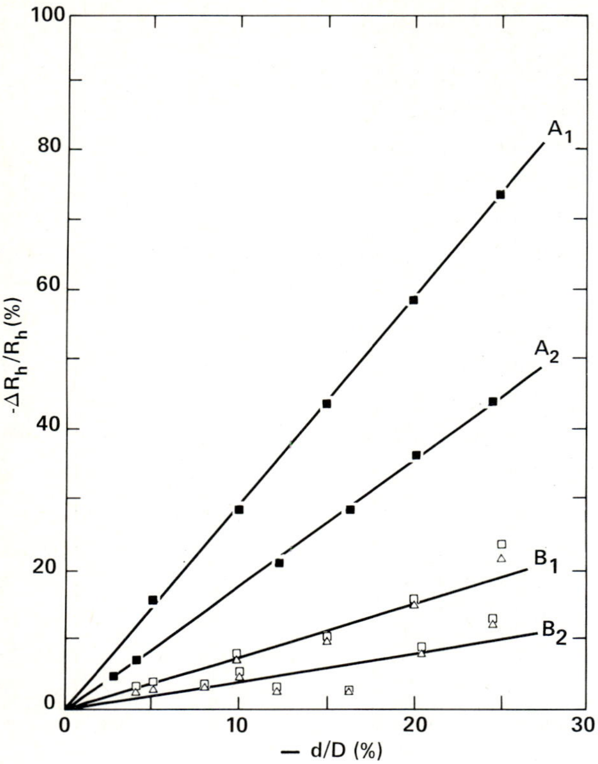

Fig. 2.11. Systematic errors in Hall coefficient $-\Delta R_h/R_h$ as a function of ratio of contact diameter d to disc diameter D. Line A_1 is characteristic of p-type Ge disc D=20.0 mm with thickness δ=0.20 mm, line A_2 is characteristic of similar disc D=24.5 mm whose thickness δ=1.3 mm. Lines B_1 and B_2 correspond to respective discs after slots were cut into them approximately midway along orthogonal diameters. (after H. J. van Daal, Philips Res. Repts., Suppl. No. 3,(1965)8]

the contacts. Experimental measurements made by the use of an electrolytic tank having an analogous contour to that of contact-equipped square specimens were used to estimate an upper bound for the errors corresponding to case (b) (for zero Hall angle). The total correction factor is about twice the value of the zero Hall angle current correction when the voltage and current contacts are identical. Their calculated Hall voltage correction factors are shown in Fig. 2.12 as a function of the ratio of the contact dimension to specimen size and as a function of the Hall angle θ. Although this specimen configuration does not yield the inherent linearity of v_h vs. B of clover-leaf or symmetrical cross-shaped structures, it provides significant advantages for an evaluation

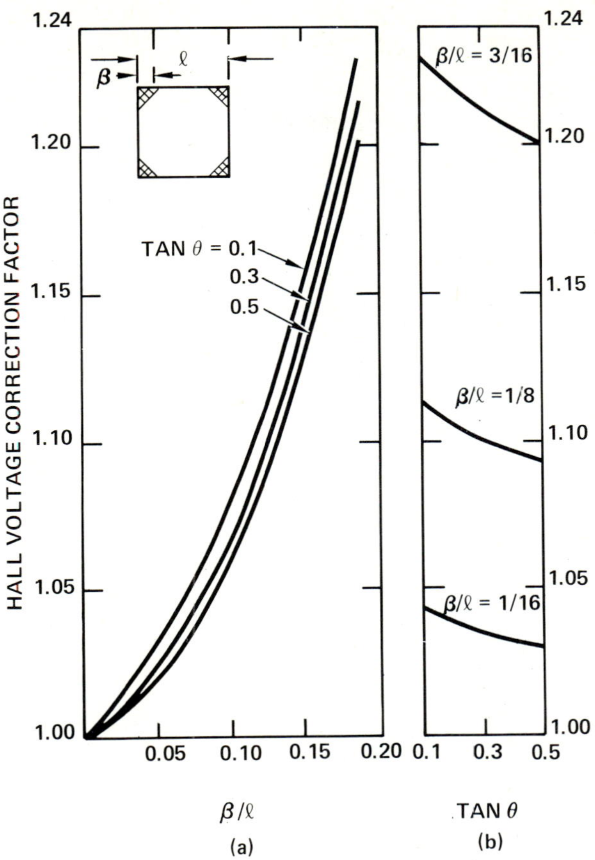

Fig. 2.12. (a) Hall voltage correction factor as a function of ratio of contact dimension β to side ℓ of square specimen with Hall tangent as running parameter; (b) correction factor as a function of tangent of Hall angle for selected values of β/ℓ. [after R. Chwang, B. J. Smith and C. R. Crowell, Solid-State Electron., 17(1974)1217]

of the conductivity and Hall coefficient of epitaxial layers grown on semi-insulating substrates without subjecting them to extensive and potentially damaging photolithographic procedures.

2.4 Apparatus and techniques

The measurements described in the previous sections of this Chapter require a stable DC current source and a high-performance galvanometer or electrometer for measuring low level Hall potentials. A simple and inexpensive current source capable of providing currents in the range from $i_0=5\times10^{-10}$ to 5×10^{-3}A with a stability of at least 0.01%, for resistive loads from $R=i_0/5$ to 0 was described by Leach (ref. 38) and is shown schematically in Fig. 2.13. The current source is isolated from ground and shielded by mounting it on teflon pillars in a Faraday cage. Isolation of the current source from ground by an impedance much larger than that of the specimen to be measured is considered desirable so that if an electrometer is used to measure v_h then its guard can be grounded. The power source consists of two 9V batteries drawing a quiescent current of $\sim$ 7 mA. The output current is sensed across a resistor selected by the switch SW and fed to one of the input ports of the integrated-circuit differential amplifier 741. The other input port has a reference signal generated by the Zener diodes applied to it and the error signal is amplified and the servo loop is closed between the amplifier and transistor FET-3. Transistors FET-1 and FET-2 should be selected as a matched pair and the $10^4\Omega$ potentiometer is used to provide a continuously variable output current.

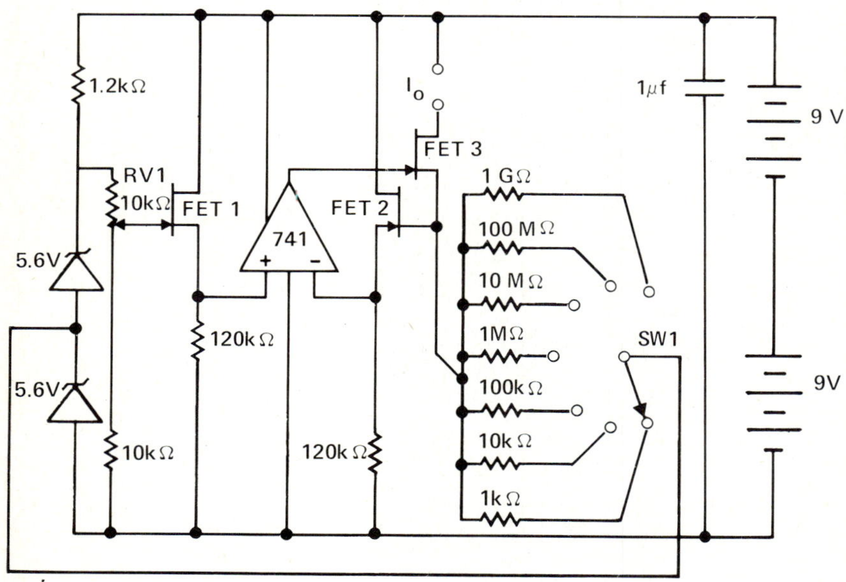

Fig. 2.13. A constant current generator; all field effect transistors are type N 3819. [after M. F. Leach, J. Phys. E., Sci. Instr., 10(1977)879]

Potentiometric methods are used for the measurement of small Hall voltages. A potentiometer can be used to produce a potential which matches the Hall voltage. The difference between them is then amplified for ease of observation. Such a potentiometer is shown in Fig. 1.3. Similar apparatus with a DC output can be coupled either to a mechanical chopper amplifier or a photoelectric galvanometer amplifier with a precision of $\sim 10^{-9}$V. Clark and Ficket (ref. 39) have described apparatus suitable for the detection of low level Hall voltages with a precision of $\sim$ 25 pV. A photoelectric galvanometer amplifier with internal feedback and a current gain of $\sim 10^5$ is used; after further amplification the output signal is converted to a frequency by a voltage-to-frequency converter and then measured by an electronic counter. Inherent advantages include the signal averaging properties of the oil-damped galvanometer, the precision inherent in voltage-to-frequency conversion, and an adjustable counting period. A mechanical mirror adjustment on the galvanometer allows measurements to be made in the presence of spurious voltages of the order of μV and allows switching of the input signal, since the galvanometer deflections are essentially linear for small deflections. An internal nanovolt source is used as a reference and for calibration purposes. A major advantage of this Hall voltage sensor system is that it can be operated in the presence of large, constant, or constantly drifting thermally-induced error voltages.

Superconductive galvanometers or superconductive parametric amplifiers (ref. 40) can be used to achieve Hall voltage sensitivities of 10^{-14}V for low resistance specimens such as metals measured at cryogenic temperatures. Even higher sensitivities can be obtained (ref. 41) with galvanometers using Josephson tunneling junctions. Hall effect measurements can also be made, on occasion, more advantageously than with DC by using a synchronous time variation of the measurement parameters. Such measurements can be made by means of alternating current supplied to a specimen with a static magnetic field, or, as an alternative, DC supplied to the specimen and a transverse time-variable magnetic field or a double AC technique in which both the specimen current and magnetic field are time variables. Techniques based on AC specimen currents and static magnetic fields were investigated and used by Donoghue and Eatherly (ref. 42) and Lavine (ref. 43). The latter employed this technique to perform Hall measurements on ferromagnetic metals and semiconductors as a function of temperature between 77°K and 900°K with a Hall voltage resolution of one part in 10^5. The use of AC provides some advantages in terms of the apparatus used for the detection, amplification, and for the use of noise rejection techniques such as sampling, synchronous detection and "lock-in" techniques. However, if the misalignment voltage of a specimen is too large then the AC amplifier may be driven into saturation. It is necessary to reduce the misalignment voltage to a minimum by external circuit compensation. A temperature-dependent drift of the misalignment voltage can complicate the compensation procedure. Mechanical vibration of the specimen or its leads can induce spurious voltages in the Hall circuit; in order to minimize these it is essential that the specimen, its mounting and leads be mechanically stable.

One of the principal advantages of the alternative method, that of a DC input to the specimen and an AC-driven magnetic field, is that the misalignment voltage for this situation is a DC parameter while the Hall voltage appears at the frequency of the magnetic field. Narrow bandwidth detection or synchronous detection of the Hall voltage can be used to eliminate error signals thus improving the resolution of the apparatus. However, the AC magnetic field is coupled inductively to the leads at the current input and Hall voltage output of a specimen. For a magnetic field of $\sim$ 0.5T, a frequency of 1 Hz and a Hall voltage loop of $\sim$ 0.3 mm^2 an error signal of $\sim$ 1 μV is obtained in the Hall voltage circuit. The elimination of such inductive pick-up is, as a rule, a complex process which requires attention to detail in design of the experiment.

A bridge circuit suitable for the AC measurement of the Hall coefficient of specimens having large thermoelectric and thermomagnetic error voltages has been described by Parodo and Zedda (ref. 44). The circuit shown schematically in Fig. 2.14 is a modification of Goldsmid's bridge (ref. 45) developed specifically

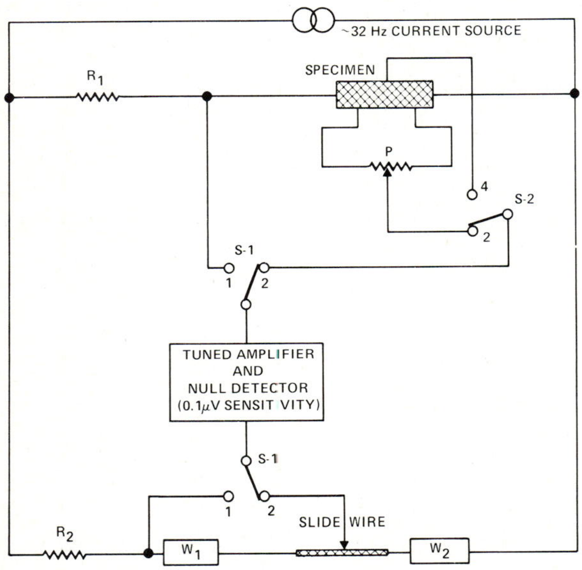

Fig. 2.14. AC-bridge circuit for Hall measurements. A DC magnet is transverse to the specimen whose misalignment voltage is minimized by potentiometer P. W_1 and W_2 are resistance decade boxes adjustable in 1 ohm steps to 10^6 ohms; slide-wire is adjustable in 10^{-3} ohm steps. [after A. Parodo and A. Zedda, J. Sci. Instr., J. Phys. E. Ser. 2,2(1969)748]

for the measurement of the resistivity of materials in the presence of large contact resistances. The switch S-1 connects the tuned amplifier and null detector, used to balance the bridge, first to terminals 1 of the circuit. A preliminary bridge adjustment is then made by means of the decade resistance boxes W_1 and W_2 in order to bring $W_1/W_2 \simeq 1$. Thereafter, with the switch S-2 connected to terminals 2 of the circuit the precision slidewire is adjusted and its position is noted. Switch S-2 is then turned to position 4 and the slidewire is readjusted for balance. The difference of the two positions of the slidewire is defined as ΔS. With the static transverse magnetic field B acting on the specimen its Hall coefficient is $R_h = \delta R_1 \cdot \Delta S \cdot (BR_2)^{-1}$. The precision of measurement of this bridge circuit was estimated (ref. 44) to be $\sim$ 1 part in 10^6.

A time-variable magnetic field and DC specimen input current were used by Lomas *et al.* (ref. 46) for Hall effect measurements. The magnetic field was provided by an electromagnet capable of producing magnetic fields up to 0.5T in a 2.54 cm air gap. It was driven by a variable 1 to 3 sec./Hz symmetrical square wave current with a polarity reversing switch so arranged that the magnet armature is short-circuited before the power supply polarity is reversed. To prevent any damage to the microvoltmeter used to measure the Hall voltage, due to switching transients, arrangements were made to short-circuit it during the switching cycle by a relay triggered from a pulse derived from the switching circuit; after polarity reversal is completed the relay contacts are opened and Hall voltage is recorded. Inductive pick-up in the current loop and Hall voltage circuit is circumvented by gating the Hall sensor circuit in synchronism with the magnetic field and drift of the misalignment voltage is not significant due to the synchronous time-variable character of the Hall voltage. This apparatus (ref. 46) was reported to be capable of resolving Hall voltages of the order of 10^{-7}V.

In the double AC method a sinusoidal current with an angular frequency ω_1 is used to drive the specimen and a different sinusoidal current of angular frequency ω_2 is used to provide the magnetic field. In consequence, the Hall voltage may be expressed by

$$v_h = R_h \frac{i_{max} B_{max}}{2\delta} [\cos(\omega_1 - \omega_2)t - \cos(\omega_1 + \omega_2)t] \tag{2.28}$$

where the specimen current is $i = i_{max} \sin\omega_1 t$ and the magnetic field $B = B_{max} \sin\omega_2 t$ with δ the thickness of the specimen. The Hall specimen behaves as a suppressed carrier modulator and the Hall voltage has two sidebands: $(\omega_1 - \omega_2)$ and $(\omega_1 + \omega_2)$ with each of them representing half the amplitude of the equivalent DC Hall voltage (ref. 48).

Russell and Wahlig (ref. 48) used a 70 Hz Hall current and a 60 Hz magnetic field to perform double AC Hall measurements with the Hall voltage detected at ν=10 Hz, the difference frequency. It is desirable to have the two sidebands far apart in the frequency domain in order to minimize intermodulation errors. The frequency used for the magnetic field is limited by the rise of the impedance of the magnet armature with frequency. This, in turn, makes the peak attainable magnetic field a function of frequency. Lupu *et al.* (ref. 49) developed a variation of the double AC Hall measurement method in which sequential double phase-sensitive detectors are used. These are synchronized with the input current frequency and also with the frequency of the magnetic field. The relative phase of the Hall voltage can be determined with respect to either of the two excitation signals. This permits a differentiation between spurious components and the actual Hall voltage. The latter appears at the frequency of the input current. With the magnetic field varied at 2 Hz and the input current with a frequency of 510 Hz the Hall voltage is mixed with a reference signal of 510 Hz and the signal is then phase-detected against a reference signal of 2 Hz thus extracting the Hall voltage. A detailed analysis of the spurious signals which can arise in this type of double AC Hall measurement have been given by McKinzie and Tannhauser (ref. 50). They also provide a repertory of diagnostic techniques for distinguishing between the real Hall voltage and the various error signals. A versatile double AC Hall and resistivity van der Pauw-type measurement apparatus suitable for use with large specimen current and low magnetic fields of $\sim$ 0.02T(rms) and with a large sideband separation between the frequencies of the current and magnetic fields was given by McLevige *et al.* (ref. 51). Its components are shown in the block diagram of Fig. 2.15. The input current frequency, f_1=1 kHz and that of the magnetic field, f_2=250 Hz provide a large enough separation of the sidebands for suppression of spurious intermodulation products. The Hall voltage is detected at the difference frequency

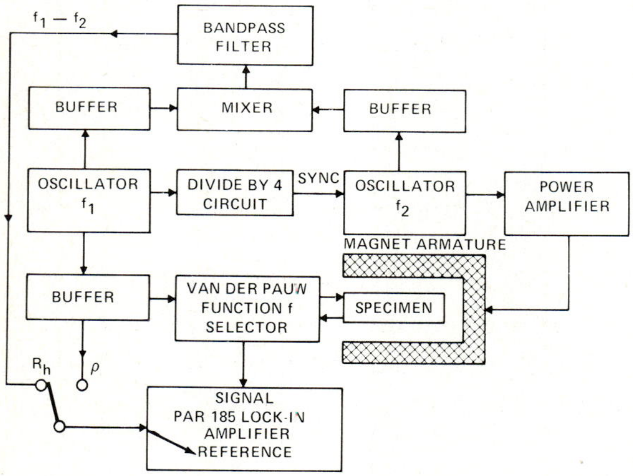

Fig. 2.15. Schematic diagram of double AC Hall measurement apparatus using van der Pauw's method. [after W. V. McLevige, P. K. Chaterjee and B. G. Streetman, J. Phys. E. Sci. Instr., 10(1977)335]

f_1-f_2=750 Hz. A substantial reduction in noise is obtained by the use of buffer stages between the mixer and the oscillators, shown in Fig. 2.15, by control of the spectral purity of the reference waveform, by careful shielding of all leads, and by the use of 3 stages of tuned active filters to generate the 750 Hz reference signal. The high dynamic range of the PAR 186A "lock-in" amplifier used to detect the Hall voltage allows the use of relatively large input currents to the specimen without saturating its input channel.

2.5 Microwave Hall effect

The classical theory of the interaction of high frequency and microwave electromagnetic radiation with matter treats a material in which there are drifting charge carriers as a medium with a high frequency conductivity, a dielectric permittivity ε and a magnetic permeability μ_m. In the presence of transverse electric and magnetic vectors the tensor conductivity can be expressed in the form (ref. 52)

$$\hat{\sigma} = \frac{\sigma}{(\omega_c\tau_r)^2+(1+i\omega\tau_r)^2}\begin{vmatrix} (1+i\omega\tau_r) & \omega_c\tau_r \\ -\omega_c\tau_r & (1+i\omega\tau_r) \end{vmatrix} \qquad (2.29)$$

where ω_c is the radian cyclotron frequency, $\omega_c=(e/m^*)\,B$; here m^* is the effective mass of the charge carriers, and B is the transverse magnetic field; the imaginary, $i=\sqrt{-1}$, and τ_r is an isotropic free charge carrier relaxation time.

Suppose that the electric field of a plane-polarized wave, in which a specimen is immersed, generates a collinear current density vector in it. Then this current may interact with a DC transverse magnetic field to generate a Hall electric field in a mutually perpendicular direction of the same frequency as that of the initial electric field. The Hall field when combined vectorially with the applied electric field causes a rotation of the plane of polarization of the net electric vector. There is a progressive increase in this angle of rotation as the wave propagates through the specimen. The effect of the Hall field can be treated in a manner similar to that of Faraday rotation of a gyromagnetic medium in which right- and left-handed circularly-polarized components of the electromagnetic wave propagates through a material with different velocities thus giving it an equivalent tensor permittivity.

The experimental investigations of Barlow and Kataoka (ref. 53) made on n-type Ge with a resistivity of 11 ohm-cm and on a p-type Ge wafer with a resistivity of 15 ohm-cm and a relative permittivity of 16 at 0.36 GHz, using the apparatus shown schematically in Fig. 2.16, indicated that the calculated microwave Hall coefficients are of the same order of magnitude as those determined at DC and that the displacement current contribution to the Hall effect is, to first order, negligible.

The microwave Hall effect can also be measured by means of a bimodal cavity (refs. 54 to 58) such as shown in Fig. 2.17 with the specimen attached to one

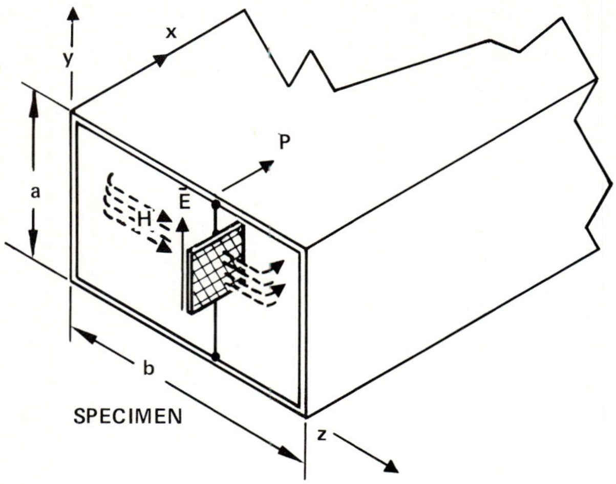

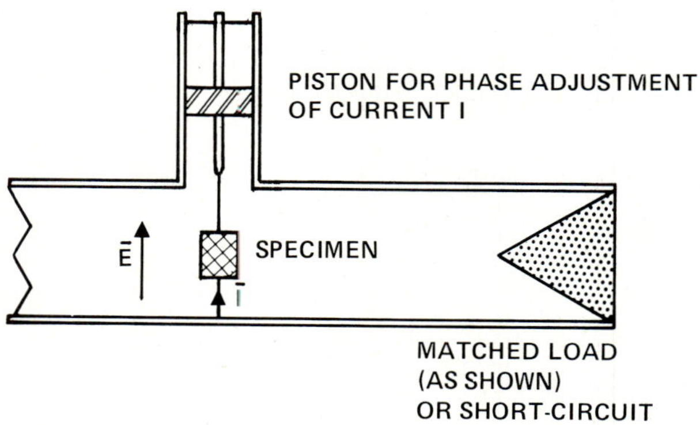

Fig. 2.16. (a) Rectangular semiconductor wafer mounted in a TE_{01} waveguide, P is the direction of power flow, E and H are, respectively, the electric field and magnetic field in the waveguide; (b) a cross-sectional view of waveguide structure.

wall. The modes of the two waveguides are isolated from each other if the externally applied magnetic field B=0. If B is set to some finite level then the modes are coupled through the tensor nature of the conductivity given in Eq. (2.29). The coupling between the modes arises from the same physical mechanism as that of the Hall effect. However, since current flow is not restricted to a direction transverse to B the behavior is best described as a magneto-Kerr effect or microwave Faraday rotation. A detailed analysis of the effect based

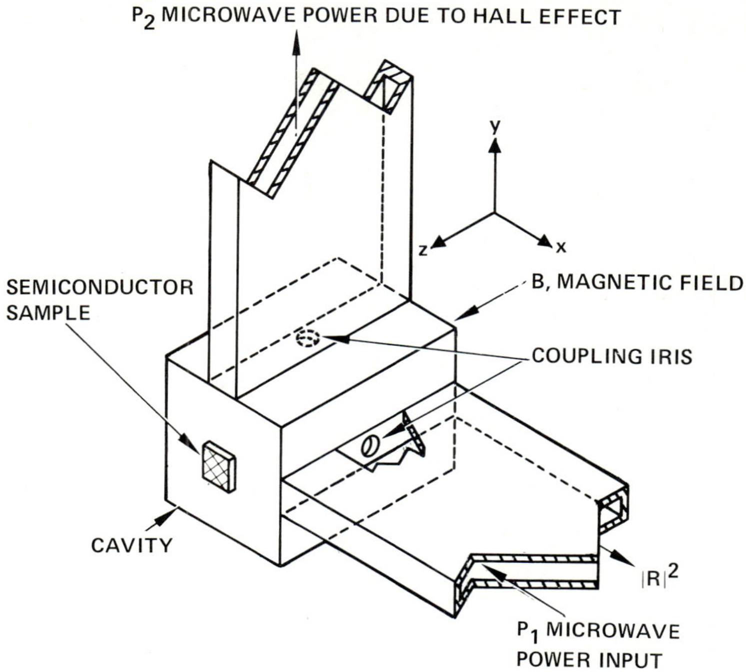

Fig. 2.17. Bimodal cavity for measuring the microwave Hall effect in a semiconductor specimen.

on the field distributions in the cavity and the wave propagation in the specimen as a function of the applied field has been derived by Liu *et al.* (ref. 55). The sensitivity of this method is limited by a signal which appears when a magnetic field is applied to the empty cavity and is due to the Hall effect of the material of which the cavity is made. In the presence of a specimen this signal is superposed on the signal from the specimen and may become the dominant contribution for materials of low mobility and high resistivity (ref. 59). A method which overcomes this problem has been described by Fletcher (ref. 58). He built a cavity with crossed-grid end walls at right angles to each other. In a grid of parallel conductors electrons are free to move in the direction of the conductors but if the grid spacing is much less than the wavelength of the incident radiation then the electrons are constrained from moving in a perpendicular direction. Thus the power coupled into the output guide was reduced by 20 dB in comparison with a full wall cavity.

2.6 Helicon waves

If an electromagnetic wave of radian frequency ω is incident on a metal or high conductivity semiconductor then the wave is either attenuated severely or it is reflected. However, if a static magnetic field is applied in the direction of propagation of the wave then a mode of propagation called a helicon wave can propagate through the material. In the plane perpendicular to the steady applied field B_0 the motion of the electron orbits is confined to a closed path; in high values of B_0 the transverse electrical conductivity approaches zero and the material appears to be a dielectric in a plane transverse to B_0 while remaining a normal conductor along B_0. This is the reason why an electromagnetic wave whose electric field is polarized in a plane perpendicular to B_0 can propagate through the material. Consider a metal or semiconductor plate oriented in a cartesian system with the steady magnetic field B_0 oriented along the z-axis with the time-dependent transverse components of the electric field E_x and E_y, current density J_x and J_y, and magnetic field, B_x and B_y represented in complex notation (with $i=\sqrt{-1}$) in the form

$$E = E_x+iE_y \,, \qquad B = B_x+iB_y \,, \qquad J = J_x+iJ_y \tag{2.30}$$

Taking z=0 as the origin of a semi-infinite material with conductivity σ and Hall coefficient R_h the effective electric field in it is

$$\overline{E} = \rho\overline{J}+iR_h(\overline{B}x\overline{J}) \tag{2.31}$$

By expressing a complex resistivity $\rho^*=\rho-iR_hB_z$, Eq. (2.31) can be written as $\overline{E}=\rho^*\overline{J}$ and the Hall angle θ between $\overline{J}$ and $\overline{E}$ is determined by

$$u = \tan\theta = R_hB_z/\rho \tag{2.32}$$

In the high field limit where $\tan\theta\approx\omega_c\tau>>1$, $\tan\theta\to\Pi/2$. If Eq. (2.31) is substituted in Maxwell's equations then the oscillatory magnetic field normal to the z-axis, $B=B_0 \exp(i\omega t-kz)$, with the differential equation

$$\frac{d^2B}{dz^2} = \frac{4\Pi i\omega B}{\rho(1+iu)} \tag{2.33}$$

has solutions for clockwise B_ℓ and counterclockwise B_r circularly-polarized electromagnetic waves of the form $B_{r,\ell}=B_0\psi \exp(\pm i\omega t)$ where $\psi=\exp(\pm ikz)$ and $k= [4\Pi\omega(i\rho-R_hB_z)^{-1}]^{1/2}$. If $\tan\theta>>1$ then B_ℓ is propagated with little attenuation while B_r decays exponentially. If the magnetic field is reversed then B_r and B_ℓ interchange roles. The reverse situation applies when holes are the charge carriers. This is the helicon mode of propagation described by Chambers and Jones (ref. 60) and by Aigrain (ref. 61); the Hall coefficient can be determined from such transmission measurements in terms of the coefficient k. A thin rectangular plate of thickness δ will behave as a resonant cavity for helicon mode propagation whenever $|k|=N\pi/\delta$ where N is an integer and the corresponding resonant frequencies are determined by

$$\omega_r = \pi(N/2\delta)^2 \cdot |\rho(1-i\tan\theta| \tag{2.34}$$

and the Q of each resonance peak is

$$Q = (1/2)\cdot(1+\tan^2\theta)^{1/2} \tag{2.35}$$

Since $\tan\theta$ is proportional to B, from Eq. (2.32), ω_r can be extrapolated to $\tan\theta\to\infty$ which, from Eq. (2.34), is

$$\left(\omega_r\right)_\infty = \pi(N/2\delta)^2 \cdot R_h B \tag{2.36}$$

Thus R_h can be determined from the dependence of ω_r on B and the resistivity ρ can also be determined from Eqs. (2.34) and (2.35). Helicon propagation has been reviewed by Fawcett (ref. 62), by Kaner and Skobov (ref. 63), by Morgan (ref. 64), and by Hurd (ref. 22). Experimental measurements using helicon mode propagation have been reported on a variety of metals at cryogenic temperatures. A specimen in the form of a thin rectangular plate has a rectangular solenoid detector wound around it. A driving coil which supplies the alternating magnetic flux is located outside the cryogenic dewar oriented along the x-axis and the detecting solenoid is oriented along the y-axis of a cartesian system with the static magnetic field B oriented along the z-axis. By measuring the detected signal along the y-axis as a function of the frequency and phase of the driving field a determination of R_h and ρ can be obtained in this manner. Helicon propagation methods of measuring R_h suffer from the errors common to other AC methods. Spurious voltages or shifts in resonant frequency can be produced by parasitic coupling to external circuits or mechanical vibration of the components. There are no problems, however, with misalignment voltages and measurements can be made in very low magnetic fields. An important source of errors are the finite specimen dimensions and the basic theory is valid only for a semi-infinite specimen. The calculation of the appropriate geometrical correction is a difficult boundary-value problem. Chambers and Jones (ref. 60) gave an approximate solution for thin rectangular samples but, even after this correction was applied, their measured Hall coefficients were still about 5% different from the theoretically-expected values, and this discrepancy could not be explained by experimental error. Legendy (ref. 65) used a different approach from that of Chambers and Jones and obtained corrections which seem to account for the discrepancy found by them. Harding and Thonemann (ref. 66) and Amundsen (ref. 67) have obtained results which are in good agreement with Legendy's theory.

REFERENCES

1 A. R. Billings, Tensor Properties of Materials, Wiley Interscience, London (1969)78.
2 R. G. Chambers, Proc. Phys. Soc. (London), A65(1952)903.
3 A. H. Wilson, The Theory of Metals, Cambridge University Press, London (1953)212.
4 C. Hilsum and A. C. Rose-Innes, Semiconducting III-V Compounds, Pergamon Press, Oxford(1961)120.
5 American Society for Testing Materials, Standard Method for Measuring Hall Mobility in Extrinsic Crystals, ASTM F76-78(1968).
6 D. Fraenkel and U. Gruen, Solid-State Electron., 12(1969)201.
7 A. Fernandez, J. Sci. Instr.(J. Phys. E), ser 2, 1(1968)782.
8 L. P. Hunter, Handbook of Semiconductor Electronics, McGraw Hill Book Co., New York, Third ed.(1970)8-14.
9 A. B. J. Sullivan, Electron. Lett., 12(1976)133.
10 A. N. Dau and R. N. Mitra, Solid-State Electron., 8(1965)697.
11 N. Braslau, J. B. Gunn and J. L. Stephens, Solid-State Electron., 10(1967)381.
12 C. J. Nuese and J. J. Gannon, J. Electrochem. Soc., 115(1968)327.
13 V. L. Rideout, Solid-State Electron., 18(1975)541.
14 W. J. Patrick, Solid-State Electron., 9(1966)203.
15 R. Jaggi, The Galvanomagnetic and Thermomagnetic Transverse Effects in "Zahlenwerte and Functionen," K. H. Hellwege and A. M. Hellwege, ed., Berlin(1959)161.
16 J. P. Jan, Solid State Phys., 5(1957)1.
17 A. C. Beer, Galvanomagnetic Effects in Semiconductors, Solid State Phys. Suppl. 4, Academic Press, New York(1963)344.
18 C. M. Hurd, J. Sci. Instr., 42(1965)465.
19 J. E. A. Alderson, T. Farrell and C. M. Hurd, Phys. Rev., 174(1968)729.
20 G. de Mey, Solid-State Electron., 20(1977)139.
21 H. Koppe and J. M. Bryan, Can. J. Phys., 29(1951)274.
22 C. M. Hurd, The Hall Effect in Metals and Alloys, Plenum Press, New York, (1972)192.
23 H. Weiss, Structure and Applications of Galvanomagnetic Devices, Pergamon Press, New York(1969)16.
24 I. Isenberg, B. R. Russel, R. F. Greene, Rev. Sci. Instr., 19(1948)685.
25 R. F. Wick, J. Appl. Phys., 25(1954)741.
26 H. J. Lippmann and F. Kuhrt, Z. Naturforsch., 13a(1958)474.
27 J. R. Drabble and R. Wolfe, J. Electron. Contr., 3(1957)259.
28 J. Haeusler, Arch. Electrotech, 52(1968)11.
29 J. R. Drabble, M. M. Kaila and H. J. Goldsmid, J. Phys. D, Appl. Phys., 8(1975)790.
30 L. J. van der Pauw, Philips Res. Repts., 13(1958)1.
31 L. J. van der Pauw, Philips Tech. Rev., 20(1958)220.
32 B. Jacobs and G. de Mey, Electron. Lett., 12(1976)4.
33 H. J. van Daal, Philips Res. Repts., Suppl. No. 3(1965)8.
34 J. Haeusler and H. J. Lippmann, Solid-State Electron., 11(1968)173.
35 G. de Mey, Arch. Elektron. Uebertragungstech., 27(1973)309.
36 G. de Mey, Y. Burvenich and M. de Molder, Phys. Stat. Sol., a23(1974)K45.
37 R. Chwang, B. J. Smith and C. R. Crowell, Solid-State Electron., 17(1974) 1217.
38 M. F. Leach, J. Phys. E, Sci. Instr., 10(1977)879.
39 A. F. Clark and F. R. Ficket, Rev. Sci. Instr., 40(1969)465.
40 R. P. Ries and C. B. Satterthwaite, Rev. Sci. Instr., 38(1967)1203.
41 J. Clarke, Phil. Mag., Ser. 8, 13(1966)15.
42 J. J. Donoghue and W. P. Eatherly, Rev. Sci. Instr., 22(1951)513.
43 J. Lavine, Rev. Sci. Instr., 29(1958)970.
44 A. Parodo and A. Zedda, J. Sci. Instr. (J. Phys. E) Ser. 2, 2(1969)748.
45 H. J. Goldsmid, Proc. Phys. Soc., 71(1958)633.

46 R. A. Lomas, M. J. Hampshire and R. D. Tomlinson, J. Phys. E, Sci. Instr., 5(1972)819.
47 H. H. Wieder, Hall Generators and Magnetoresistors, Pion Press Ltd., London (1971)72.
48 B. R. Russel and C. Wahlig, Rev. Sci. Instr., 21(1950)1028.
49 N. Z. Lupu, N. M. Tallan and D. S. Tannhauser, Rev. Sci. Instr., 38(1967) 1658.
50 H. L. McKinzie and D. S. Tannhauser, J. Appl. Phys., 40(1969)4954.
51 W. V. McLevige, P. K. Chaterjee and B. G. Streetman, J. Phys. E, Sci. Instr., 10(1977)335.
52 B. Lax and L. A. Roth, Phys. Rev., 98(1955)548.
53 H. E. M. Barlow and S. Kataoka, Proc. Inst. Electr. Eng., 105B(1958)53.
54 Y. Nishina and W. J. Spry, J. Appl. Phys., 29(1958)230.
55 S. H. Liu, Y. Nishina and R. H. Good, Rev. Sci. Instr., 32(1961)748.
56 Y. Nishina and G. C. Danielson, Rev. Sci. Instr., 32(1961)790.
57 D. W. Griffin, Proc. IREE Australia, April(1967)121.
58 J. R. Fletcher, J. Phys. E, Sci. Instr., 9(1976)481.
59 E. M. Trukhan, Radio Eng. Electron Phys., 11(1966)1097.
60 R. G. Chambers and B. K. Jones, Proc. Roy. Soc.(London) A270(1962)417.
61 P. Aigrain, Proc. Internat. Conf. Semicond. Phys., Prague(1960) Academic Press, New York, p.224.
62 E. Faucett, Adv. Phys., 13(1964)139.
63 E. A. Kaner and V. G. Skobov, Adv. Phys., 17(1968)605.
64 D. P. Morgan, Phys. Stat. Sol., 24(1967)9.
65 C. Legendy, Phys. Rev., 135(1964)1713.
66 G. N. Harding and P. C. Thonemann, Proc. Phys. Soc.(London) 85(1965)317.
67 T. Amundsen, Proc. Phys. Soc.(London) 88(1966)757.

CHAPTER 3

MAGNETORESISTANCE

3.0 Introduction

The magnetoresistance is usually defined in terms of the fractional change in resistance of a specimen produced by an applied magnetic field. Charge carriers drifting in a material under the combined action of transverse electric and magnetic vectors are subject to the Lorentz force which deflects some of them producing the Hall field described in Section 2.0. If all the charge carriers are identical and can be described by a spherical Fermi surface and by a single isotropic relaxation time then the Hall field compensates the Lorentz force, the current streamlines are invariant with magnetic field, and in consequence, the magnetoresistance, $\Delta\rho/\rho_o$ is zero. If the charge carriers are not of the same charge type, if their effective masses m* are different or their relaxation times are energy dependent then the Lorentz force is not fully compensated by the Hall field. This produces a current transverse to the electric and magnetic vectors and a change in the resistivity of a specimen which can be measured along its longitudinal direction of current flow. Although a single charge carrier conduction band defined by a spherical Fermi surface and a single isotropic relaxation time does not produce a magnetoresistance effect, two or more such conduction bands in the same material make its galvanomagnetic coefficients dependent on the magnetic field. In terms of Eqs. (2.12) and (2.13) the magnetoresistance of a material in which there are two conduction bands whose Hall coefficients and conductivities are identified by subscripts 1 and 2, respectively, (ref. 1) is

$$\frac{\Delta\rho}{\rho_o} = \frac{\alpha(R_{h_1}^2\sigma_1 + R_{h_2}^2\sigma_2) - \alpha^2(R_{h_1} + R_{h_2})^2}{1 + B^2\alpha^2(R_{h_1} + R_{h_2})^2} B^2 \tag{3.1}$$

where $1/\alpha=1/\sigma_1+1/\sigma_2$ and the resistivity in zero magnetic field, $\rho_o=(\sigma_1+\sigma_2)^{-1}$.

If the two different charge carrier bands are affected differently by the magnetic field B because they have different charge types, different effective masses or different scattering lengths then the net current density which is the vector sum of the current density components of the two bands is reduced and the effective resistance in the direction of current flow is increased by the magnetic field. The magnetoresistance in Eq. (3.1) vanishes if $R_{h_1}\sigma_1 = R_{h_2}\sigma_2$, i.e., if the mobilities of the charge carriers in the two bands $\mu_1=\mu_2$ are identical. In weak magnetic fields $\Delta\rho/\rho_0$ is proportional to B^2 and saturates in large magnetic fields. The two-band expression in Eq. (3.1) can be generalized to many different, not necessarily spherical bands, all of which contribute separately to the net conductivity of a material.

Transverse magnetoresistance phenomena are those in which an increase in resistance is observed experimentally if the applied magnetic field is perpendicular to the direction of current flow in a specimen. In materials in which the charge carriers have a nonspherical Fermi surface a magnetoresistance can also be observed when the magnetic field is oriented in a direction parallel to direction of current flow. The orientation dependence of the magnetoresistance with respect to the crystallographic axes of a single crystal specimen and the direction of current flow and magnetic field direction can be used for mapping the geometrical configuration of the Fermi surface. The main topic of this Chapter is that of transverse magnetoresistance and the term magnetoresistance used here is meant to apply to the transverse case.

Suppose that all the electrons in a metal have the same relaxation time τ and $\Delta\rho/\rho_0$ is a function only of τB. Then, since τ is inversely proportional to ρ_0,

$$(\Delta\rho/\rho_0) = F(B/\rho_0) \tag{3.2}$$

where $F(B/\rho_0)$ is a characteristic function of the metal. Equation (3.2) is known as Kohler's rule and might be interpreted qualitatively by considering the path traversed by an electron between collisions with the crystal lattice in terms of the fraction ℓ/r where ℓ is the electron mean free path, $\ell=m^*v/(e^2n\rho_0)$ and r is the radius of the electron orbit, $r=m^*v/eB$, where n is the electron density and v the electron velocity; therefore,

$$\frac{\ell}{r} = \frac{B}{ne\rho_0} \tag{3.3}$$

and B/ρ_0 would be an index of the ℓ/r ratio.

Justin (ref. 2) has determined the magnetoresistance of a large number of metals as a function of their resistivity normalized to their resistivity ρ_θ at the Debye temperature θ. In general, Kohler's rule applies; deviations are considered to be an indication of complex Fermi surfaces. However, the multi-band model, each having a spherical Fermi surface, cannot explain the magnetic field-dependent anisotropy of metals (ref. 3). While $\Delta\rho/\rho_0$ has only a slight dependence on the direction of B in weak magnetic fields large anisotropies are observed in large values of B/ρ_0. A calculation of such anisotropies must take into account the actual shape of the Fermi surface and of possible anisotropies in τ. Jones and Zener (ref. 4) have performed such calculations assuming a constant τ and weak magnetic fields. These were subsequently extended to include anisotropic τ leading to a longitudinal as well as a transverse magnetoresistance attributed to nonspherical Fermi surfaces.

Lifshitz *et al.* (ref. 5) and Lifshitz and Peschanskii (ref. 6) have shown theoretically that in high magnetic fields the magnetic field-dependence of $\Delta\rho/\rho_0$ of different metals, and that of a particular metal for different magnetic field and current orientations, is a function only of the specific geometry of its Fermi surface. The electrons move in $\bar{k}$-space along curves of constant energy perpendicular to the direction of $\bar{B}$ and the electrons can be specified in terms of their orbits in $\bar{k}$-space. For a Fermi surface that does not touch the Brillouin zone boundary the electron orbits are closed curves; for these the magnetoresistance saturates in high magnetic fields. However, if the Fermi surface intersects some of the faces of the Brillouin zone then it is defined as multiply-connected and maybe open. For such directions the magnetoresistance does not saturate but continues to increase quadratically with B. The above description concerns electron orbits in $\bar{k}$-space; in real space the trajectories of electrons are nearly always open since the component of the velocity along B is not affected by the applied magnetic field. In general, the Fermi surface of a metal is not necessarily contained in a single Brillouin zone, some electrons have enough energy to spill over into the next higher energy zone and their number will be roughly equal to the number of unoccupied states left in the zone of lower energy.

A number of theoretical and experimental investigations have shown that magnetoresistance measurements which depend on bandstructure parameters may be obscured by perturbing effects such as those related to the geometry of a specimen, its contacts, conductive or nonconductive inclusions and by surface properties. Weak field-magnetoresistance is particularly sensitive to such influences. If the band-structure parameters of a material are known, then anomalies in the magnetic field-dependence of $\Delta\rho/\rho_0$ might be interpreted in terms of such perturbations. If the bandstructure of a material is not known then the orientation dependence of $\Delta\rho/\rho_0$ must be evaluated in terms of the energy surface symmetry relations appropriate to the specific crystallographic system.

3.1 Geometrical magnetoresistance

Although the magnetoresistance of a material having a single spherical Fermi surface and an isotropic charge carrier relaxation time is zero, a magnetoresistance of "geometrical" origin can be produced in it by physically or electrostatically short-circuiting its Hall field, in accordance with Eq. (2.10).

The effect of the geometry of a specimen on its magnetoresistance has been calculated by Wick (ref. 7) and by Drabble and Wolfe (ref. 8). In a rectangular specimen the current contacts cause a partial electrostatic short-circuit of the Hall field and perturb the current streamlines. Broom (ref. 9) has demonstrated that potential probes in a Hall plate also perturb the Hall field and introduce a magnetoresistance of geometrical origin which is significant in materials with high electron mobilities. The Hall effect and magnetoresistance are complementary phenomena; a reduction in the Hall voltage of a rectangular specimen produced by a decrease in its length to width ratio, such as shown in Fig. 2.6, is accompanied by an increase in its magnetoresistance. One might argue, intuitively, that as ℓ/w decreases the Hall voltage approaches zero due to the electrostatic short-circuit of the Hall field by the current electrodes and $(\Delta\rho/\rho_0)$ approaches its maximum value for a given B.

Weiss and Welker (ref. 10) have investigated experimentally and analytically the ℓ/w dependence of the magnetoresistance of rectangular InSb specimens. They found that the magnetoresistance increases with decreasing ℓ/w approaching, as a limiting value, the magnetoresistance of a Corbino disc (ref. 11) in which the Hall field is physically short-circuited by virtue of its structure. Figure 3.1 shows such a Corbino disc equipped with coaxial center and peripheral electrodes. A current is applied between the inner and outer electrodes; the current has

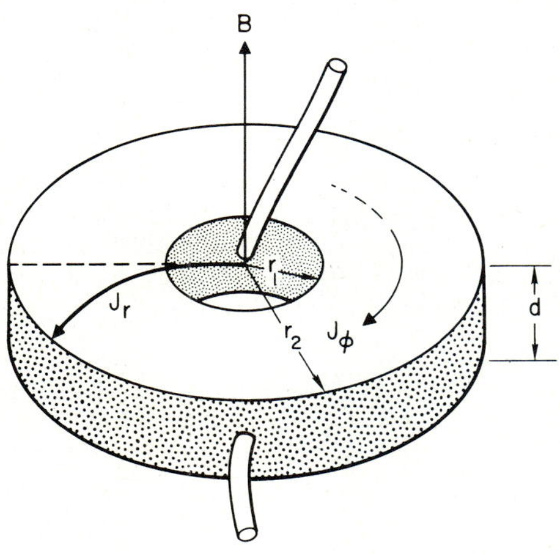

Fig. 3.1. Cylindrical Corbino disc with coaxial electrode, thickness d, radial current density J_r, circumferential Hall current J_ϕ and transverse magnetic field B.

radial symmetry producing a current density J_r. In the presence of a steady magnetic field the current streamlines become logarithmic spirals as shown in Fig. 3.2 and by virtue of the short-circuit of the Hall field the current density J_ϕ is generated in the disc. The Hall angle θ can be determined from

$J_\phi/J_r=\tan\theta$. In cylindrical coordinates the components of the electric fields in a Corbino disc are

$$E_r = \rho_o J_r - R_h J_\phi B \qquad\qquad E_\phi = \rho_o J_\phi + R_h J_r B \tag{3.4}$$

In the steady state, $\nabla\times\bar{E}=-\partial B/\partial t=0$ and the line integral of the circumferential field is

$$\oint_0^{2\pi} E_r r d\phi = 0 \tag{3.5}$$

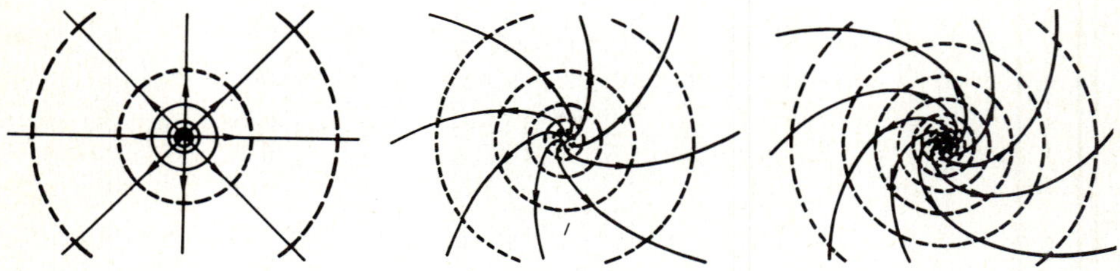

Fig. 3.2. Distribution of current streamlines and equipotentials for an ideal Corbino source as a function of Hall angle θ; (a) $\theta=0$; (b) $\theta=26.6°$; (c) $\theta=45°$. Current streamlines are shown as continuous lines, equipotentials as dashed lines. [after M. G. Buehler, Solid-State Electron., 10(1967)801]

The resistivity, $\rho(B)=E_r/J_r$ and with $E_\phi=0$ the magnetoresistance of the Corbino disc is

$$(\Delta\rho/\rho_o)_c = \tan^2\theta = \mu^2 B^2 \tag{3.6}$$

where μ is the mobility of the charge carriers. Figure 3.2 shows that the equipotentials are concentric circles and under the influence of the transverse magnetic field the current streamlines spiral outward at an angle θ with respect to the normals to the equipotential planes.

Buehler and Pearson (ref. 12) and Buehler (ref. 13) have investigated the distribution of equipotentials and streamlines in geometrical structures which cannot properly be classified either as Corbino discs or as Hall plates by the method of "Corbino images." They show, by an extension of the method of images, that in the case of a point contact source on an infinite plane the current flow in the presence of a magnetic field is the same as that of a Corbino disc shown in Fig. 3.2. For a thin layer of thickness δ whose area is circumscribed by a peripheral electrode with a point contact at its center, the current density and electric field in the plane are

$$\begin{aligned} J_c &= J_r \cdot \overline{r} + J_\phi \cdot \overline{\phi} = (i/2\Pi r\delta)(\overline{r} + \tan\theta\, \overline{\phi}) \\ E &= (\rho i/2\Pi r\delta)(1 + \tan^2\theta)\overline{r} \end{aligned} \tag{3.7}$$

where $\overline{r}$ and $\overline{\phi}$ are unit vectors in cylindrical coordinates. A comparison between the magnetic field dependence of a Hall plate and a Corbino disc, presented in Table 3.1, illustrates the complementarity between them. The parameters which the experimenter can measure are E, E_h, J and B. The quantities to be determined are $R_h(b)$ and $\rho(B)$. In a Corbino disc the ratio E/J is the Corbino resistivity $\rho_c = \rho(B)(1+\tan^2\theta)$. In a Hall plate, $E/J = \rho(B)$ while in a Corbino disc $\rho(B)$ can be determined only for B=0. Furthermore, in a Hall plate E_h/J is directly proportional to $R_h(B)$ while in a Corbino disc $E_h/J=0$.

TABLE 3.1

Magnetic field-dependence of Hall plate and Corbino disc.

Structure	Equipotentials	Current streamlines	E/J	E_h/J
Hall plate	Dependent	Independent	$\rho(B)$	$R_h(B)\cdot B$
Corbino disc	Independent	Dependent	$\rho(B)(1+\tan^2\theta)$	0

Equation (3.6) may also be calculated (ref. 14) in terms of Fig. 3.1 by equating the work $W = i^2\Delta R_r$ produced in effecting a change ΔR_r in the radial resistance $R_r = (\rho/2\Pi d)\ell n(r_2/r_1)$ of the Corbino disc with the Joule heat $Q = v_\phi{}^2/R_\phi$, where $v_\phi = R_h i_r B/d$ is the line integral of the circumferential Hall voltage and $R_\phi = (2\Pi\rho/d)[\ell n(r_2/r_1)]^{-1}$. For rectangular specimens having no magretoresistance of "physical" origin Lippmann and Kuhrt (ref. 15) have evaluated $\Delta\rho/\rho_o$ as a function of the Hall tangent and their ℓ/w ratio by a conformal representation. Their calculations, represented in Fig. 3.3, are in good agreement with the experimental measurements of Welker and Weiss (ref. 10) made on InSb wafers. Conformal mapping does not provide, in closed form, analytical solutions over the entire range of $\tan\theta$. For small Hall angles and small ℓ/w ratios the magnetoresistance can be approximated by $(\Delta\rho/\rho_o)_G = (\Delta\rho/\rho_o)_c \cdot (1-0.543\ell/w)$ where $(\Delta\rho/\rho_o)_c$ is the magnetoresistance of a Corbino disc of the same material. For large Hall angles the pertinent integrals have to be evaluated by numerical methods. For small Hall angles, $\mu B \leq 0.45$ the magnetoresistance is quadratic in B and for $\theta \to \Pi/2$ the magnetoresistance becomes linear in B. Green (refs. 16, 17) has demonstrated a variety of other geometrical structures whose magnetoresistance is equivalent to a Corbino disc. These include one electrode circumscribed by another, two electrodes circumscribed by a third, floating electrode, and a structure which consists of essentially two point-contacts with the specimen boundaries effectively at infinity.

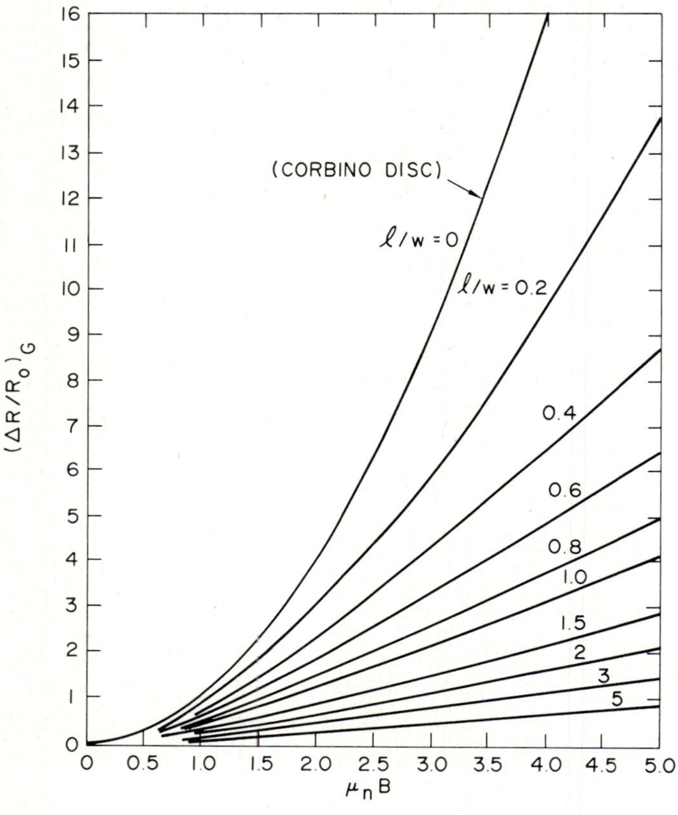

Fig. 3.3. Geometrical magnetoresistance $(\Delta R/R)_G$ of a rectangular plate as a function of the Hall tangent, $\tan\theta=\mu_n B$ and its length to width ratio, ℓ/w. [after H. J. Lippmann and F. Kuhrt, Z. Naturf. 13a(1958)474]

3.2 Perturbations and inhomogeneities

Equation (3.6) indicates that $(\Delta\rho/\rho_o)_c$ can be calculated if the mobility of the single conduction band and spherical Fermi surface of electrons or holes is known; alternatively, the mobility can be determined from a measurement of $(\Delta\rho/\rho_o)_c$ vs. B of a Corbino disc or a rectangular plate with $\ell/w\to 0$. In practice a number of perturbations may produce an anomalous dependence of $(\Delta\rho/\rho_o)_c$ on B.

These include inhomogeneities in the form of conductive or nonconductive inclusions, the presence of more than one band of charge carriers and admixtures of "physical" and geometrical magnetoresistances (ref. 18). If the charge carrier relaxation time of a single conduction band of charge carriers is energy dependent then Eq. (3.6) becomes

$$(\Delta\rho/\rho_o)_c = (\xi\mu B)^2 \tag{3.8}$$

In an intrinsic semiconductor whose electron-to-hole mobility ratio $b>>1$ and dominant mobility limiting mechanism is acoustic phonon scattering, the Corbino magnetoresistance (ref. 20) in the weak field region ($\mu_n B<<1$) is

$$(\Delta\rho/\rho_o)_c = (9\Pi/16)\cdot(\mu_n B)^2 \tag{3.9}$$

In a stronger field region such that electrons are in the range $\mu_n B \geq 1$ and holes are still in the weak field region

$$(\Delta\rho/\rho_o)_c \simeq (9\Pi/16) - (\mu_p B)^2 + b \tag{3.10}$$

while in the very strong field range, $\mu_p B>>1$ the Corbino magnetoresistance is

$$(\Delta\rho/\rho_o)_c \simeq (9\Pi/32)\mu_n\mu_p B^2 \tag{3.11}$$

Thus $(\Delta\rho/\rho_o)_c$ follows a B^2 dependence in low fields, drops below the B^2 curve in higher fields and eventually returns to a B^2 dependence in very high fields.

The Corbino magnetoresistance of a semiconductor with two isotropic conduction bands (defined by subscripts 1 and 2) with energy independent relaxation times (ref. 19) and with $\sigma_0=\sigma_1+\sigma_2$, is

$$\left(\frac{\Delta\rho}{\rho_o}\right)_c = \frac{[\mu_1{}^2(\sigma_1/\sigma_0) + \mu^2{}_2(\sigma_2/\sigma_0)]B^2 + (\mu_1\mu_2B^2)^2}{1 + B^2[\mu_1{}^2\cdot(\sigma_2/\sigma_0) + \mu_2{}^2\cdot(\sigma_1/\sigma_0)]} \tag{3.12}$$

Impurity concentration gradients can have a strong effect on magnetoresistance measurements. Herring (ref. 21) has described a model which takes into account the important role of random inhomogeneities, macroscopic inclusions and stratified, laminar variations in doping density. The problem of gross inhomogeneities was examined by Bate and Beer (ref. 22) and explicit solutions for the boundary value problems were obtained for an expotential variation of the carrier concentration longitudinally along a rectangular specimen and also for step function variations in conductivity (ref. 23). Experimentally, the magnetoresistance produced by the inhomogeneities may predominate over the macroscopic "physical" or "geometrical" magnetoresistance. This is illustrated by the directional magnetoresistance measurements made by Rupprecht *et al.* (ref. 24) on Te-doped InSb in which Te is incorporated preferentially in a crystal during its growth from a Te-doped melt. It is also illustrated by the large ($\Delta\rho/\rho_o$) of two-phase structures such as those of InSb containing preferentially oriented NiSb eutectic needle-like inclusions described by Weiss and Wilhelm (ref. 25) and thin film dendritic or raster pattern magnetoresistors (refs. 26-28).

The effect of inhomogeneities on the magnetoresistance of rectangular specimens has been evaluated by Jervis and Johnson (ref. 29) by assuming a first order one-dimensional variation of the conductivity and Hall tangents in a parallelipiped. If the resistivity varies along its length, in the direction of current flow, then its effect on the measured $\Delta\rho/\rho_o$ is large. On the other hand, if the gradient in ρ is transverse to the current then its effect on $\Delta\rho/\rho_o$ is small. In a thin specimen such that $\ell/w \to 0$ transverse gradients in ρ cause large anomalies in the magnetoresistance while longitudinal gradients lead to relatively minor perturbations. For small Hall angles and small deviations from homogeneity a perturbation calculation (ref. 29) can be made by assuming the conductivity to have a small sinusoidal fluctuation of the form

$$\sigma(y,B) = \sigma_0 B(1 + m \cdot \cos Ky); \qquad m << 1 \tag{3.13}$$

Then to first order in m and to second order in θ the magnetoresistance is

$$\Delta\rho/\rho_o = \theta^2 - m(\theta + \theta^2) \tag{3.14}$$

The most significant term is the linear variation of the Hall angle which is a linear function of the magnetic field. Reversing the direction of the magnetic field will exhibit the presence of such inhomogeneities and averaging the measurements for two opposite orientations of B will eliminate it.

Magnetoresistance measurements can be used for a first order evaluation of the properties of semiconducting epitaxial GaAs layers grown on an n^+ GaAs substrate, covered with a second n^+ layer or coated with a suitable alloy (Table 2.2) ohmic contact. Such structures are used for the fabrication of transferred electron oscillators or amplifiers (Gunn effect) whose small size precludes conventional conductivity and Hall effect measurements. Their geometrical configuration is similar to that of a rectangular semiconductor plate whose $\ell/w \to 0$ in which the Hall field is effectively short-circuited. A constant current may be applied to the specimen and the potential drop v measured across its terminals for B=0. If the effective mobility μ_b measured in a magnetic field B is expressed (ref. 30) in terms of the drift mobility μ_o (in B=0), then

$$\mu_b = \mu_o(1 + \mu_o^2 B^2)^{-1} \tag{3.15}$$

In a transverse magnetic field the potential drop across the specimen terminals changes by an amount Δv; in consequence, μ_o can be determined from

$$\mu_o = (1/B)(\Delta v/v)^{1/2} \tag{3.16}$$

Typical measurements made on GaAs specimens indicate that for B=0.5T, $\Delta v/v=0.0625$ if $\mu_o=5\times10^3$ cm^2/V-sec. A first order determination of the average carrier concentration $\langle n \rangle$ of the epitaxial layer can also be made from

$$\frac{1}{\langle n \rangle} = \frac{1}{\ell}\int_0^{\ell} \frac{dx}{n(x)} = \frac{e\,\mu_o A}{\ell} R \tag{3.17}$$

where ℓ is the length of the specimen between its contacts, A is its cross-sectional area and R is the measured two-terminal resistance between its contacts. The above presumes that both contacts to the epitaxial layer are ohmic in character. However, a common problem encountered in making such devices is that a thin high resistance layer forms at the semiconductor-contact interface. Since the mobility is not a strong function of electron density the mobility of this interfacial region may be the same as that of the epilayer, however, a significant gradient in electron density may be present and its quantitative evaluation is necessary. If the correct drift mobility can be determined by other methods then the contact resistance can be calculated (ref. 30) from

$$R_c/R_b = (\mu_c/\mu_o)^2 - 1 \tag{3.18}$$

where R_c is the contact resistance, R_b the resistance of the epitaxial layer, μ_c is the correct drift mobility and μ_o is the mobility value determined from a magnetoresistance measurement using Eq. (3.16). After R_c has been determined it should be subtracted from the total specimen resistance R before calculating $\langle n \rangle$ from Eq. (3.17). Gurney (ref. 31) indicated that a distinction should be made between the degraded mobility (mobility notch) and reduced carrier concentration (concentration notch) of the interfacial high resistance region of a GaAs transferred electron device made during the production of the device or following its thermally-induced degradation. In general, such problems are encountered more often with metal-contacted semiconducting epitaxial layers probably because of interfacial metallurgical interactions between them. Evaluation of the geometrical magnetoresistance of such structures made at room temperature and 77°K before and after degradation suggests that the dominant degradation mechanism is a decrease in mobility in the interfacial region.

Gutai and Mojzes (ref. 32) have described a method for determining the interfacial contact resistance between a metal and a semiconductor or between two semiconductors having a strong conductivity gradient at their interface. They propose that the angular dependence of the magnetoresistance be measured in a constant magnetic field intensity, $B(\psi)$. If an epitaxial semiconductor layer is slightly anisotropic so that it has a small but not negligible longitudinal magnetoresistance then the magnetic field-dependent resistance, $R_B(B,\psi)$ can be expressed as

$$R_B = R_o\left(\frac{1 + \mu^2B^2}{1 + \mu^2B^2\cos^2\psi}\right) + R_c \tag{3.19}$$

where R_o is the resistance of the semiconductor epilayer in B=0 and R_c is the interfacial contact resistance. Experimentally, (ref. 32) the method yields R_c to within 0.5% of the total resistance which is usually of the order of 1 to 2 ohms and is best used in combination with Hall effect measurements required in order to determine the electron mobility μ.

Upadhyayula *et al.* (ref. 33) found that the average carrier concentration of an epitaxial layer determined from magnetoresistance measurements made on a Gunn-effect type structure can differ by an order of magnitude from that measured on a simultaneously-grown epitaxial layer used for Hall effect measurements. In these, the electron transport is parallel to the semi-insulating substrate whereas for magnetoresistance measurements electron transport is perpendicular to the n^+ substrate. A thin high-resistivity interfacial layer of thickness ℓ_2 and conductivity σ_2 is effectively an extension of the semi-insulating substrate on which the epilayer of thickness ℓ_1, and conductivity σ_1, used for Hall measurements, is deposited. It produces only a slight perturbation and Hall measurements provide essentially the value of n_1. For magnetoresistance measurements the resistance of a composite two-layer structure, in terms of the parameters of the two layers, is

$$R = \frac{\ell_1}{n_1e\mu_1A} + \frac{\ell_2}{n_2e\mu_2A} = \frac{\ell}{\langle n\rangle e\mu A} \tag{3.20}$$

Assuming that $\mu_1 \simeq \mu_2$ then Eq. (3.20) can be expressed as

$$\langle n\rangle = \frac{n_1(1 + \ell_2/\ell_1)}{1 + (\ell_2/\ell_1)(n_1/n_2)} \tag{3.21}$$

A thin region of high electron density, $n_1/n_2 \ll 1$ does not alter appreciably the value calculated from Eq. (3.17) using magnetoresistance measurements; the presence of n_2 can be ignored. However, if a high resistivity interfacial layer is present then the electron density will differ substantially from n_1 and will depend strongly on ℓ_2 and n_2. In fact, if $n_1/n_2 \gg \ell_1/\ell_2$ then $\langle n\rangle \simeq n_2(1+\ell_1/\ell_2)$. For all values of $\ell_2/\ell_1 > 0.1$ the effective carrier concentration determined from magnetoresistance measurements in the manner described above approaches n_2. The validity of this simple two-layer interpretation of interfacial inhomogeneities and their effect on magnetoresistance measurements was confirmed experimentally (ref. 33) on GaAs and InP specimens.

If the interfacial region has a much lower mobility than that of the epitaxial layer such that $(\mu_2 B)^2 \ll 1$ then the equivalent Corbino magnetoresistance (ref. 19) is

$$\frac{\Delta\rho}{\rho_0} = \frac{\mu_1^2 B^2}{1 + k(1 + \mu_1^2 B^2)} \tag{3.22}$$

where $k=\sigma_2 \ell_2/\sigma_1 \ell_1$. Equation (3.22) indicates that a saturation of $\Delta\rho/\rho_0$ occurs in high magnetic fields and that the magnetic field required for saturation can be determined by taking the second derivative of Eq. (3.21) with respect to B and solving for B_s,

$$B_s = (1/\mu_1)[(1 + k)/3k]^{1/2} \tag{3.23}$$

Experimental measurements made on an InSb Corbino disc were found to be in fair agreement with such a two-layer model in which the mobility of an accumulated surface layer is much smaller than that of the bulk. A further and more detailed treatment of surface-induced and bulk inhomogeneities and their effect on galvanomagnetic measurements is provided in Chapter 5.

3.3 Physical magnetoresistance

In the two preceding Sections the emphasis has been placed on the magnetoresistance of geometrical origin and perturbations encountered in its use for determining fundamental charge carrier transport coefficients. The need arises for measuring the "physical" magnetoresistance in order to determine the details of Fermi surfaces, the specific coefficients of the various conduction bands and the nature of their interactions. It is highly desirable that specimens prepared for resistivity and Hall effect measurements and the apparatus and techniques used for these purposes should also be amenable for physical magnetoresistance measurements. In choosing specimens and measurement techniques adequate care must be used to eliminate the role of geometrical magnetoresistance contributions and the perturbations introduced by spatially ordered or randomly distributed inhomogeneities. The influence of geometrical effects can be eliminated to first order by employing a specimen cut in the form of a parallelipiped whose length to width ratio, $\ell/w \gg 1$, and by placing the potential probes, such as shown in Fig. 1.1, far from the current contacts in order to minimize the electrostatic short-circuit induced in their immediate vicinity. Specimens which contain impurity gradients whose presence is determined either from Hall effect measurements, from different values of ρ measured by the displacement of the potential probes or from different values of $\Delta\rho/\rho_0$, obtained by reversing the direction of the magnetic field, should be viewed with caution and may have to be discarded. A more reliable method for the identification of impurity gradients is the measurement of $\Delta\rho/\rho_o$ as a function of B for crystallographically equivalent directions which lie at different angles with respect to the growth direction of a crystal (ref. 34). Two-band conduction in which, to first order, the conduction bands can be considered spherical are represented by the measurements of Hilsum and Barrie (ref. 35) made on p-type InSb. In the temperature range dominated by the intrinsically generated charge carriers $\Delta\rho/\rho_o \simeq 0.55$ and decreases with increasing electron density to $\Delta\rho/\rho_o < 0.01$ for $n = 4\times10^{17}/cm^3$. Longitudinal magnetoresistance is not observed at room temperature. In the temperature range in which mixed intrinsic and p-type conductivity is present $\Delta\rho/\rho_o$ rises to a peak, $\Delta\rho/\rho_o \simeq 1.6$ and then decreases monotonically.

In multi-valley semiconductors such as silicon and germanium magnetoresistance measurements must be made along different crystallographic orientations in order to determine the elements of the resistivity tensor and the shape of the Fermi surface. Seitz (ref. 36) has given a general equation for the constitutive Ohm-Hall equations which include the transverse and longitudinal magnetoresistance phenomena. These may be expressed in an inverted form (ref. 37) by

$$\bar{E} = \rho\bar{J} + R_h(\bar{B}\times\bar{J}) + \rho[bB^2\bar{J} + c\bar{B}(\bar{J}\cdot\bar{B}) + d\hat{A}\bar{J}\ ] \tag{3.24}$$

where ρ, R_h, b, c, and d are parameters which depend on the bandstructure and charge carrier relaxation processes and A is a diagonal tensor whose elements are,

$$\hat{A} = \begin{vmatrix} B_1^2 & 0 & 0 \\ 0 & B_2^2 & 0 \\ 0 & 0 & B_3^2 \end{vmatrix} \tag{3.25}$$

where the B subscripts refer to a coordinate system which is coincident with the crystallographic axes and the magnetoresistance coefficients b, c and d are functions of the symmetry of the energy surfaces in accordance with Table 3.2 below.

TABLE 3.2

Dependence of magnetoresistance coefficients on symmetry of energy surface. [after M. Glicksman, Progr. Semicond., 3(1958)3]

Energy surface symmetry	Relationship
spherical	$b+c=0$; $d=0$
[111] ellipsoids	$b+c=0$; $d>0$
[100] ellipsoids	$b+c=-d$; $d<0$
[110] ellipsoids	$b+c=d$; $d>0$

The magnetoresistance coefficients can be determined experimentally for the case of weak field magnetoresistance ($\mu^2B^2<<1$)

$$\frac{\Delta\rho}{\rho_0} = b\mu^2B^2 + c\left(\sum_j i_j n_j\right)^2 + d\sum_j i_j^2 n_j^2 \tag{3.26}$$

where i_j and n_j are direction cosines of the electric field and current density, respectively. For a [100]-oriented epitaxial layer of a cubic symmetry crystal Eq. (3.24) can be expressed as

$$E = \rho\bar{J} + R_h(\bar{J}x\bar{B}) + \rho bB^2\bar{J} \tag{3.27}$$

Suppose that an arbitrary contour specimen of such an [100]-oriented epilayer with peripheral line-electrodes, such as described in Section 1.5, is used for van der Pauw-type resistivity measurements. With the current derived from a constant current source applied to terminals 1 and 2 and the potential $(v_{34})_0$ measured in B=0 and $(v_{34})_B$, measured in a transverse magnetic field, between terminals 3 and 4,

$$v_{34} = -(1 + bB^2)\int_3^4 \rho\bar{J}\cdot d\bar{s} - R_h\int_3^4 (\bar{J}x\bar{B})\ d\bar{s} \tag{3.28}$$

The path of integration is taken along the periphery of the specimen where the integrand of the second integral is zero and a measurement of $(v_{34})_0$ yields the value of the first integral. If the current density vector is considered to be invariant with B then the value of the first integral in Eq. (3.28) is invariant in B. Therefore a measurement of $(v_{34})_B$ yields the resistance $(R_{12,34})_B=(v_{34})_B/i$ and from Eq. (3.28)

$$(R_{12,34})_B = (R_{12,34})_0\cdot(1 + bB^2) \tag{3.29}$$

in consequence, the magnetoresistance coefficient

$$b = (1/B^2)\cdot(\Delta v/v) \tag{3.30}$$

For the derivation of Eq. (3.30) the projections of the peripheral line-contacts are presumed to be point-contacts on the [100]-plane of the specimen; no account has been taken of the correction factors required for finite-size contacts. This was done by de Mey *et al.* (ref. 38) who calculated the magnetoresistance of a Greek cross, such as shown in Fig. 1.20. If a current is supplied from a constant current source to two adjacent arms of such a device then a voltage v measured between the contacts of the other two adjacent arms (in B=0) is proportional to its resistivity. If a transverse magnetic field is applied to the specimen then the voltage v will change by an amount Δv, proportional to the magnetoresistance, provided that the current density vector $\bar{J}$ is invariant in B. Finite size contacts affect the boundary conditions which determine the electric field and current density distribution in the cross. De Mey *et al.* (ref. 38) have calculated this effect by a conformal transformation of the square into a circle and obtained a solution expressed in the form of a Taylor series in μB,

$$\frac{\Delta V}{V} = b\mu^2B^2 + A_1\mu B + A_2\mu^2B^2 \tag{3.31}$$

where $A_1 \simeq 0.593 \exp(-\pi A/S)$ and $A_2 \simeq 1.147 \exp(-2\pi A/S)$. For a perfect symmetrical cross the errors due to $A_1\mu B$ cancel; in practice, a slight asymmetry introduces a slight error of this type; the error due to $A_2\mu^2B^2$ is usually negligible.

The scattering parameter ξ in Eq. (3.8) is the ratio of the experimentally-determined mobility μ_m derived from low field magnetoresistance measurements and of the Hall mobility μ_B of a semiconductor with a single isotropic conduction band as shown by Jervis and Johnson (ref. 29)

$$\xi = \mu_m/\mu_B \tag{3.32}$$

The scattering parameter is related to the charge carrier relaxation time τ (ref. 39) by

$$\xi = \langle\tau^3\rangle\langle\tau\rangle\cdot\langle\tau^2\rangle^{-2} \tag{3.33}$$

These equations provide the experimental basis for the determination of ξ in GaAs. For Hall angles $\theta<0.3$ radians and $\ell/w<0.3$ the magnetoresistance was found to be quadratic in B and the Hall mobility was determined from van der Pauw-type measurements on bulk-grown GaAs at 300°K to be in the range between 4040 and 6100 cm^2/V-sec. The average value of $\xi=1.03\pm0.07$. This corresponds within experimental error to optical polar-mode phonon scattering (ref. 40). Further investigations on the scattering parameter of GaAs were made by Blood and Tree (ref. 41) in the temperature range between 77°K and 400°K in magnetic fields up to 0.7T. They found that ξ is approximately unity throughout this temperature range and at 300°K it is 0.96 ± 0.08, in good agreement with the results obtained by Jervis and Johnson. The scattering parameter is related directly to the Hall factor discussed in Chapter 4 Section 4.0.

3.4 Microwave magnetoresistance

A simple experimental set-up for measuring the microwave magnetoresistance of semiconductors is shown schematically in Fig. 3.4. The calculations of $\Delta\rho/\rho_o$ from microwave transmission and reflection measurements of the specimen-loaded waveguide are based on the theoretical treatment of Jacobs *et al.* (ref. 42) developed for an experimental and analytical evaluation of the microwave conductivity of semiconductors. These are extended to include, under appropriate boundary conditions, an evaluation of the magnetic-field dependence of $\Delta\rho/\rho_o$. The arrangement shown in Fig. 3.4 is limited to thin layers whose thicknesses are much smaller than their skin depth and whose microwave resistivity is large.

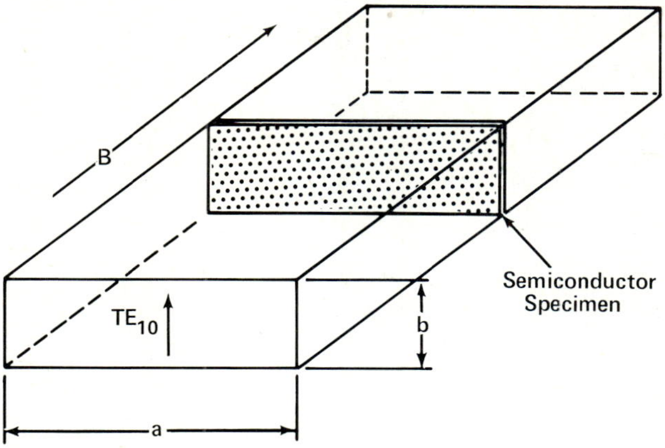

Fig. 3.4. A thin semiconducting specimen mounted in a rectangular waveguide driven in the TE_{10} mode; arrow shows direction of E-field in the guide and external arrow shows orientation of DC magnetic field.

The solution of Maxwell's equations leads to the propagation coefficient $\gamma=\alpha+i\beta$ where

$$\gamma = (\Pi^2/a^2 - \omega^2\varepsilon^*/c^2)^{1/2} \tag{3.34}$$

with the complex dielectric constant ε^* expressed in terms of the conductivity, σ_0 the lattice dielectric constant, ε_L and the radian frequency, ω with

$$\varepsilon^* = \varepsilon_L - i\sigma_0/\omega\varepsilon_0 \tag{3.35}$$

For a semiconductor whose lattice dielectric constant, $\varepsilon_L \approx 16$ and with the guide dimension a=1.067 cm the phase and attenuation components α and β are essentially equal (ref. 43);

$$\alpha = \beta = (1/c)(\sigma_0\omega/2\varepsilon_0)^{1/2} \tag{3.36}$$

For $\alpha >> \beta_0$ where β_0 is the phase coefficient of the air-filled portion of the waveguide the absolute value of the transmission coefficient is

$$|T| = [\beta_0/(\beta_0 + \alpha^2 d)](1 - 2\alpha d + 2\alpha^2 d^2)^{1/2} \tag{3.37}$$

where d is the thickness of the specimen. For $\alpha d << 1$ Eq. (3.36) leads to

$$\alpha = [\beta_0(1 - |T|)/d|T|]^{1/2} \tag{3.38}$$

consequently, the conductivity expressed in terms of the transmission coefficient by means of Eqs. (3.36) and (3.38), is

$$\sigma_0 = (2\varepsilon_0 c^2/\omega)[\beta_0(1 - |T_0|)/d|T_0|] \tag{3.39}$$

where $|T_0|$ represents the transmission coefficient in zero magnetic field. The effective conductivity σ_0 can be determined in terms of the known values of ω and β_0, the measured $|T_0|$ and d. Assuming, in analogy, that in the presence of a magnetic field the attenuation coefficient $\alpha(B)$ is dependent on $\rho(B)$ then

$$\alpha(B) = [\omega/2\varepsilon_0 c^2 \rho(B)]^{1/2} \tag{3.40}$$

The magnetoresistance can then be expressed as

$$(\Delta\rho/\rho_0) = (|T| - |T_0|)\cdot\Big[|T_0|\cdot(1 - |T|)\Big]^{-1} \tag{3.41}$$

where $|T|$ is the absolute value of the transmission coefficient. The microwave magnetoresistance of the specimen can thus be determined in terms of measurements of $|T|$ vs. B.

A block diagram of the apparatus used for microwave magnetoresistance measurements is shown in Fig. 3.5 and the details of the specimen mounting in a specimen

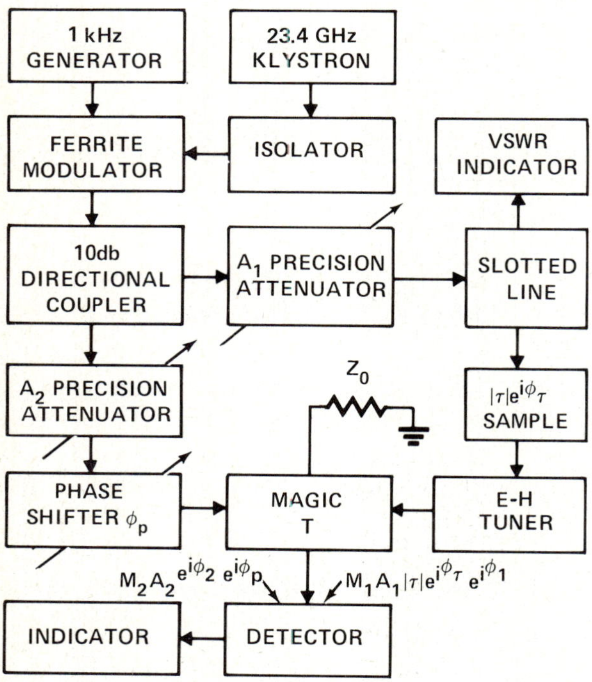

Fig. 3.5. K-band microwave bridge circuit used for measuring microwave resistivity and magnetoresistance of InSb films as a function of magnetic field. [after H. H. Wieder and D. J. White, J. Appl. Phys., 39(1968)2401]

holder and its position within an electromagnet equipped with hollow pole pieces are shown in Fig. 3.6. The specimen extends over the entire inner cross-section of the K-band waveguide with an overlap of $\sim$0.1 cm on each side. In view of the geometry of the waveguide and its ℓ/w ratio a significant portion of the Hall field in the specimen is short-circuited. The apparatus assembled in Fig. 3.5 is a microwave K-band bridge circuit. With the attenuators A_1 and A_2 and the phase adjusted for balance (ref. 43) the absolute value of the transmission coefficient is

$$|T| = A_{10}A_2/A_{20}A_1 \tag{3.42}$$

where A_{10} and A_{20} correspond to the positions of attenuators A_1 and A_2 with the specimen removed and the bridge rebalanced. The magnetoresistance can now be

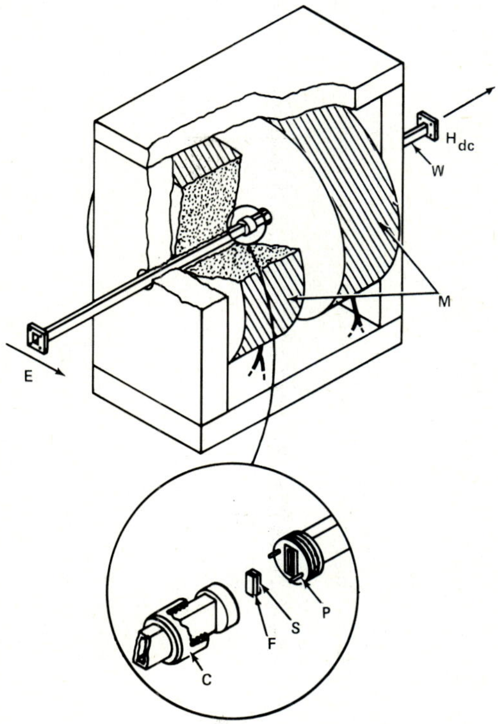

Fig. 3.6. Film specimen mounted in its sample holder made of K-band rectangular TE_{01} waveguide W. Medallion shows specimen S, held in place by flange C and guide-pins P. The direction of the rf electric field E and that of the magnetic field H_{dc} generated by electromagnet M are also indicated. [after H. H. Wieder and D. J. White, J. Appl. Phys., 39(1968)2401]

calculated by means of Eqs. (3.39), (3.41) and (3.42). Figure 3.7 compares measurements made at DC and 23.4 GHz on an InSb film grown on a glass substrate (ref. 43). The DC measurements were made for two orientations, with electrodes along the a and b directions (with respect to the waveguide), $(\Delta\rho/\rho_o)_a$ and $(\Delta\rho/\rho_o)_b$. The good agreement between the microwave magnetoresistance and

$(\Delta\rho/\rho_o)_a$ is due to the fact that the TE_{01} mode microwave E-field is directed along the shorter dimension of the waveguide and the specimen.

Koike and Barlow (ref. 44) measured the microwave magnetoresistance of an InSb thin film grown on an insulating mica substrate placed along the longitudinal axis of a waveguide, as shown in Fig. 3.8, with a variable magnetic field B, up

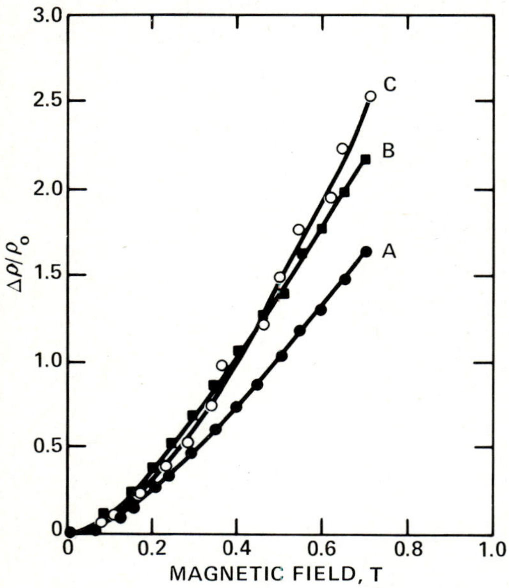

Fig. 3.7. Magnetic field-dependence of magnetoresistance of an InSb film; (a) measurements made at DC with electrodes across b edges (Fig. 3.4); (b) DC magnetoresistance with electrodes across a edges; (c) magnetoresistance calculated from transmittance measurements made at 23.4 GHz. [after H. H. Wieder and D. H. White, Appl. Phys., 39(1968)2401]

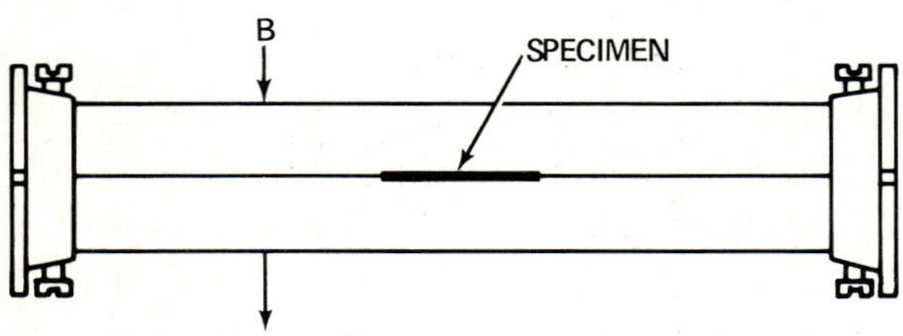

Fig. 3.8. Longitudinal orientation of specimen and magnetic field for magnetoresistance measurements in a waveguide. [after R. Koike and H. E. M. Barlow, Proc. IEE, 109(1962)137]

to 0.45 T, oriented in a direction transverse to the waveguide. A microwave impedance bridge circuit was used to determine, at 9.478 GHz the change in attenuation and phase along the guide as a result of the applied magnetic field acting on the specimen. A precision attenuator, having a $\cos^2\theta$ calibration curve with an error of ±2%, and a precision phase shifter were used to compensate for the properties of the specimen whose attenuation coefficient depends linearly on the conductivity: $\alpha \simeq c_1[\sigma(B)]^{\eta}$ where $\eta \simeq 0.891$ and c_1 is a constant of proportionality. The fractional change in resistance and the corresponding change in attenuation is $\Delta R/R=(1/\eta)\cdot\Delta\alpha/\alpha$. A similar experimental configuration was used by Sun (ref. 45) to measure the 10 GHz magnetoresistance of InSb wafers with a room temperature electron mobility of $\sim$ 7xk0^4 cm^2/V-sec. With an external constant magnetic field applied such as shown in Fig. 3.8 the resistance of the specimen can be related to the oscillatory magnetic field H in the waveguide by a linear relation of the form

$$R = R_o + mH\cdot\cos(\omega t - \beta z) \tag{3.43}$$

where $m=(dR/dH)_B$ is a coefficient proportional to the slope of the magnetoresistance curve at the biasing value of the constant magnetic field B and R_o is the initial resistance of the semiconductor with only the steady-state magnetic field B applied to it. Since the conductivity of InSb is essentially the same at DC and at 10 GHz the voltage drop across the specimen is a function of R and has a DC component which is a function of the microwave power absorbed by the specimen. This DC voltage produced by the oscillatory component of the magnetoresistance is shown in Fig. 3.9 to be in good agreement with theoretical calculations and the coefficient m was found to be of the same order of magnitude as that measured at DC.

Kataoka and Fujisada (ref. 46) used a rectangular cavity such as shown in Fig. 3.10 to determine, at room temperature, the magnetic field dependence of the magnetoresistance of single crystal InSb in magnetic field up to 1.5T. One wall of the cavity shown in Fig. 3.10 is removable and can be replaced by a rectangular tight-fitting InSb wafer whose thickness, $\sim$ 1 mm, is considerably greater than the effective skin depth of $\sim$ 20 μm at the frequency of measurement, 34 GHz. A variable DC magnetic field B was applied parallel to the specimen surface. The magnetoresistance was determined by measuring the Q factor of the cavity under various conditions. If the loaded Q of the cavity in which the specimen is mounted is Q_ℓ (for B=0), that of the cavity when a metal plate replaces the specimen is Q_r, and that of the metal plate itself is Q_m then the $Q(B=0)=Q_o$ of the specimen can be calculated from

$$1/Q_o = 1/Q_\ell - 1/Q_r + 1/Q_m \tag{3.44}$$

In the presence of a magnetic field B the characteristic Q of the specimen,

$Q(B)=Q_B$ and the microwave magnetoresistance is

$$\Delta\rho/\rho_o = (Q_o/Q_B)^2 - 1 \tag{3.45}$$

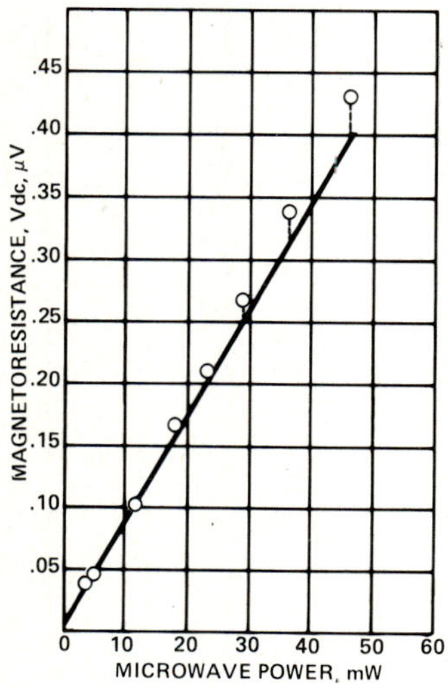

Fig. 3.9. Magnetoresistance voltage V_{dc} in relation to total input microwave power at a constant bias magnetic field. The small circles are measured points and the straight line is calculated from theory. [after S. F. Sun, J. Appl. Phys., Phys., 35(1964)211]

Their experimental magnetoresistance measurements made in terms of Eq. (3.45) as a function of B determined the "physical" $\Delta\rho/\rho_o$ due to two-band conduction since at room temperature undoped InSb is essentially in the intrinsic regime and both, electrons and holes, contribute to the conductivity. They found that the microwave magnetoresistance is smaller than the corresponding DC value and the discrepancy between them increases with B. The magnetoresistance is expected to decrease with increasing frequency of measurement as it approaches the charge carrier relaxation frequencies. Kataoka and Fujisada calculated the ratio of the DC magnetoresistance to its 34 GHz value as γ=1.15 using the high frequency conductivity tensor such as that given in Eq. (2.29) and respective electron and hole masses of $m^*_n=0.013\ m_o$ and $m^*_p=0.6\ m_o$. They found fair agreement between their experimental measurements and the calculated γ=1.15 for $B\geq 1$ T.

Similar investigations were made by Clark and Assenheim (ref. 47) on bismuth using a resonant coaxial line to measure the magnetoresistance of Bi at frequencies of 1.1, 1.5, and 9.2 GHz. A portion of the center conductor of the coaxial cavity was made of Bi and $\Delta\rho/\rho_o$ vs. B was calculated from the change in

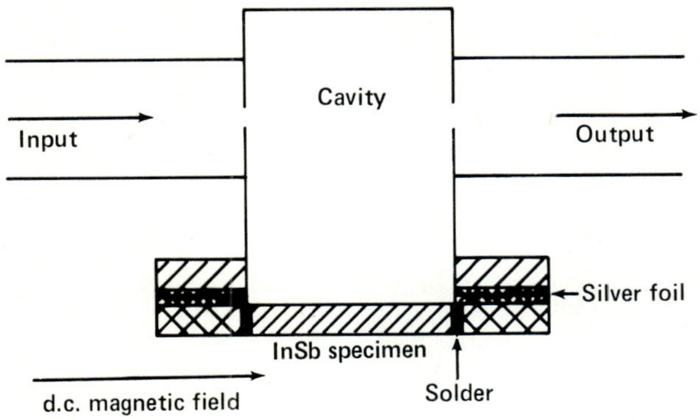

Fig. 3.10. Rectangular microwave cavity used for measuring the magnetoresistance of semiconducting specimens whose thickness is much greater than their skin depth at 34 GHz. [after S. Kataoka and H. Fujisada, Solid-State Electron., 11(1968)163]

the cavity Q with magnetic field. Their experimental measurements were found to be in fair agreement with the model of Donovan (ref. 48), in which electrons occupy two over-lapping partially-filled bands and the magnetoresistance falls off sharply above their relaxation frequency, and the experimental measurements of Heaps (ref. 49) in which the magnetoresistance measured at 3 GHz is half of its DC value.

REFERENCES

1 R. A. Smith, Semiconductors, Cambridge Univ. Press (1968)114.
2 E. Justi, Z. Tech. Physik, 21(1940)315.
3 R. G. Chambers, "Magnetoresistance" The Fermi Surface, W. A. Harrison and M. B. Webb, ed's, Wiley (1961)100.
4 H. Jones and C. Zener, Proc. Roy. Soc.(London)A145(1934)268.
5 I. M. Lifshitz, M. Ya. Azbel and M. I. Kaganov, Sov. Phys. JETP, 3(1956)143.
6 I. M. Lifshitz and V. G. Peschanskii, Sov. Phys. JETP, 8(1958)875.
7 R. F. Wick, J. Appl. Phys., 25(1954)741.
8 J. R. Drabble and R. Wolfe, J. Electron. Control, 3(1957)259.
9 R. F. Broom, Proc. Phys. Soc.(London) 71(1958)500.
10 H. Weiss and H. Welker, Z. Physik, 138(1957)322.
11 O. M. Corbino, Nuovo Cimento, 1(1911)397.
12 M. G. Buehler and G. L. Pearson, Solid-State Electron., 9(1966)395.
13 M. G. Buehler, Solid-State Electron., 10(1967)801.
14 L. Halpern and K. M. Koch, Acta Phys. Austria., 5(1951)129.
15 H. J. Lippmann and F. Kuhrt, Z. Naturforsch., 13a(1958)462.
16 M. Green, Solid-State Electron., 3(1961)314.
17 M. Green, Solid-State Electron., 6(1963)63.
18 A. C. Beer, J. Appl. Phys., 32(1961)2107.

19 H. H. Wieder, J. Appl. Phys,. 40(1969)3320.
20 A. C. Beer, Solid State Phys., 4(1963)159.
21 C. Herring, J. Appl. Phys., 31(1960)1939.
22 R. T. Bate and A. C. Beer, J. Appl. Phys., 32(1961)800.
23 R. T. Bate, J. C. Bell and A. C. Beer, J. Appl. Phys., 32(1961)806.
24 H. Rupprecht, Z. Naturforsch., 16a(1961)395.
25 H. Weiss and M. Wilhelm, Z. Physik, 176(1963)399.
26 H. H. Wieder and D. A. Collins, Appl. Phys. Lett., 8(1966)239.
27 H. H. Wieder and D. A. Collins, Solid-State Electron., 11(1968)1093.
28 G. S. Nadkarni, A. Simoni and J. G. Simmons, Solid-State Electron., 18(1975)393.
29 T. R. Jervis and E. F. Johnson, Solid-State Electron., 13(1970)181.
30 R. D. Larrabee, W. A. Hicinbothem, Jr. and M. C. Steele, Trans. Electron. Dev., ED-17(1970)271.
31 W. S. C. Gurney, Electron. Lett., 7(1971)711.
32 L. Gutai and I. Mojzes, Appl. Phys. Lett., 26(1975)325.
33 L. C. Upadhyayula, R. E. Enstrom, C. J. Nuese, J. F. Wilhelm and J. R. Appert, IEEE Trans. Electron Dev. ED-23(1976)1348.
34 H. Weiss in "Semiconductors and Semimetals," R. K. Willardson and A. C. Beer, ed. Physics of III-V Compounds, Academic Press, 1(1966)333.
35 C. Hilsum and R. Barrie, Proc. Phys. Soc.(London) 71(1958)676.
36 F. Seitz, Phys. Rev., 79(1950)372.
37 M. Glicksman, Progr. Semicond., 3(1958)3.
38 G. de Mey, X. Burvenish and M. de Molder, Phys. Stat. Sol., a23(1974)K45.
39 R. A. Smith, Semiconductors, Cambridge Univ. Press (1968)Ch. V.
40 D. J. Howarth and E. H. Sondheimer, Proc. Phys. Soc., A219(1953)53.
41 P. Blood and R. J. Tree, J. Phys. D, Appl. Phys., 4(1971)L29.
42 H. Jacobs, F. A. Brand, J. D. Meindl, M. Benanti and R. Benjamin, Proc. IRE, 49(1961)928.
43 H. H. Wieder and D. J. White, J. Appl. Phys., 39(1968)2401.
44 R. Koike and H. E. M. Barlow, Proc. IEE, 109B(1962)137.
45 S. F. Sun, J. Appl. Phys., 35(1964)211.
46 S. Kataoka and H. Fujisada, Solid-State Electron., 11(1968)163.
47 D. E. Clark and J. G. Assenheim, Brit. J. Appl. Phys., 11(1960)35.
48 B. Donovan, Proc. Phys. Soc., A67(1954)305.
49 C. W. Heaps, Phys. Rev., 80(1950)892.

CHAPTER 4

CHARGE CARRIER TRANSPORT PARAMETERS

4.0 Introduction

The type of charge carriers, their densities, their mobilities, their effective masses and scattering parameters can be determined by means of electrical and galvanomagnetic measurements such as those described in Chapter 1 through 3. The carrier concentration of a metal or a degenerate semiconductor with a single conduction band, a spherical Fermi surface and an isotropic energy independent relaxation time can be determined from a measurement of its Hall coefficient

$$n = -1/eR_h \qquad\qquad p = 1/eR_h \tag{4.1}$$

where e is the electronic charge and the Hall coefficient R_h, following the usual sign convention, is negative for electrons in a specimen with the orientation of the applied current and magnetic field shown in Fig. 2.1 and it is positive if holes are the charge carriers. The electron concentration n or the hole concentration p are usually expressed in cm^{-3} with R_h expressed in cm^3/C. The measured conductivity $\sigma=en\mu_n$ or $\sigma=ep\mu_p$ can be used, together with the measured Hall coefficient $\pm R_h$ to calculate the electron mobility μ_n or the hole mobility μ_p from

$$\mu_n = R_h\sigma \qquad\qquad \mu_p = R_h\sigma \tag{4.2}$$

where μ_n and μ_p are expressed in cm^2/V-sec. with σ expressed in $(ohm\text{-}cm)^{-1}$. The mobility can be expressed in terms of an effective mass $m^*_{n,p}$ and relaxation time $\tau_{n,p}$ as

$$\mu_n = e\tau_n/m^*_n \qquad\qquad \mu_p = e\tau_p/m^*_p \tag{4.3}$$

If τ is a function of the energy of the charge carriers then it can be expressed as an energy-averaged relaxation time, $<\tau>$. The Hall coefficient of a material containing a single, non-degenerate, spherically symmetric, conduction band can then be expressed in terms of a scattering factor or Hall factor, r_B (ref. 1) as $R_h=r_B/en$; the value of r_B depends on the characteristics of the scattering process (ref. 2) in accordance with Table 4.1

$$r_B = <\tau^2>/<\tau>^2 \tag{4.4}$$

The energy-averaged relaxation time also makes the Hall mobility different from the drift mobility (measured in B=0),

TABLE 4.1
Dependence of Hall factor r_B on scattering process.

Scattering process	r_B
Acoustic lattice scattering (non-polar)	$3\pi/8$
Polar lattice scattering (piezoelectric)	1.10
Optical lattice scattering (non-polar)	$\sim$1.00
Optical lattice scattering (polar mode)	1.10-1.34
Ionized impurity scattering	1.18-1.70
Lattice defect scattering	1.34-1.40

$$\mu_B = r_B \mu \tag{4.5}$$

If r_B is not known or its value is in doubt then the assumption that r_B=1.0 does not introduce any gross errors. For polar mode scattering (refs. 3-6) for which the relaxation time approximation is invalid r_B is greater than unity and for non-spherical energy surfaces r_B depends on the anisotropy of the energy surfaces and on the anisotropy of charge carrier scattering (refs. 7-9). The effective mass m*, considered as a constant of proportionality which relates the energy E of an electron to the square of its wave vector $\bar{k}$ defines the curvature of the E vs. $\bar{k}$ relation; the smaller the effective mass the faster does the energy increase as $\bar{k}$ increases. In general, m* depends on the direction of the wave vector in a crystal and is, therefore, a tensor quantity. For the special case of spherical constant-energy surfaces the effective mass is a scalar. If the Fermi surface is not spherical with a single energy minimum at $\bar{k}$=0 then a choice of coordinate axes can be made such that the effective mass tensor is diagonal; in that case, the mobility is also a diagonal tensor whose components may be described by $\mu_i = e\langle\tau\rangle/(m^*)_i$ where i=1,2,3 and m_1^*, m_2^*, and m_3^* are the diagonal elements of the effective mass tensor. If there are N equivalent minima in the conduction band then there are n/N electrons per unit volume in each of them. For silicon, there are six equivalent minima and the constant-energy surfaces are ellipsoids. The conductivity tensor components are $\sigma_i = (1/3)ne\mu_i$ (with i=1,2,3) and the components of the effective mass tensor are

$$\frac{1}{m_n^*} = \frac{1}{3}\left(\frac{1}{m_x^*} + \frac{1}{m_y^*} + \frac{1}{m_z^*}\right) \tag{4.6}$$

Figure 4.1 shows the first Brillouin zone for electrons in silicon. The six equivalent Δ_{1c} minima are located along <100> directions; two of the effective

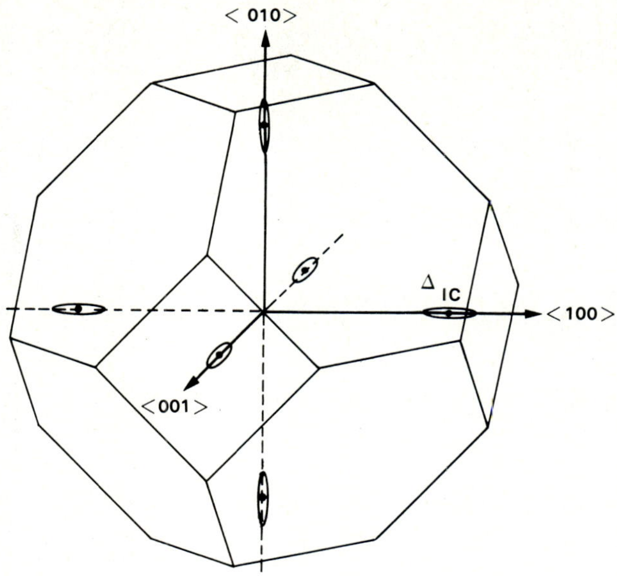

Fig. 4.1. First Brillouin zone for electrons in Si. The six Δ_{1c} minima located along <100> are ellipsoidal constant-energy surfaces.

masses, m_y^* and m_z^* are equal and may be represented by an equivalent transverse effective mass, m_T^* while the longitudinal effective mass $m_L^*=m_x^*$. Thus Eq. (4.6) can be written in the form,

$$\frac{1}{m_n^*} = \frac{1}{3}\left(\frac{1}{m_L^*} + \frac{2}{m_T^*}\right) \tag{4.7}$$

The Hall factor r_B is also dependent on the effective mass anisotropy and can be expressed for ellipsoidal energy surfaces as

$$r_B = \frac{\langle\tau^2\rangle}{\langle\tau\rangle^2} \cdot \frac{3\gamma(\gamma + 2)}{(2\gamma + 1)^2} \tag{4.8}$$

where $\gamma=m_L^*/m_T^*$.

4.1 Intrinsic and near-intrinsic conduction

Electrical and galvanomagnetic measurements made in order to determine fundamental charge carrier transport parameters should be made over a wide temperature range in order to identify the specific attributes of intrinsic and extrinsic conduction processes. A semiconductor exhibits intrinsic conduction when thermally-excited electrons and holes dominate charge carrier transport in a particular temperature region. An energy bandgap, $E_g=E_c-E_v$, separates the lowest conduction band minimum from the uppermost energy level of the valence band. At

T=0°K the valence band is completely filled and the conduction band is empty. At some finite temperature electrons in the valence band may gain sufficient energy from the lattice to make a transition into an allowed state in the conduction band; in the effective mass approximation, the holes in the valence band behave as positive charge carriers with a positive mass and positive charge. For a non-degenerate semiconductor with a parabolic conduction band and valence band the electrons and holes can be described by Maxwell-Boltzmann statistics (ref. 10) with

$$n = N_c \exp[(E_F - E_c)/kT] \qquad (4.9)$$
$$p = N_v \exp[(E_v - E_F)/kT]$$

where k is Boltzmann's constant, T the absolute temperature and E_F is the Fermi energy which is expressed as

$$E_F = -\frac{E_g}{2} + \frac{3}{4}\left(kT \cdot \ln \frac{m_n^*}{m_p^*}\right) \qquad (4.10)$$

and the conduction band and valence band density of states functions, N_c and N_v, are, respectively,

$$N_c = 2\left(\frac{2\pi m_n^* kT}{h^2}\right)^{3/2} \qquad N_v = 2\left(\frac{2\pi m_p^* kT}{h^2}\right)^{3/2} \qquad (4.11)$$

where h is Planck's constant. In view of the mass action law, $np=n_i^2$, the density of intrinsic charge carriers is

$$n_i = (N_c N_v)^{1/2} \exp[-(E_c - E_v)/2kT] \qquad (4.12)$$

$n_i(T)$ is a material characteristic and is independent of the Fermi energy. If ionized impurities present in a material contribute electrons or holes then the term extrinsic refers to the temperature range in which these are the dominant charge carriers. In an extrinsic semiconductor, if N_d is the density of donor atoms per unit volume and N_a the density of acceptor centers per unit volume, n_d the electron density per unit volume bound to donor sites and p_a the hole density per unit volume bound to acceptor sites, then for electrical neutrality

$$n + N_a - p_a = p + N_d - n_d \qquad (4.13)$$

Now $N_d - n_d$ is the density of donor atoms which have lost their extra electrons and $N_a - p_a$ is the density of acceptor atoms which have lost their trapped electrons. In an n-type semiconductor in which $N_a=0$ for all temperatures for which the donors are fully ionized so that $n_d \ll N_d$, Eq. (4.13) becomes $n=p+N_d$ and by means of Eq. (4.12)

$$n = \frac{N_d}{2} + \left[\left(\frac{N_d}{2}\right)^2 + n_i^2\right]^{1/2} \tag{4.14}$$

As the temperature rises from absolute zero, electrons or holes may be activated from donor or acceptor sites. At some particular temperature the impurities are fully ionized. As the temperature rises further thermally-generated electrons and holes contribute to the conduction process. At sufficiently high temperatures $n_i >> N_d$, N_a and the conduction process is essentially intrinsic.

In the intrinsic temperature region, the weak field (B→0) Hall coefficient and resistivity are

$$R_h = \frac{p - b^2 n}{e(p + bn)^2} \qquad \rho^{-1} = e\mu_p(p + bn) \tag{4.15}$$

where b is the electron-to-hole mobility ratio. Dunlap (ref. 11) has investigated the dependence of the Hall mobility, $\mu = R_h/\rho$ on ρ in terms of the relative contributions of intrinsic and extrinsic processes. By algebraic manipulation Eq. (4.15) can be expressed as

$$\left[\frac{\mu + (1/2)(\mu_n - \mu_p)}{(1/2)(\mu_n + \mu_p)}\right]^2 + \rho^2(4e^2\mu_n\mu_p np)^{-1} = 1 \tag{4.16}$$

A functional representation of μ vs. ρ in Eq. (4.16) is the ellipse of Fig. 4.2 (ref. 12); its semi-minor axis is half the sum of the hole and electron mobilities; its semi-major axis is the resistivity of the specimen. The maximum resistivity can be greater than the intrinsic resistivity of a pure specimen. For materials whose mobility ratios b>1 the maximum resistivity ρ_{max} occurs for a slightly p-type material even though their Hall coefficients, may still be negative. The Hall coefficients can also be greater than those of the intrinsic Hall coefficients R_{hi}; the locus of points in the plane of Fig. 4.2 which represents a fixed value of R_{hi} is determined by an equilateral hyperbola which intersects the ellipse at points A and I. If the mobility ratio b is independent of temperature then a single ellipse and its associated hyperbola represents all specimens of the same mobility ratio for all temperatures, as shown by Hunter (ref. 12), who described, in detail, the resistivity and Hall coefficient overshoot for various mobility ratios.

A correct evaluation of the overshoot is necessary if the energy gap of a semiconductor is to be determined from the slope of the intrinsic Hall coefficient or resistivity as a function of temperature, in terms of Eq. (4.12). If b is known then an ellipse may be constructed and the overshoot determined. The true extrapolated intrinsic line may then be plotted from the data using the proper overshoot. The applicable assumptions include: (a) an infinitesimal

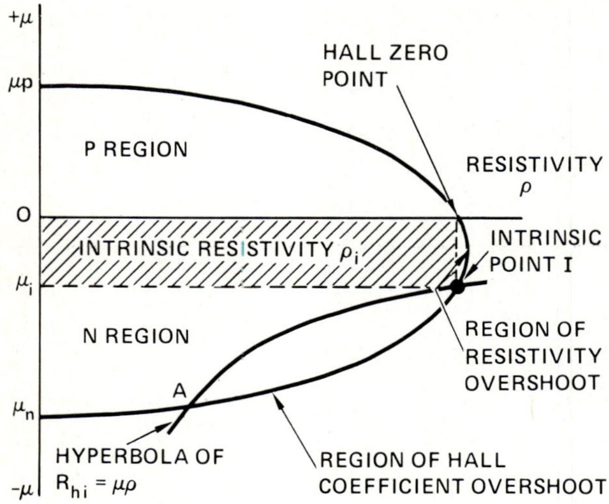

Fig. 4.2. Dependence of the Hall mobility, $\mu=R_h/\rho$, on the resistivity ρ for a mobility ratio b=1.5. [after L. P. Hunter, Phys. Rev., 94(1954)1157]

lifetime of the charge carriers, (b) a mobility ratio b which is not dependent on temperature, and (c) a mobility ratio which is invariant with magnetic field. Condition (a) can usually be met by chemically or physically treating the surfaces of a semiconductor in order to increase the carrier surface recombination velocities. Condition (b) is often assumed to be the case but requires justification; it is usually true only over a limited range of temperature. Condition (c) has been discussed by Hunter (ref. 12) who showed that even if the drift mobilities differ from the Hall mobilities the μ vs. ρ functional dependence is still that of an ellipse. However the R_h=0 point and the intrinsic point no longer occupy positions of equal resistivity value on the ellipse. Only if b=b(B) will the resistivity crossover take place at the same temperature as for R_h=0. The Hall coefficient crossover and overshoot are, in general, functions of both b and b(B) while the resistivity overshoot is a function of b alone.

The effect of finite charge carrier lifetime on the transport parameters was analyzed by Landauer and Swanson (ref. 13). Figure 4.2 was derived under the tacit assumption of an infinitesimal charge carrier lifetime. For a finite lifetime the μ vs. ρ ellipse is distorted (ref. 12) in such a way that the R_h=0 point moves into the region of resistivity overshoot; for an essentially infinite lifetime the R_h=0 point coincides with the maximum overshoot.

Hunter (ref. 14) has proposed a method for determining the mobility ratio from resistivity measurements alone without recourse to Hall measurements by the use of resistivity vs. reciprocal temperature curves measured experimentally, such as the one shown in Fig. 4.3. For p-type specimens the mobility ratio $b=(r-1)^{-1}-r$ and for n-type specimens $b^{-1}=(r-1)^{-1}-r$ where r is the ratio of the slopes of the extrinsic to the intrinsic ρ vs. 1/T lines shown in Fig. 4.3. Using such a

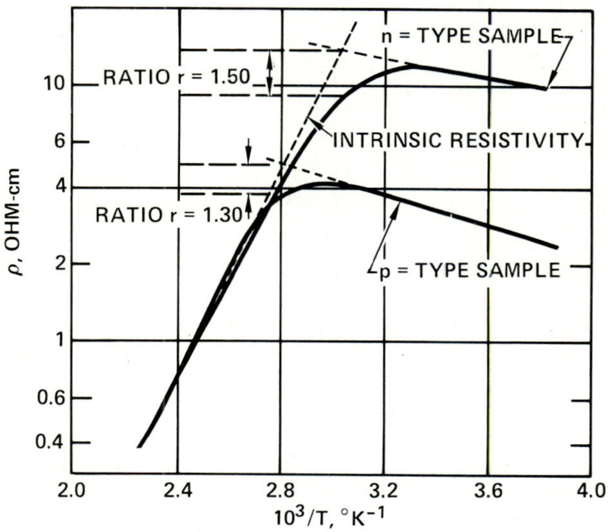

Fig. 4.3. Resistivity of n- and p-type Ge as a function of temperature; extrapolation of the intrinsic and extrinsic line provides the data for the evaluation of the mobility ratio. [after L. P. Hunter, Phys. Rev., 94(1954)1157]

procedure, Hunter determined a value of b=2.1 for germanium in good agreement with measurements made by means of other methods.

If the expression for the weak-field Hall coefficient in Eq. (4.15) is combined with that of the mass action law $n_i^2=np$ where n_i is the intrinsic charge carrier density then a quartic equation is obtained whose solution leads to four sets of values of n and p; two sets are negative and two are positive. Robinson *et al.* (ref. 15) developed a graphical method for calculating the positive values from a single measurement of R_h if n_i and b are known or determined from other data. From Eq. (4.15) solving for p and introducing the Hall factor r_B,

$$p = \frac{r_B}{2}\left\{\frac{1}{R_h} - \frac{2bn}{r_B} \pm \left[\frac{1}{R_h^2} - \frac{4bn(1+b)}{R_h r_B}\right]^{1/2}\right\} \tag{4.17}$$

Multiplying Eq. (4.17) by R_h and writing n^* and p^* for nR_h and pR_h, respectively, leads to

$$p^* = \frac{r_B}{2}\left\{1 - \frac{2bn^*}{r_B} \pm \left[1 - \frac{4bn^*(1+b)}{r_B}\right]^{1/2}\right\} \qquad (4.18)$$

Thus p^* is a function only of n^* and a graphical construction such as shown in Fig. 4.4 is valid for all values of R_h. The continuous curves represent values

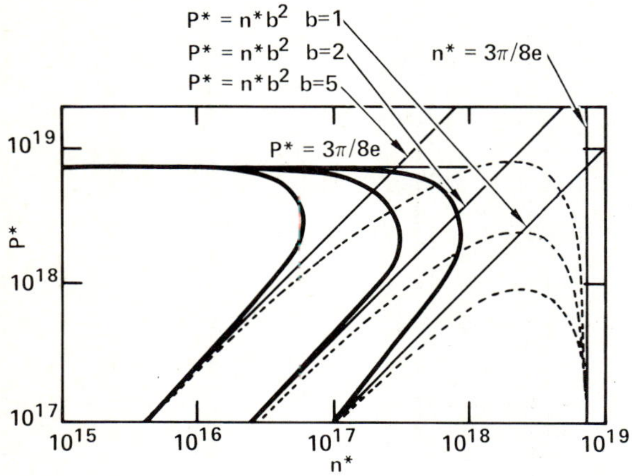

Fig. 4.4. Dependence of p^* on n^* from Eq. (4.18) with b as a running parameter. [after T. S. Robinson, M. Smolett and R. G. Pratt, Brit. J. Appl. Phys., 8(1957) 130]

of n^* and p^* for positive R_h and the dashed curves correspond to a negative R_h. Both negative and positive curves tend to the straight lines (on the logarithmic plot) of $n^*, p^*=r_B$. Each curve tends towards $p^*=n^*b^2$ corresponding to the particular values of b shown here. The mass action law may also be expressed as $n^*p^*=(R_h n_i)^2$. This equation corresponds to a hyperbola which becomes a straight line on the logarithmic scale in Fig. 4.4 and it intersects every curve in two specific points each of which corresponds to a pair of values of n^* and p^*. Using these, together with the measured conductivity, the corresponding values of the mobilities can be calculated rejecting the "non-physical" values.

A plot of $\ell n\, n_i$ vs. $1/T$ from Eq. (4.12) made in order to determine the intrinsic energy bandgap, $E_g=E_v-E_c$ is not quite linear because the density of states functions in Eq. (4.11) are themselves functions of $T^{3/2}$. It is, therefore, often preferred to construct graphs of $\ell n\,(n_i T^{3/2})$ vs. $1/T$ (ref. 10). Care is required in the quantitative evaluation of E_g, usually expressed as a linear relation of the form $E_g=E_{g_0}-\alpha T$, where α is a coefficient which represents the change with temperature of the fundamental bandgap and E_{g_0} is its value at absolute zero,

$$(n_i T^{3/2}) = 2\left(\frac{4\pi^2 m_n^* m_p^* k^2}{h^4}\right)^{3/4} \cdot \exp(\alpha/2k) \cdot \exp(-E_{g_0}/2kT) \tag{4.19}$$

A plot of $\ell n(n_i T^{3/2})$ vs. 1/T using Eq. (4.19) still yields a linear relation but the slope is indicative of E_{g_0}. Further complications arise from the presence of non-parabolic conduction bands or intrinsic degenerate narrow bandgap semiconductors. These issues have been treated by Blakemore (ref. 10) and the reader is referred to his monograph for details.

4.2 Extrinsic and compensated materials

A semiconductor containing only donor impurities, N_d characterized by a single energy level E_d may have a temperature-dependent fraction, N_{di} present in ionized form, with each ionized donor atom presumed to contribute a conduction electron and the residue of the donors, $N_{dn}=N_d-N_{di}$ present as neutral centers. For such a non-degenerate semiconductor the electron concentration in the presence of a negligible intrinsic charge carrier contribution is

$$n = 2N_d \left\{1 + \left[1 + \frac{4N_d}{\beta N_c} \cdot \exp\left(\frac{E_d}{kT}\right)\right]^{1/2}\right\}^{-1} \tag{4.20}$$

where N_c is the density of states in the conduction band in Eq. (4.11); however the effective mass m_n^* which appears in N_c is not, as a rule, the same as the conductivity effective mass in Eq. (4.6). For materials with high symmetry conduction bands it is a function of m_1*, m_2*, and m_3* of the form $m_n^*=(m_1^* m_2^* m_3^*)^{1/3}$ and an analogous relation applies to the effective masses of valence bands. The parameter β is the electron spin level degeneracy. It is usually considered to have a value of 1/2 for donor impurities and $\beta=2$ for acceptor impurities; E_d is the ionization energy of the donor impurities. At low temperatures such that $kT<<E_d$ and $n<<N_d$ Eq. (4.20) can be approximated by

$$n \simeq (\beta N_c N_d)^{1/2} \cdot \exp -\left(\frac{E_d}{2kT}\right) \tag{4.21}$$

Therefore, a plot of ℓn n vs. 1/T, made at low temperatures, yields the value of E_d from the slope of the curve provided that due account is taken of the temperature dependence of N_c.

It is extremely difficult to prepare a material with a single donor level or one which does not also contain acceptor impurities (ref. 16). Such a material is usually described as compensated since the number of free electrons is less than the number of impurity centers. The total donor density, $N_d=n+N_{dn}+N_a$. If $N_a<<n<<N_d$ then Eq. (4.21) is still applicable. At low temperatures, where $n\leq N_a$ the electron density is

$$n = \beta N_c \left(\frac{N_d - N_a}{N_a} \right) \exp \left(- \frac{E_d}{kT} \right) \qquad (4.22)$$

Care is required in differentiating between Eqs. (4.21) and (4.22) in the evaluation of experimental measurements presented as ℓn n vs. 1/T; the ionization energy might appear to be half its actual value if the material is considered to be uncompensated. For heavy compensation

$$n = (N_d - N_a) \left[1 + \frac{N_a}{\beta N_c} \exp \left(\frac{E_d}{kT} \right) \right]^{-1} \qquad (4.23)$$

More than one donor level may be present in an n-type material, as shown schematically in Fig. 4.5 for two donor levels N_{1d} and N_{2d}, with corresponding

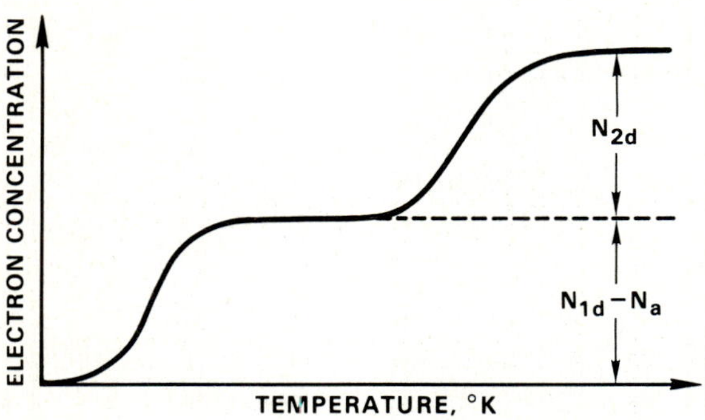

Fig. 4.5. Example of electron density as a function of temperature for a compensated semiconductor with two different donor impurities centers.

energies E_{1d} and E_{2d}. If $N_a < N_{1d}$ then the states corresponding to E_{2d} are full and $(N_{1d}-N_a)$ electrons will occupy the E_{1d} levels at 0°K. As the temperature rises the N_{1d} donors are activated providing conduction electrons. When all the N_{1d} levels are exhausted the electron density reaches the plateau shown in Fig. 4.5 and, as the temperature increases further the electron concentration rises again until all the N_{2d} levels are ionized. The qualitative arguments used here for an n-type material in which acceptors compensate the donors apply, by analogy, to p-type materials in which the donors act as compensators. In that case p replaces n, N_v replaces N_c and N_a is interchanged with N_d in the appropriate equations.

An interpretation of experimentally measured data, such as the plot of ℓn n vs. 1/T of a partially compensated non-degenerate n-type semiconductor shown, schematically, in Fig. 4.6 may be made by means (ref. 10) of the equation

$$\frac{n(n + N_a)}{N_d - N_a - n} = \beta N_c \bullet \exp\left(- \frac{E_d}{kT}\right) \tag{4.24}$$

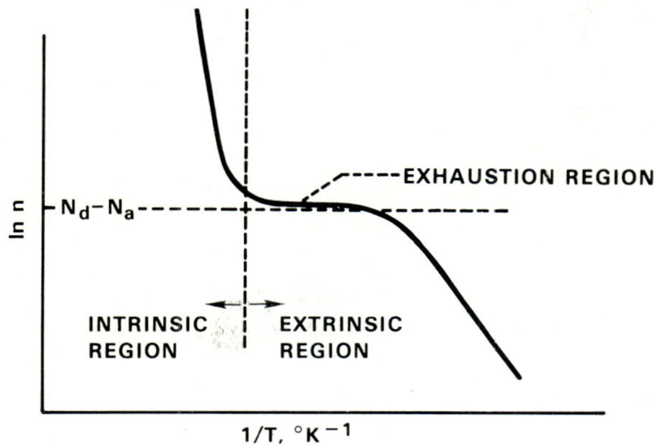

Fig. 4.6. Typical temperature dependence of the electron density of a partly compensated n-type semiconductor.

A first order estimate is made of the parameters N_a, N_d, N_c, E_d, and β and then, by an iterative procedure such as a least squares method of analysis, the parameters are adjusted until the temperature dependence of the calculated carrier concentration matches, as closely as required, the experimentally measured n(T). The density of states function can be determined as a function of temperature if the effective mass m_n^* and its temperature dependence is available from other, independent measurements such as those based on cyclotron resonance, optical or magneto-optic measurements. N_d and N_a can be evaluated from a plot such as that shown in Fig. 4.7 (refs. 16,17). The low temperature portion of the curve is extrapolated using Eq. (4.22) until it intercepts the exhaustion line. At the temperature which corresponds to their intersection the left hand side of Eq. (4.24) is N_a and the carrier concentration in the exhaustion region, measured experimentally, is equal to N_d-N_a. The donor ionization energy can be obtained from the slope of the ℓn n vs. 1/T curve at low temperatures, as described earlier and β can be fitted by iterative procedures to low temperature measurements of n vs. T using Eqs. (4.22) or (4.23).

The energy level E_d is a ground state for electrons of the donor impurity atoms; an analogy can be drawn between the energy level structure of atoms and those associated with donor impurities. If r=1 is the ground state and r=2 the first excited state then each state has a spin degeneracy β_r^{-1} with an energy $E_{1r} \pm kTE_{1r}$ above the ground state. The measurements made by Eddolls (ref. 18) on the temperature dependence of the electrical properties of n-type epitaxial GaAs layers required the consideration of such excited states by a modification of Eq. (4.24) in the form

$$\frac{n(n + N_a)}{N_d - N_a - n} = \frac{N_c}{\beta_1} \cdot \frac{\exp(- E_d/kT)}{(1 + F)} \qquad (4.25)$$

$$F = \sum_{r=2}^{m} (\beta_r/\beta_1)\exp(- E_r/kT)$$

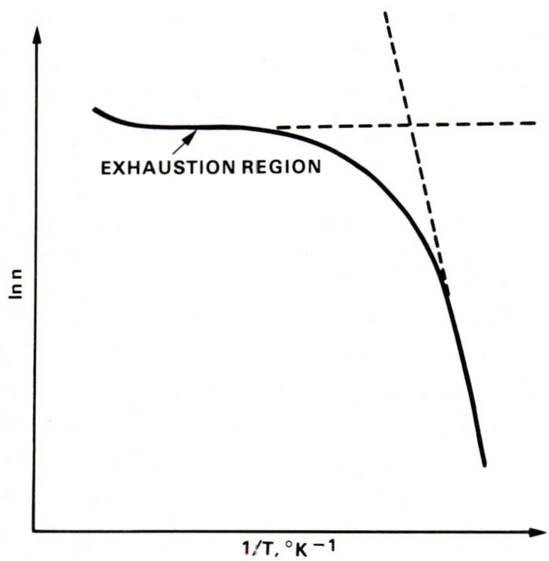

Fig. 4.7. A plot of carrier concentration as a function of reciprocal temperature used for determining the donor and acceptor densities of a compensated semiconductor.

Equations (4.24) and (4.25) coincide at low temperatures where $F \ll 1$ and at high temperatures where $n \approx N_d - N_a$. By assuming that the excited donor states resemble those of a Bohr atom re-normalized to take into account the proper effective mass and dielectric constant,

$$F = 4\exp(- 3E_d/4kT) + 9\exp(- 8E_d/9kT) \qquad (4.26)$$

$$(1 + F) = \sum_{r=1}^{q} r^2\exp[E_d(1 - r^{-2})/kT]$$

It is only necessary to consider the first few terms of this series because for a finite donor density the wave functions of the higher excited states overlap and form part of the conduction band. By making the simplest assumption of one excited state Eddolls (ref. 18) found that with q=2 in Eq. (4.26), used in conjunction with Eq. (4.25), the computed values of $R_h(T)$ differs from experimentally measured values by less than 2% in the temperature range between 8 and 200°K; in contrast, if Eq. (4.24) is used for such calculations then the discrepancy between the theoretically calculated and experimental curves is ∿30%.

4.3 Degeneracy

In the previous Sections of this Chapter the explicit or implicit assumption has been made that the distribution function of the charge carriers is non-degenerate and can be represented by an exponential, in accordance with Maxwell-Boltzmann statistics. In metals where the electron density is of the order of $10^{22}/cm^3$ the Fermi energy is considerably greater than kT and lies at such depth in the conduction band that a simple "metallic approximation" (ref. 19) is sufficient for describing the dependence of the Fermi energy on electron concentration. In general, the electron density in a parabolic conduction band is $n=N_c\mathcal{F}_{1/2}(\eta)$ where $\mathcal{F}_{1/2}(\eta)$ denotes the integral

$$\mathcal{F}_{1/2}(\eta) = \frac{2}{\pi^{1/2}} \int_0^\infty \frac{\varepsilon^{1/2} d\varepsilon}{1 + \exp(\varepsilon - \eta)} \tag{4.27}$$

with $\varepsilon=(E-E_c)/kT$ and $\eta=(E_F-E_c)/kT$. In order to determine whether a distribution is degenerate or non-degenerate it is necessary to determine whether $\exp(\varepsilon-\eta)$ is larger or smaller than unity. When $\exp(\varepsilon-\eta)<<1$ the system is degenerate (ref. 20). The Fermi-Dirac integrals of Eq. (4.27), which have been tabulated by Blakemore (ref. 21) and others, must be used to calculate the position of the Fermi level. Such calculations can be made if the density of states is known and if the electron concentration has been determined experimentally as a function of temperature. Pankove and Annavedder (ref. 22) have prepared the nomogram shown in Fig. 4.8 which allows the determination of any one of the four parameters n,m,*E_F or T if the other three are known. For example, if n and m* are known and are joined by a straight line and then from their intercept on the x-axis a straight line is drawn to the desired temperature T then the corresponding value of η may be found on its scale. Thereafter, transferring this value to the right-hand η scale another straight line to the corresponding temperature on the T scale provides the Fermi level E_F. If η is negative then this implies that E_F is within the energy bandgap. The nomograph is not applicable in the temperature region where carrier freeze-out occurs. A neglect of nonparabolicity which causes an increase of m* away from the band edges (ref. 23) leads to errors; however, the value of E_F determined from the nomogram in this manner is an upper bound. Heavy doping which brings about degeneracy causes a deformation of the lower edge of the parabolic conduction bands by tailing of the density of states function into the bandgap also leads to a value of E_F which is an upper bound.

Intrinsic degeneracy occurs in narrow bandgap semiconductors such that the Fermi level is close to either or both band edges. Using a model which consists of a parabolic density of states, acoustic phonon scattering and a linear temperature dependence of the bandgap, Dalven (ref. 24) found that the temperature dependences of R_h and σ of materials with degenerate charge carrier distributions differ significantly from those of the same non-degenerate materials. Plots of $\ell n|R_h|T^{3/2}$ and of $\ell n\,\sigma$ vs. 1/T are nonlinear in the degenerate intrinsic case and do not have the constant slope $\pm E_g/2k$ obtained by means of Eq. (4.19) using classical statistics; the deviation increases with increasing degeneracy leading to an underestimate of the value of E_g.

4.4. Interconduction and intervalence band transfer of charge carriers

In normal semiconductors the Hall coefficient decreases with increasing temperature because n increases with T, in accordance with Fig. 4.6. There are, however, a considerable number of extrinsic semiconductors in which, over a given temperature region, R_h increases with T. Such behavior is encountered in GaAs (ref. 25); the extrinsic Hall coefficient increases with T reaching a pronounced maximum just before its typical decrease with T at the onset of intrinsic conduction. Aukerman and Willardson (ref. 26) showed that a simple two-band conduction process such as described in Chapter 2, Section 2.0 can be used, to a first approximation, to explain these experimental observations in terms of the conduction band structure of GaAs (ref. 27). At temperatures above $\sim$500°K electrons present in the conduction band minimum at $\Gamma(000)$ in the center of the Brillouin zone are thermally-excited into subsidiary L minima where their effective masses are larger and, therefore, their mobilities are much smaller. They showed that R_h depends only on the ratio of mobilities and effective masses of the two bands and on the energy separation ΔE between these minima. Ikoma and Wang (ref. 28) have measured the Hall coefficient of GaAs specimens with electron concentrations in the range between 10^{14} and $10^{17}/cm^3$. They found that only for $n \geq 10^{16}/cm^3$ is the effect of such interband electron transfer appreciable. If the Hall coefficient in the exhaustion region, where all the electrons are at the Γ minimum of the conduction band is defined as R_{ho} and the measured Hall coefficient in a temperature range in which there is an appreciable population of electrons in the subsidiary L-minimum is R_{hm}, then

$$\frac{R_{hm} - R_{ho}}{R_{ho}} = \left(\frac{b-1}{b}\right)^2 \cdot \left(\frac{m_2^*}{m_1^*}\right)^{3/2} \cdot \exp\left[-\frac{\Delta E(T)}{kT}\right] \qquad (4.28)$$

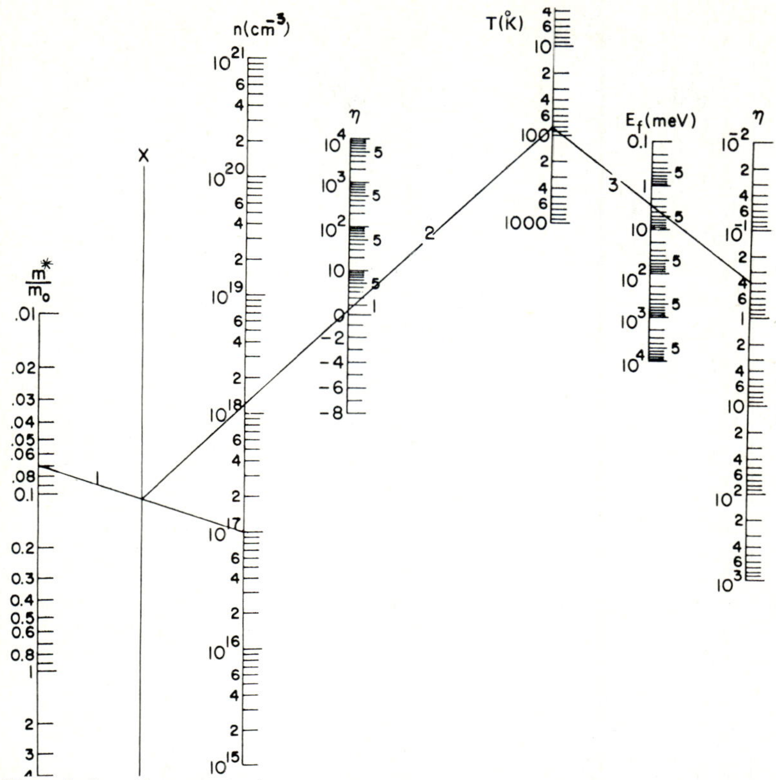

Fig. 4.8. Nomograph for determining any one of the four parameters if the other three are known with the carrier concentration calculated by means of the Fermi integral. [after J. I. Pankove and E. K. Annavedder, J. Appl. Phys., 36(1965) 3948]

where b is the mobility ratio of the electrons at the Γ-minimum, μ_1 with respect to μ_2, that of the electrons at the L-point; m_1* and m_2* are the corresponding effective masses and $\Delta E(T)$ is the temperature dependent energy separation of the two minima $\Delta E(T)=\Delta E(0)+\gamma T$. At room temperature, $\mu_1 \simeq 6\times10^3$ cm^2/V-sec, $\mu_2 \simeq 150$ cm^2/V-sec, b≃40, $m_1^*=0.072m_o$ and $m_2^*=1.2m_o$. Ikoma and Wang (ref. 28) used these values to calculate ΔE from experimental measurements of $\ell n[(R_{hm}-R_{ho})/(R_{ho}]$ vs. 1/T. They found $\Delta E(0)=0.36$ eV and $\gamma=3\times10^{-5}$ eV/°K in good agreement with the calculations of Ehrenreich (ref. 29).

The analysis of electron transfer between conduction band minima of Aukerman and Willardson (ref. 26) using classical statistics was generalized by Allgaier (ref. 30) to include hole transfer between valence band minima as well as degenerate charge carrier distributions in the sub-bands. He found that the experimentally measured $R_h(T)$ of n-type GaAs and GaSb and on p-type GeTe, SnTe, PbTe and Bi_2Te_3 conform with theoretical predictions of the model.

Gallium antimonide is another intermetallic III-V semiconducting compound in which the increase in Hall coefficient with temperature (refs. 31,32) is related to electron transfer from the center (000) of the Brillouin zone to a set of four minima at the zone boundary in the <111> directions. An increase in resistivity is associated with the rise in R_h with temperature indicating a decrease in the average mobility when the electrons are excited into the subsidiary conduction band minima. A two-band method of analysis combined with Faraday rotation measurements made at 77°K and 273°K was used by Liang *et al.* (ref. 33) and Liang (ref. 34) to investigate the electron transfer process in Te-doped n-type GaSb. With the weak-field Hall coefficient expressed as

$$R_h = \frac{1}{e}\frac{(n_1\mu_1^2 + n_2\mu_2^2)}{(n_1\mu_1 + n_2\mu_2)^2} \tag{4.29}$$

where the subscripts 1 and 2 refer, respectively, to the (000) and (111) minima of the conduction band. The mobility ratio, $b=\mu_1/\mu_2$ was considered to be sufficiently large so that terms containing $(\mu_1\mu_2)^2$ are negligible. Therefore the electron concentration in the (111) band, from Eq. (4.29), is

$$n_2 = b/(2eR_h) - bn_1/2 \tag{4.30}$$

They calculated b=9.0 at 77°K and b=5.59 at 273°K. Assuming that the effective mass of the electrons in the (000) band is $m^*_{1n}=0.05m_o$ (ref. 35,36) the density of the electrons in the (111) band was determined to be

$$n_2 = 4\pi(2m^*_t kT/h^2)^{3/2}\cdot 4\Phi^{1/2}\cdot \mathcal{F}_{1/2}\left[\frac{E_F - \Delta E(T)}{kT}\right] \tag{4.31}$$

where $\Phi=m^*_\ell/m^*_t$; here m^*_t and m^*_ℓ are, respectively, the transverse and longitudinal effective masses of the (111) band. Using experimentally-measured parameters and Eq. (4.31) they found that $\Delta E(0)=0.084$ eV and $\gamma=5.9\times10^{-5}$ eV/°K for Te-doped GaSb. However, the investigations of Strauss (ref. 37) suggest that the type of donor impurities may also have a significant role in determining the electron transfer process in GaSb.

Electron transfer from the $\Gamma(000)$ conduction band minimum to the subsidiary L(111) minima at the edges of the Brillouin zone has been suggested as the reason for the experimentally observed increase with increasing temperature of the Hall coefficient of InSb doped with donor impurities in excess of 10^{18} cm^{-3}. The data and calculations of Filipchenko *et al.* (ref. 38) suggest that the energy interval between the Γ and L minima $\Delta E\approx0.56$ eV and that the effective mass ratio of electrons in the L valley varies in a non-systematic manner, $0.20\leqq m_2^*\leqq0.47$ within the donor impurity density range 1.68×10^{18} cm$^{-3}\leqq N_d\leqq1.2\times10^{19}$ cm^{-3}; moreover, both ΔE and m_2^* were found to be dependent on the type of donor impurity (Se,Te) used as a dopant. These calculations were based on the experimentally measured temperature dependence of the Hall coefficient in low magnetic fields. Above room temperature, the increase of the Hall coefficient R_h with temperature was interpreted in terms of electron transfer from the Fermi level to the L valleys. The implicit assumption was made that the mobility ratio $b=\mu_1/\mu_2$ is very much greater than 1, where μ_1 and μ_2 are the electron mobilities in the Γ and L valleys respectively. The total electron density n_t in the conduction band is equal to n_1+n_2, where n_2 is the electron density of the L valleys; on taking n_0 to be the impurity-activated electron density and n_i the intrinsically generated electron concentration, $n_t=n_0+n_i$. Therefore, the temperature dependence of n_2 was given as

$$n_2(T) = n_0 + n_i(T) - n_1(T) \tag{4.32}$$

Moreover, n_1 is considered here to be degenerate in a non-parabolic conduction band, while n_2 is presumed to follow classical Maxwell-Boltzmann statistics so that

$$n_2 = N_c T^{3/2} \exp\left(-\frac{1}{k}\frac{\partial\Delta E}{\partial T}\right)\exp\left(\frac{E_F - \Delta E_0}{kT}\right) \tag{4.33}$$

where the density of states function $N_c=2(2\pi m_0 m_2{}^*k/h^2)^{3/2}$ with m_0 the electron rest mass, k and h, respectively, Boltzmann's and Planck's constants, E_F the Fermi level and ΔE_0 the Γ to L valley energy interval at absolute zero. From Eq. (4.33), the slope of $\ell n(n_2/T^{3/2})$ vs. T^{-1} is $(E_F-\Delta E_0)/k$; ΔE_0 can thus be obtained in a straightforward manner by neglecting the temperature dependence of ΔE, provided that E_F can be determined from independent measurements or theory, and $m_2{}^*$ is then calculated by fitting the experimental data to Eqs. (4.32) and (4.33). Wieder and Collins (ref. 39) have investigated such electron transfer in degenerate S-doped and Te-doped InSb layers. For $N_d \geq 10^{18}/cm^3$, with the square of the mobility ratio, $b^2 >> 1$ and with $n_2 \leq 0.2 n_1$ Eq. (4.29) becomes

$$n_2 \simeq \frac{b}{2}\cdot\left(\frac{1}{eR_h} - 1\right) \tag{4.34}$$

At low temperatures all the electrons are expected to be in the Γ valley and the measured low magnetic field value of the Hall coefficient is

$$R_{h_0} = \frac{1}{e}\left(\frac{1}{n_1 + n_2}\right) \tag{4.35}$$

Therefore, by means of Eqs. (4.34) and (4.35)

$$n_2 = \left(\frac{1}{1 - 2/b}\right)\left(\frac{1}{eR_{h_0}} - \frac{1}{eR_h}\right) \tag{4.36}$$

A simplification can be achieved by restricting the temperature region under consideration to that portion in which the thermally-induced depletion of the Γ

valley by electron transfer is slight and no energy states are available for any intrinsically-generated carriers. Furthermore, (1-2/b) may be assumed to be temperature independent and ΔE_0 can be determined from the experimental data and from an evaluation of the slope of $\ell n(n_2/T^{3/2})$ vs. 1/T. In this manner they found the energy separation to be $0.40<\Delta E_0<0.46$ eV.

It is interesting to compare the above-described investigations with data obtained by electric field-activated electron transfer from the Γ to the L valley of pure InSb using high electric fields applied for a time short enough to inhibit the onset of impact ionization. McGroddy *et al.* (ref. 40) have shown that hydrostatic pressure applied to bulk single-crystal high purity InSb can decrease ΔE and increase the fundamental bandgap E_g sufficiently for bulk negative differential conductivity to be observed in high electric fields. Their experimental data and calculations and the subsequent results of Lorenz *et al.* (ref. 41) and Porowski *et al.* (ref. 42) suggest that $\Delta E_0 \simeq 0.45$ eV. This is supported by the Monte Carlo calculations of electron velocity vs. electric field applied to n-type high purity InSb at 77°K by Fawcett and Ruch (ref. 43), who assumed that $m_1^*=0.014$ and $m_2^*=0.2$. The low field electron mobility was calculated to be $\mu_1 \simeq 1.7\times10^6$ cm^2/V-sec. and the negative differential mobility $\mu_2 \simeq 5\times10^4$ cm^2/V-sec; thus b=34. With a threshold field for electron transfer of 525 V/cm, a field of approximately 0.7 kV/cm is required in order to reduce the electron population in the Γ valley by about 10%. These calculations are also supported by the experimental measurements of Neukermans and Kino (ref. 44) using a time-of-flight technique to determine the electron velocity-electric field characteristic and the negative differential mobility of p-type InSb at 77°K.

Hole transfer between sub-bands of the valence band of $Pb_{1-x}SnTe$ takes place between 77°K and 300°K; this process was investigated by Farinre and Zemel (ref. 45). In this temperature range, the Hall coefficient of p-type epitaxially grown layers increases with temperature from 38% to 55% over their 77°K value. The analysis of their experimental data was made in terms of the model shown schematically in Fig. 4.9. A rise in temperature transfers holes from sub-band 1 to sub-band 2 depleting the high mobility sub-band 1 and, in analogy with conduction band electron transfer, making the measured Hall coefficient rise with temperature. They found that the hole transfer process, expressed in terms of the fractional change in R_{hm} with respect to R_{ho}, that of the Hall coefficient measured at 77°K, is

$$\frac{\Delta R_h}{R_{ho}} = \left(\frac{b-1}{b}\right)^2 \left(\frac{m_2^*}{m_1^*}\right)^{3/2} \cdot \frac{F_{1/2}(\eta_F - \Delta\eta)}{F_{1/2}(\eta_F)} \qquad (4.37)$$

where $F_{1/2}(x)$ are the appropriate Fermi integrals, m_1^* and m_2^* are the effective masses which correspond to valence bands 1 and 2 in Fig. 4.9, the normalized

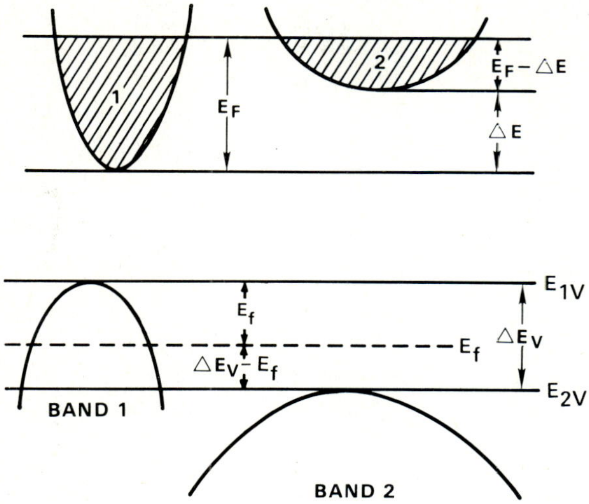

Fig. 4.9. (a) Electron transfer between subsidiary minima of conduction band separated by energy interval ΔE; electrons are transferred from sub-band 1 to sub-band 2 as a function of Fermi level E_F. (b) Hole transfer between sub-bands of valence band separated by energy interval ΔE_V.

Fermi level, $\eta_F=E_F/kT$ and the normalized sub-band separation $\Delta\eta=\Delta E_V/kT$. They found good agreement between theoretical calculations and experimental data for different alloy compositions with ΔE_V decreasing with temperature in the PbTe and SnTe-rich layers; at 0°K, ΔE_V=0.197 eV for $Pb_{0.77}Sn_{0.23}Te$ and ΔE_V=0.373 eV for $Pb_{0.295}Sn_{0.705}Te$.

4.5 Mobility of charge carriers

The mobility of charge carriers is one of the most important transport coefficients used for the characterization of electronic materials; it is often cited (without adequate justification) as an index of their purity or crystalline perfection. The mobility μ is defined as the average drift velocity per unit electric field. In the case of a constant relaxation time and a parabolic conduction band $\mu=e\tau/m^*$ with the relaxation time τ considered to be the mean time between scattering events or as the time constant with which the electron population returns to equilibrium after being subjected to a time-dependent perturbation.

The harmonic vibrations of atoms about their equilibrium positions can be resolved into standing plane waves called the normal modes of the crystal lattice and the interaction of an electron with lattice vibrations can result in the absorption or emission of phonons from the normal modes. In the case of a crystal with N unit cells each containing two atoms there are 6N normal modes. When both atoms vibrate in phase the vibration is called acoustical; when they move in phase opposition then the mode is called optical; these vibrations can be further subdivided into transverse and longitudinal modes.

In a crystal structure with a center of symmetry the acoustical modes produce scattering of electrons due to the disturbance of the periodic potential of the lattice. The energy of the electrons is altered by a periodic dilation of the lattice induced by longitudinal acoustic vibrations. In the deformation potential theory of Bardeen and Shockley (ref. 46) a local elastic strain is assumed to shift the minimum of the conduction band in proportion to the strain and the coefficient of proportionality is defined as the deformation potential, E_{ac}. The longitudinal acoustic lattice scattering mobility μ_{ac} (in cm^2/V-sec.) is

$$\mu_{ac} = 3.1727\text{x}10^{-5} \frac{\rho_d \mu_\ell^2}{(m^*/m_o)^{5/2}\ E_{ac}^2 T^{3/2}} \tag{4.38}$$

where ρ_d is the density of the crystal in g/cm^3, μ_ℓ is the velocity of longitudinal acoustic waves in it, in cm/sec. and E_{ac} is in eV.

In a crystal in which the unit cell consists of electrically-charged atoms and the structure is not centrosymmetric, acoustic modes also produce a polar form of scattering called piezoelectric scattering (refs. 47,48). The mobility (in cm^2/V-sec.) may be expressed in the form

$$\mu_{pz} = K(m_o/m^*)^{3/2}\ (300/T)^{1/2} \tag{4.39}$$

where the coefficient K is related to the electromechanical coupling coefficients; $K \simeq 160$ for ZnO and $K \simeq 400$ for CdS (ref. 48).

Any perturbation of the periodicity of a crystal lattice caused by optical vibrations produces optical mode scattering. If the temperature $T \ll \Theta$ where Θ is a characteristic temperature of the optical phonon scattering process, then the change in energy of a charge carrier produced by the absorption or emission of a phonon is so large that the Boltzmann transport equation cannot be solved in terms of a relaxation time (ref. 49) because the scattering process is inelastic. For temperatures $T > \Theta$ an optical deformation potential, E_{opt} such as introduced by Seitz (ref. 50) may be postulated with the square of the lattice strain proportional to the energy of the optical mode. For the non-polar optical phonon mobility (in cm^2/V-sec.),

$$\mu_{npo} = 6.345\text{x}10^{-5} \cdot \frac{\rho_d \mu_\ell^2}{(m^*/m_o)^{5/2}\ E_{opt}^2 T^{3/2}} \tag{4.40}$$

For T<Θ combined acoustic and non-polar optical scattering lead to a temperature dependence of the form T^{-x} where x has a value in the range between 1.8 and 2.4. Figure 4.10 shows a comparison of the experimentally-measured Hall mobility of

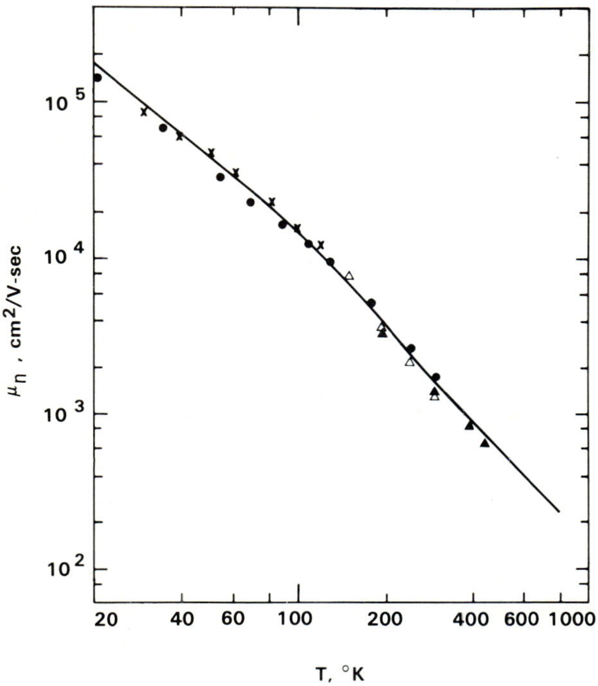

Fig. 4.10. Electron drift mobility of silicon as a function of temperature; Hall measurements indicate the material to have $N_d=1.1x10^{14}/cm^3$ and $N_a=4x10^{12}/cm^3$.

silicon against theoretically-calculated values as a function of temperature. Acoustic mode scattering is dominant at T<80°K. At higher temperatures inter-valley scattering (i.e., scattering between the minima of the conduction band) produced by non-polar optical as well as acoustic scattering contribute to the temperature dependence of the mobility. If the atoms in a unit cell are electrically charged then optical mode polar scattering arises from the interaction of charge carriers with the polarization associated with this mode. This type of scattering mechanism is dominant in direct bandgap semiconductors. A theoretical analysis of polar optical mode scattering has been made by Howarth and Sondheimer (ref. 49) by solving the Boltzmann transport equation using a variational method. Ehrenreich (ref. 5) has derived a comprehensive theoretical treatment of polar optical scattering and Hilsum (ref. 51) has calculated the corresponding mobilities of III-V compounds. A detailed description of many theoretical aspects of polar optical mode scattering is available in Beer's monograph (ref. 52). Figure 4.11 represents the temperature dependence of the experimentally-measured Hall

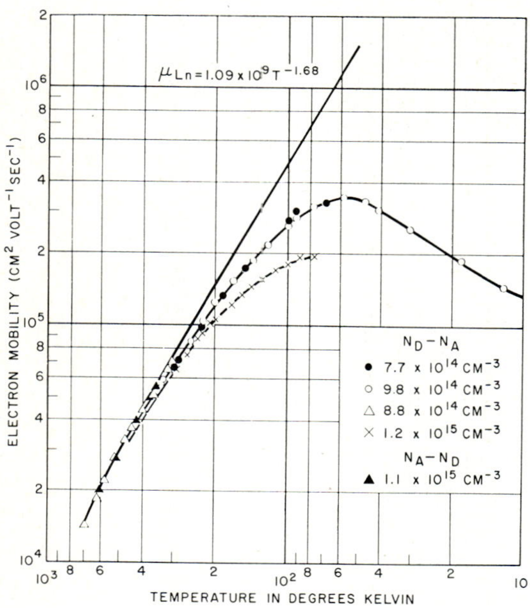

Fig. 4.11, Temperature dependence of electron mobility of InSb. [after H. J. Hrostowski, F. J. Morin, T. H. Geballe and G. H. Wheatley, Phys. Rev., 100(1955)1672]

mobility of InSb specimens having various impurity concentrations. In the intrinsic temperature region above 300°K polar optical modes scattering is dominant and the mobility is proportional to $T^{-1.68}$. Other deviations from periodicity of the crystal lattice, such as produced by charged point defects, give rise to impurity scattering. When electrons are scattered by ionized impurities their mobility (in cm²/V-sec.) can be expressed in terms of the Brooks-Herring equation (refs. 54,55),

$$\mu_i = 3.284\times10^{15}\left[\frac{\kappa_s{}^2 T^{3/2}}{N_i(m^*/m_o)^{1/2}}\right]\left[\ell n(1+x) - \frac{x}{1+x}\right]^{-1} \qquad (4.41)$$

where $x = 1.294\times10^{14} m^* \kappa_s T^2/m_o n°$ with the density of impurity scattering centers, $N_i = n + 2N_a$, κ_s is the dimensionless static dielectric constant and $n° = n + (N_d - N_a - n)(n + N_a)N_d^{-1}$ with N_i and n° in cm^{-3}. Figure 4.12 illustrates the effect of ionized impurities on the low temperature electron mobility of GaAs and the good agreement with the Brooks-Herring expression from Eq. (4.41). Other mobility limiting mechanisms include neutral impurity scattering analyzed by

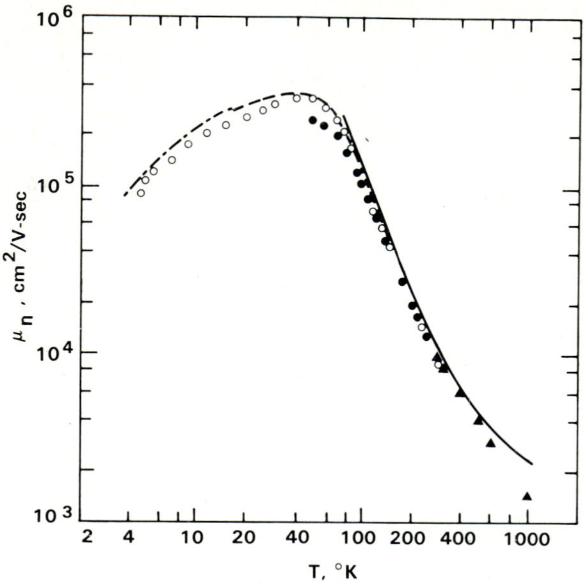

Fig. 4.12. Theoretically-calculated and experimental measurements of Hall mobility of electrons in GaAs as a function of temperature (solid line); dashed curve is calculated impurity scattering mobility using Brooks-Herring theory. [after D. L. Rode (in Semiconductors and Semimetals) Academic Press, New York 10(1975)69]

Erginsoy (ref. 56) and modified by Sclar (ref. 57) whose derivation suggests that the mobility due to neutral impurity scattering varies as $T^{1/2}$ while the derivation of Erginsoy predicts a temperature-independent mobility which is inversely proportional to the density of the scattering centers. The expression for the ionized impurity scattering mobility in Eq. (4.41) neglects contributions of electron-electron scattering which tend to randomize the manner in which the total momentum is distributed between electrons with different energies. The effect of electron-electron scattering is to reduce the lattice scattering mobility (ref. 59). Li (ref. 59) has calculated the dependence of the electron mobility of Si on the donor concentration using an equation which includes a combination of scattering mechanisms. In the impurity density range $2x10^{17}<N_d<10^{19}/cm^3$ the experimental data indicates that the full effect of electron-electron scattering on lattice and impurity scattering must be included. In the intermediate doping range, $2x10^{16}<N_d\leq10^{17}/cm^3$ the effect of electron-electron scattering is considered to increase gradually with increasing, impurity concentration and it is negligible for $N_d<10^{15}/cm^3$.

Compensation can also lead to a reduction in charge carrier mobilities as a function of doping, as shown in Fig. 4.13. Wolfe and Stillman (ref. 60) have

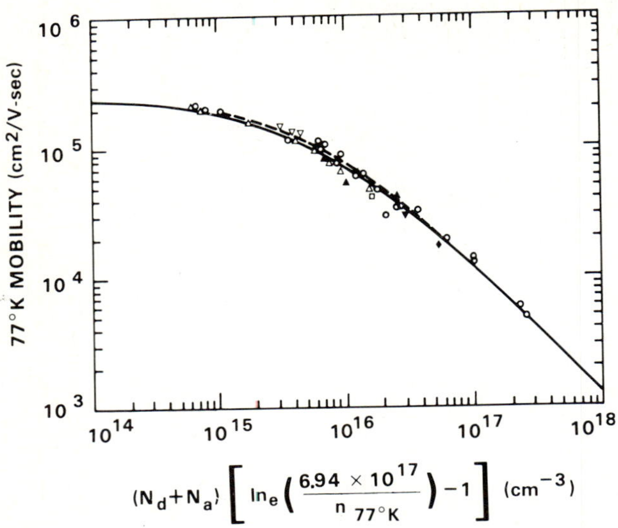

Fig. 4.13. Electron mobility of GaAs at 77°K as a function of total impurity concentration (N_d+N_a) multiplied by Brooks-Herring screening factor used to normalize the data. [after C. M. Wolfe, G. E. Stillman and J. O. Dimmock, J. Appl Phys., 41(1970)504]

shown that all common donors in high purity GaAs are compensated by vacancy-impurity complex acceptors. Figure 4.13 includes theoretical and experimental measurements of a number of workers using different methods in order to determine N_d and N_a in GaAs. The dashed line corresponds to two different theoretical analyses (refs. 61,62) which overlap. The main curve represents the 77°K mobility as a function of (N_d+N_a) multiplied by the Brooks-Herring screening factor used to normalize the data; it is an empirically-derived curve (ref. 63) in remarkably good agreement with experimental measurements. The reduction in mobility with increasing (N_d+N_a) applies to amphoteric column IV impurities as well as to column VI impurities which are normally donors localized on As sites and are compensated by acceptor centers which are donor impurity-vacancy complexes. For a comprehensive treatment of theoretical and experimental aspects of electron and hole mobilities in solids the reader is referred to the reviews by Rode and Wiley (ref. 64), the monograph on galvanomagnetic effects in semiconductors by Beer (ref. 2) and the book on the physics of electronic conduction in solids by Blatt (ref. 3).

REFERENCES

1 R. A. Smith, Semiconductors, Cambridge Univ. Press (1968)102.
2 A. C. Beer, Solid-State Phys., Suppl. 4(1963)23.
3 F. J. Blatt, Physics of Electronic Conduction in Solids, McGraw Hill Book Co., New York (1968)130.
4 C. Hilsum and A. C. Rose-Innes, Semiconducting III-V Compounds, Pergamon Press, Oxford (1961)116.
5 H. Ehrenreich, J. Phys. Chem. Solids, 2(1957)131.
6 H. Ehrenreich, J. Phys. Chem. Solids, 8(1959)130.
7 R. S. Allgaier, Phys. Rev., 158(1967)699.
8 R. S. Allgaier, Phys. Rev., 165(1968)775.
9 R. S. Allgaier and R. Perl, Phys. Rev., B2(1970)877.
10 J. S. Blakemore, Semiconductor Statistics, Pergamon Press (1962)94.
11 W. C. Dunlap, Phys. Rev., 79(1950)286.
12 L. P. Hunter, Phys. Rev., 94(1954)1157.
13 R. Landauer and J. Swanson, Phys. Rev., 91(1953)555.
14 L. P. Hunter, Phys. Rev., 91(1953)579.
15 T. S. Robinson, M. Smollett and R. G. Pratt, Brit. J. Appl. Phys., 8(1957)130.
16 E. H. Putley, The Hall Effect and Related Phenomena, Butterworth, London (1960)125.
17 P. A. Lee, Brit. J. Appl. Phys., 8(1957)340.
18 D. V. Eddolls, Phys. Stat. Sol., 17(1966)67.
19 A. H. Wilson, The Theory of Metals, Cambridge Univ. Press (1958)16.
20 R. A. Smith, The Theory of Metals, Cambridge Univ. Press (1958)p.74.
21 J. S. Blakemore, The Theory of Metals, Cambridge Univ. Press (1958)p.354.
22 J. I. Pankove and E. K. Annavedder, J. Appl. Phys., 36(1965)3948.
23 E. O. Kane, J. Phys. Chem. Solids, 1(1956)82.
24 R. Dalven, J. Appl. Phys., 38(1967)1732.
25 O. Folberth and H. Weiss, Z. Naturforsch., 10a(1955)615.
26 L. Aukerman and R. Willardson, J. Appl. Phys., 31(1960)939.
27 J. Callaway, J. Electron., 2(1957)330.
28 H. Ikoma and S. S. Wang, J. Phys. Soc. Japan, 25(1968)1739.
29 H. Ehrenreich, Phys. Rev., 120(1960)1951.
30 R. S. Allgaier, J. Appl. Phys., 36(1965)2429.
31 A. Sagar. Phys. Rev., 117(1960)93.
32 A. Sagar and R. C. Miller, J. Appl. Phys., 32(1961)2073.
33 C. Y. Liang, H. Piller and D. L. Stierwalt, Appl. Phys. Lett., 12(1968)49.
34 C. Y. Liang, J. Appl. Phys., 39(1968)3866.
35. H. B. Harland and J. C. Wooley, Can. J. Phys., 44(1966)2715.
36 H. Piller, J. Phys. Chem. Solids, 24(1963)425.
37 A. J. Strauss, Phys. Rev., 121(1961)1087.
38 A. S. Filipchenko, I. P. Molodian, D. N. Nasledov, V. G. Sidorov and O. V. Emelyanenko, Phys. Stat. Solidi, 14(1966)K195.
39 H. H. Wieder and D. C. Collins, Thin Solid Films, 20(1974)201.
40 J. C. McGroddy, M. I. Nathan, W. Paul, S. Porowski, J. E. Smith, Jr., and W. P. Dumke, IBM J. Res. Develop., 13(1969)580.
41 M. R. Lorenz, J. C. McGroddy, T. S. Plaskett and S. Porowski, IBM J. Res. Develop., 13(1969)583.
42 S. Porowski, W. Paul, J. C. McGroddy, M. I. Nathan and J. E. Smith, Jr., Solid State Commun., 7(1969)905.
43 W. Fawcett and J. G. Ruch, Appl. Phys. Lett., 15(1969)368.
44 A. Neukermans and G. S. Kino, Appl. Phys. Lett., 17(1970)102.
45 T. O. Farinre and J. N. Zemel, Thin Solid Films, 7(1971)341.
46 J. Bardeen and W. Shockley, Phys. Rev., 80(1950)72.
47 H. J. G. Meyer and D. Polder, Physica, 19(1953)255.
48 A. R. Hutson, J. Appl. Phys., 32 Suppl. (1961)2287.
49 D. J. Howarth and E. H. Sondheimer, Proc. Royal Soc.(London), A219(1953)53.
50 F. Seitz, Phys. Rev., 73(1948)549.
51 C. Hilsum, Proc. Phys. Soc.(London) 76(1960)414.

52 A. C. Beer, Proc. Phys. Soc.(London) 76(1960)289.
53 D. L. Rode, Phys. Rev., 3B(1971)3287.
54 H. Brooks, Adv. Electron. Electron. Phys., 7(1955)85.
55 C. Herring and E. Vogt, Phys. Rev., 101(1956)944.
56 C. Erginsoy, Phys. Rev., 79(1950)1013.
57 N. Sclar, Phys. Rev., 104(1956)1559.
58 P. P. Debye and E. M. Conwell, Phys. Rev., 93(1954)693.
59 S. S. Li, National Bureau of Std's Special Publ. 400-33(1977)10.
60 C. M. Wolfe and G. E. Stillman, Appl. Phys. Lett., 27(1975)564.
61 C. M. Wolfe, G. E. Stillman and W. T. Lindley, J. Appl. Phys., 41(1970)3088.
62 D. L. Rode and S. Knight, Phys. Rev., 3B(1971)2534.
63 C. M. Wolfe, G. E. Stillman and J. O. Dimmock, J. Appl. Phys., 41(1970)504.
64 D. L. Rode and J. D. Wiley in"Semiconductors and Semimetals," R. K. Willardson and A. C. Beer, ed., 10(1975)pp 1-171.

CHAPTER 5

INHOMOGENEOUS AND ANISOTROPIC MATERIALS

5.0 Introduction

All natural or synthetic materials deviate to a greater or lesser extent from perfect homogeneity. Electrical and galvanomagnetic measurements are used to characterize a macroscopic spatial average which, it is hoped, represents the specific properties of a homogeneous material in which stoichiometric defects, crystallographic variations, precipitates or inclusions are minor perturbations.

A considerable effort is required to circumvent the formation of impurity fluctuations even in crystallographically "perfect" crystals; quite often the analysis of electrical and galvanomagnetic measurements of crystalline specimens is complicated by the presence of impurities introduced accidentally, during synthesis, or deliberately by diffusion or ion implantation. Witt and Gatos (ref. 1) have shown by means of interference contrast microscopy that Te-doped InSb crystals grown from the melt, along the <111> direction, have six different types of impurity striations in {211} planes parallel to the growth axes. The density and distribution of these striations are functions of the Te concentrations, in the melt, temperature gradients, rate of pulling and other parameters of the crystal growth process. Previously Mueller and Wilhelm (ref. 2) have shown that periodic microfluctuations in the impurity concentration of InSb crystals grown from the melt or by zone melting procedures are produced by temperature fluctuations prevailing at the solid-liquid interface during growth. Convective heat flow during crystal growth can also produce radial inhomogeneities in crystal boules. Mechanical vibrations of the melt or seed during crystal growth (ref. 3) affect the impurity distribution in InSb crystals by introducing conductivity striations whose density and distribution are a function of the frequencies of vibration.

One-dimensional inhomogeneities are generated in chemical vapor phase-grown homoepitaxial GaAs layers by a high-resistance interfacial region between the substrates and their epilayers (refs. 4,5). Such high-resistance interfacial regions are also formed between liquid phase epitaxially-grown n^+GaAs layers and their n-type substrates (ref. 6) and between homoepitaxially-grown n-type GaAs layers and semi-insulating substrates (ref. 7). A different type of one-dimensional inhomogeneity is that produced by surface accumulation in thin high mobility epilayers or by heteromorphic interfacial layers such as produced by the heteroepitaxial pyrolytic deposition of silicon on sapphire.

Three-dimensional inhomogeneities are present in large bandgap polycrystalline anisotropic materials and in random distribution polyphase materials. One-, two-, and three-dimensional inhomogeneities affect the electrical and galvanomagnetic measurements made on a specimen in which they are present. This Chapter sets forth methods and techniques for experimental measurements made on such inhomogeneous materials and analytical procedures used for their interpretation. The one-dimensional inhomogeneities induced in epitaxial layers by surface or interfacial accumulation or depletion are a function of the space charge parameters and surface barrier phenomena in semiconductors; they are described in Chapter 6 in the context of capacitance vs. voltage and gate-controlled galvanomagnetic measurements.

Wolfe and Stillman (ref. 8) have suggested a division of inhomogeneities into three categories: (A) microscopic inhomogeneities in which the size of inhomogeneities is comparable to that of a charge carrier mean free path; (B) intermediate inhomogeneities, large compared to a mean free path but small compared to the size of a specimen; (C) macroscopic inhomogeneities, whose size is such that due account must be taken of the specimen boundaries. An alternative classification might be made in terms of (a) a one-dimensional or polydimensional distribution of inhomogeneities in a specimen; (b) spatially ordered or random distributions of inhomogeneities in single crystal specimens; (c) polycrystalline specimens and the inhomogeneous perturbations associated with grain boundaries; (d) polyphase and polymorphic materials and the corresponding microscopic and macroscopic inhomogeneities. Not all of these are analytically tractable. However, their effect on electrical and galvanomagnetic measurements is considerable.

5.1 Diffusion and ion implantation

Consider an inhomogeneous rectangular wafer, oriented along cartesian axes, in accordance with Fig. 2.1. Let it be homogeneous in the xy-plane with a z-axis-dependent electron concentration and electron mobility, respectively, $n(z)$ and $\mu(z)$. In the steady-state, $\nabla \times \bar{J}=0$, $\nabla \times \bar{E}=0$ and, except at the boundaries of the wafer, $\nabla \cdot \bar{E}=0$. The average current densities in the x and y directions are

$$\langle J_x \rangle = \frac{1}{d}\int_0^d J(x)dz \qquad\qquad \langle J_y \rangle = \frac{1}{d}\int_0^d J_y(z)dz = 0 \tag{5.1}$$

where the bracket represents a spatial average and d is the thickness of the wafer. Suppose that the wafer is subdivided into n laminae of infinitesimal thickness. Each of them is subjected to the same electric field in the z-direction; therefore

$$E_{x1} = E_{x2} = \ldots . E_{xn} = \langle E_x \rangle \tag{5.2}$$

However, in the presence of a magnetic field oriented along the z-axis the Hall field of each lamira is different from that of its neighbor; $E_{y1} \neq E_{y2} \cdot \cdot \cdot \neq E_{yn}$. Consequently the Hall angle θ is z-dependent and

$$\tan\theta(z) = E_y(z)/E_x = \mu(z) \cdot B \tag{5.3}$$

The z-dependent Hall current density is

$$J_y(z) = \sigma(z)\left[\langle E_x \rangle \cdot \tan\theta(z) - \langle E_y \rangle\right] \tag{5.4}$$

Here $\langle E_y \rangle$ denotes the average Hall field produced by the interaction between all the laminae with the circulating Hall current J(z) flowing between them. From Eqs. (5.3) and (5.4)

$$\langle J_y \rangle = \frac{1}{d}\int_0^d \sigma(z)\left[\langle E_x \rangle \left(B\mu(z)\right) - \langle E_y \rangle\right] dz = 0 \tag{5.5}$$

The average Hall field is, therefore,

$$\langle E_y \rangle = \frac{\int_0^d \langle E_x \rangle B \cdot \sigma(z) \cdot \mu(z) dz}{\int_0^d \sigma(z) dz} \tag{5.6}$$

The average Hall coefficient $\langle R_h \rangle = \langle E_y \rangle / (\langle J_x \rangle B)$ and the average conductivity $\langle \sigma \rangle = \langle J_x \rangle / \langle E_x \rangle$ are, respectively,

$$\langle \sigma \rangle = \frac{1}{d}\int_0^d \sigma(z) dz \qquad \langle R_h \rangle = \frac{d\int_0^d \sigma(z) \cdot \mu(z) dz}{\left[\int_0^d \sigma(z) dz\right]^2} \tag{5.7}$$

and the average mobility $\langle \mu \rangle = \langle R_h \rangle \langle \sigma \rangle$; in view of $\sigma(z) = en(z)\mu(z)$, it is

$$\langle \mu \rangle = \frac{\int_0^d n(z)\mu^2(z) dz}{\int_0^d n(z)\mu(z) dz} \tag{5.8}$$

These equations were first derived by Petritz, and by Zemel and Petritz (refs. 9, 10) and used in order to account for the effects and the contributions of surface charge carrier transport to the measured electrical and galvanomagnetic properties of Ge single crystal specimens.

In terms of Eq. (5.7) the average sheet Hall coefficient and sheet conductivity are, respectively, $\langle R_{hs} \rangle = \langle R_h \rangle / d$ and $\langle \sigma_s \rangle = \langle \sigma \rangle d$. Suppose that the conductivity of a wafer such as shown in Fig. 2.1 is measured by procedures such as described in Chapter 1, then,

$$<\sigma_{1s}> = <\sigma_1>d = \int_0^d \sigma(z)dz \tag{5.9}$$

If a lamina of thickness δ is removed from the wafer and its sheet conductivity is remeasured then

$$<\sigma_{s2}> = <\sigma_2>(d - \delta) = \int_\delta^d \sigma(z)dz \tag{5.10}$$

The average conductivity of the lamina $<\sigma_\delta>$ can be obtained from the two measurements,

$$<\sigma_\delta> = \frac{<\sigma_{s1}> - <\sigma_{s2}>}{\delta} = \frac{1}{\delta}\int_0^\delta \sigma(z)dz \tag{5.11}$$

In a similar manner the first Hall measurement on the specimen yields

$$<R_{hs1}><\sigma_{s1}>^2 = <R_{h1}><\sigma_1>^2 d = \int_0^d \sigma(z)\cdot\mu(z)dz \tag{5.12}$$

and the second Hall measurements with the lamina of thickness δ removed leads to

$$<R_{hs2}><\sigma_{s2}>^2 = <R_{h2}><\sigma_2>^2(d - \delta) = \int_\delta^d \sigma(z)\bullet\mu(z)dz \tag{5.13}$$

The average Hall coefficient of the lamina can be obtained by means of Eqs. (5.7), (5.12) and (5.13) in the form

$$<R_{h\delta}> = \delta \bullet \frac{<R_{hs1}><\sigma_{s1}>^2 - <R_{hs2}><\sigma_{s2}>^2}{(<\sigma_{s1}> - <\sigma_{s2}>)^2} \tag{5.14}$$

with $(<\sigma_{s1}>-<\sigma_{s2}>)$ obtained from Eq. (5.11). The average mobility of the lamina

$$<\mu_\delta> = \frac{<R_{hs1}><\sigma_{s1}>^2 - <R_{hs2}><\sigma_{s2}>^2}{<\sigma_{s1}> - <\sigma_{s2}>} \tag{5.15}$$

By removing laminae, sequentially, between sheet conductivity and sheet Hall measurements on the remaining portion of a wafer the average conductivity and Hall coefficient of each removed layer can be determined by means of Eqs. (5.11) and (5.14). Therefore the average electron density and mobility can be determined and their z-dependence can be reconstructed from such, essentially destructive measurements.

An analysis of a Zn-diffused p-layer in n-type InAs using such techniques was made by Buehler (ref. 11). A square four-point probe was used for performing the measurements. Laminae were removed by means of a precision mechanical lapping machine and the thickness of each removed lamina was determined from the weight differential of the wafer using a Mettler "Micro Gram-atic" balance. The analysis started with the diffused p-layer, traversed the p-n junction, and continued into the n-wafer. Figure 5.1 shows the results of the measurements made at 77°K. It is assumed that the p-layer is electrically isolated from the body of the

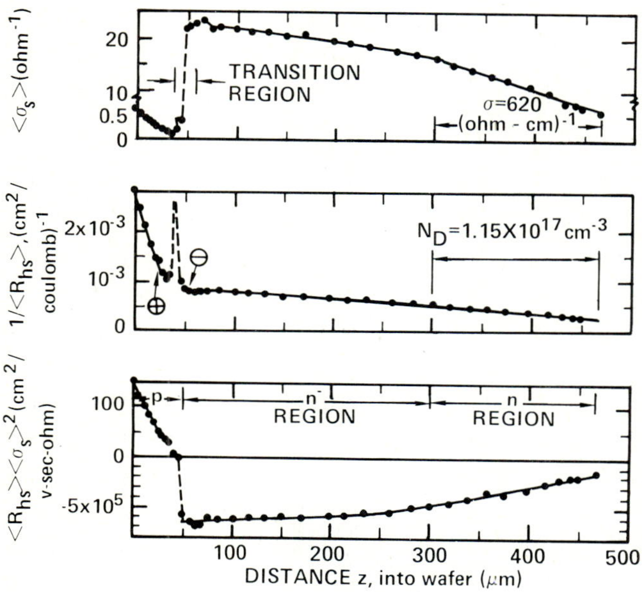

Fig. 5.1. Experimental data of sheet conductivity, reciprocal sheet Hall coefficient and parameter $<R_{hs}><\sigma_s>^2$ as a function of distance into a Zn-diffused InAs wafer measured at 77°K using Eqs. (5.11) to (5.15). [after M. G. Buehler, Stanford Electr. Lab.'s Tech. Rept. 5101(1966)45]

wafer so that the slope of the σ_s curve, shown in Fig. 5.1, should be negative with a discontinuity at the junction. The $1/<R_{hs}>$ curve is expected to change sign in going from a p-region to an n-region and its slope in the uniformly doped region, beyond 300μm, yields the background doping density, $N_d=1.15\times10^{17}/cm^3$. The $<R_{hs}><\sigma_s>^2$ curve has a negative slope in the p-region and a positive slope in the n-region. The p-n junction is located at ∿50μm into the wafer and an n-n$^-$ junction is located at ∿300μm into the wafer. The analysis is complicated by three problems: (a) near the junction both electrons and holes are present and two-band conduction has to be taken into account; (b) a non-planar junction or non-parallel lapping may cause a deviation from theory described by Eqs. (5.11) to (5.14); (c) junction leakage may occur through the junction or along its edge thus reducing the isolation of the p-layer.

Similar measurements have been made by Baron *et al.*(ref. 12) on group III and group V ion-implanted dopants in silicon. They determined the concentration and effective mobility of charge carriers in ion-implanted layers as a function of depth from the surface. In each case the implanted species produced a conductivity opposite to that of the substrate producing a p-n junction and, therefore, isolating the implanted layer. They pointed out that the carrier density calculated from Eq. (5.7) and the mobility calculated from Eq. (5.8) should be interpreted with caution because they represent weighted averages over n and μ rather than the actual averages,

$$n_{ave.} = \frac{1}{d}\int_0^d n(z)dz \qquad \mu_{ave.} = \frac{\int_0^d \mu(z)n(z)dz}{\int_0^d n(z)dz} \tag{5.16}$$

and the weighted average can differ significantly from the average conductivity mobility. Sufficiently large variations in mobility can cause the value of the effective carrier concentration calculated from Eq. (5.7) to be completely dominated by a small fraction of the carriers present in a high mobility region. They demonstrated that because of the depth dependence of ion-implantation density the mobility is a function of depth because of the dependence of the mobility on ionized impurity scattering. Baron *et al.*(ref. 12) indicated the advantages of a combination of van der Pauw-type sheet conductivity and Hall measurements coupled with anodic stripping operations. Specimens may be anodized in a buffered solution of boric acid and sodium tetraborate. The area to be stripped is defined by a rubber gasket. The solution is recirculated and directed onto the specimen by means of a jet immersed in the solution ∿1 cm above the specimen. A constant current between 1 and 10 mA is used to grow the oxide with growth stopped when a preselected change in voltage is reached (ref. 13). The oxide is dissolved in HF; in high-resistivity material the surface is illuminated to reduce the voltage drop across the specimen. The thickness of the stripped oxide can be determined from differential weight measurements or multiple beam interferometric methods. A tendency toward thicker oxides per volt in the anodization bath was noted for depth less than 0.04μm and for implanted dosages greater than $10^{14}/cm^2$.

Baron *et al.*(ref. 12) have also used a different type of anodic oxidation-layer stripping. They obtained reproducible results with a 0.04 M solution of potassium nitrate in N-metylacetimide (ref. 14). A pumped jet of the solution is driven against the face of the Si specimen in the presence of intense illumination. A variety of different anodization procedures using the apparatus shown in Fig. 5.2 were described by Johansson *et al.*(ref. 15). Actual depth values were established optically with an interference microscope using sodium illumination. Minimum steps of ~0.01μm can be obtained with this method. Equations (5.11) to (5.15) were used, in conjunction with the anodization and oxide

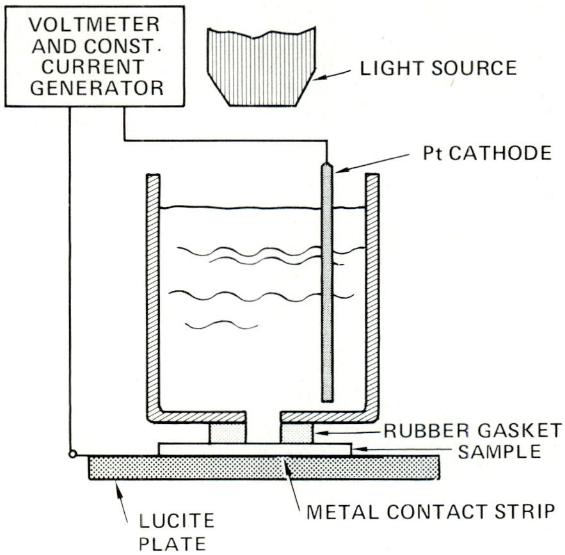

Fig. 5.2. Schematic of an arrangement used to grow anodic oxide films on ion-implanted layers in silicon. Not shown is a stirrer used to prevent the adhesion of bubbles to specimen surface. [after N. G. E. Johansson, J. W. Mayer and J. O. Marsh, Solid-State Electron., 13(1970)317]

stripping operations, to determine n(z) and μ(z). Baron *et al.*(ref. 12) found that for well-annealed ion-implanted specimens the dependence of the mobility on the free carrier concentration follows closely that determined on uniformly doped uncompensated material except in the immediate surface region of the layers where the dependence is anomalous.

Johansson *et al.*(ref. 15) have pointed out two major errors of the above described technique for determining an one-dimensional inhomogeneity induced by ion-implantation: the uncertainty in the thickness of the removed layer and the degradation of the junction characteristics. It is necessary to remove predictable thicknesses of the material without degrading the contact areas or the junction characteristics. This is made difficult by the small amounts of material removed and the difficulty of making accurate measurements by multiple beam interferometric techniques. Degradation of the junction occurs because as successive layers are removed from an ion-implanted specimen the resistance of the residual wafer increases; it is also difficult to maintain low surface leakage currents after many anodization-stripping operations.

Johnson (ref. 16) has analyzed the one-dimensional inhomogeneity of diffused or ion-implanted layers in silicon by assuming a Gaussian distribution of the carrier concentration n(z) and defining a correction factor which compares the measured values of the carrier concentration and mobility expressed in terms of Eqs. (5.7) and (5.8) against the true average mobility and carrier concentration determined by means of Eqs. (5.16), of the form,

$$f = \frac{n_{ave.}}{\langle n\rangle} = \frac{\langle \mu\rangle}{\mu_{ave.}} = \frac{\int_0^d \mu^2(z)n(z)dz \int_0^d n(z)dz}{\left[\int_0^d \mu(z)n(z)dz\right]^2} \tag{5.17}$$

The factor f is equal to unity if the mobility is independent of depth, no matter what the depth dependence of the carrier concentration. In all other cases this factor is greater than one. Johnson (ref. 16) solved Eqs. (5.7) and (5.8) by numerical integration using the empirical equation of Caughey and Thomas (ref. 17) to express the carrier mobility dependence on free carrier concentration, n, of the form

$$\mu = \frac{\mu_1}{1 + (n/n_r)^{\alpha}} + \mu_2 \tag{5.18}$$

where the constants μ_1 and μ_2 are expressed in cm^2/V-sec. and n_r is expressed in cm^{-3}. For silicon the applicable values for electrons and holes are given in Table 5.1.

TABLE 5.1.

Fixed parameters of Eq. (5.18). [after W. S. Johnson, Solid-State Electron., 13 (1970)951]

carrier type	μ_1	μ_2	n_r	α
holes	447.3	47.7	6.3×10^{16}	0.76
electrons	1265.0	65.0	8.5×10^{16}	0.72

Figure 5.3 shows his calculated correction factor dependence on carrier concentration. The average mobility corresponds closely to that occurring at one standard deviation from the peak of the Gaussian profile. The Gaussian distribution has been shown to have the same sheet carrier density and mobility as a uniform profile with a carrier density ∿55% of the Gaussian peak and a width of ∿4.5 standard deviation. Qualitative estimates may also be made about Ge, where the impurity concentration dependence of the mobility is weaker for electrons

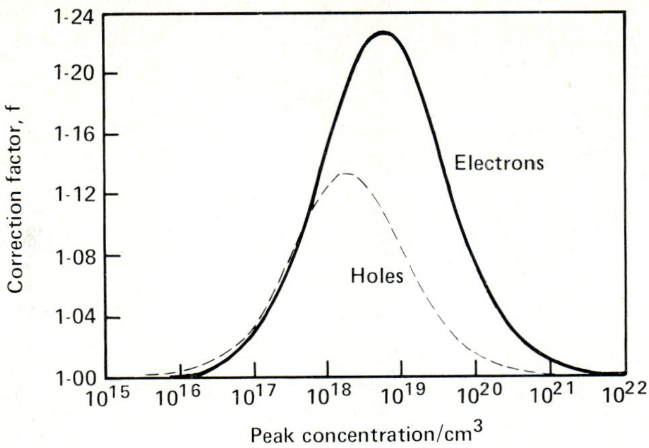

Fig. 5.3. Numerical correction factor f used for correction of measured carrier concentration and mobility of silicon to true average values provided that distribution of carrier concentration has a Gaussian profile. [after W. S. Johnson, Solid-State Electron., 13(1970)951]

and stronger for holes than for either type of carrier in Si. Therefore correction factors should be small for electrons and large for holes. For GaAs it is expected that the correction factors should be small for either type of carrier.

Anodization and etch-stripping techniques were also used by Mueller *et al.* (ref. 18) to evaluate electrical carrier concentration profiles and mobility of Cr-doped semi-insulating GaAs implanted with sulfur and tellurium. An anodic oxidation technique for both p- and n-type GaAs (with n in the range between 10^{17} and $10^{18}/cm^3$) was developed for use with a constant current source. The electrolyte was N- methylacetamide adjusted to a pH value of 8.3 by the addition of NH_4OH. Typical growth rates of the oxide were found to be 2×10^{-3} μm/V and the amount of GaAs consumed in the oxide was estimated to be 1.34×10^{-3} A/V.

One-dimensional inhomogeneities in a direction normal to a wafer surface and parallel to a magnetic field B(z) have been investigated by Subashiev and Poltinnikov (ref. 19), Tufte (ref. 20), Kravchenko *et al.*(ref. 21) and Pavlov (ref. 22). Hlasnik (ref. 23) has analyzed the dependence of the weighted average Hall coefficient $\langle R_h(B)\rangle$ and of the weighted average resistivity $\langle\rho(B)\rangle$ for a z-dependent carrier concentration and mobility; for an arbitrary magnetic field and an inhomogeneous specimen of thickness, d, these coefficients are

$$\langle R_h(B)\rangle = \frac{1}{\alpha d}\cdot\int_0^d \frac{\sigma\mu dz}{1+\mu^2B^2} \qquad \langle\rho(B)\rangle = \frac{1}{\alpha d}\cdot\int_0^d \frac{\sigma dz}{1+\mu^2B^2} \tag{5.19}$$

where

$$\alpha = \left[\frac{1}{d}\cdot\int_0^d \frac{\sigma dz}{1+\mu^2B^2}\right]^2 + B^2\left[\frac{1}{d}\cdot\int_0^d \frac{\sigma\mu dz}{1+\mu^2B^2}\right]^2 \tag{5.20}$$

The weighted average mobility of the specimen is

$$\langle\mu(B)\rangle = \int_0^d \frac{\sigma\mu dz}{1+\mu^2B^2}\cdot\left[\int_0^d \frac{\sigma dz}{1+\mu^2B^2}\right]^{-1} \tag{5.21}$$

Hlasnik (ref. 23) has obtained good experimental confirmation of these equations, between 0 and 1T, for the magnetic field dependence of the Hall coefficient and of the conductivity of an n-type diffused layer of InSb and of a copper-plated wafer of InSb measured at 77°K.

5.2 Ordered inclusions and conductivity striations

Inclusions or precipitates embedded in a homogeneous material can have an appreciable effect on its measured electrical and galvanomagnetic coefficients if the carrier concentrations and mobilities of the inclusions are much larger or much smaller than those of the matrix.

Thin films or thin layers may contain pinholes or voids assumed to have an essentially spherical symmetry. Juretschke and Landauer (ref. 24) have compared analytically the measured weak magnetic field ρ and R_h of such inhomogeneous layers to those of the same but homogeneous materials and Goldin and Juretschke (ref. 25) have confirmed experimentally their theoretical analysis.

If f is the fractional volume occupied by the inclusions in a specimen then the measured resistivity, ρ_m and the measured Hall coefficient, R_{hm} are

$$\rho_m = \frac{1+f/2}{1-f}\,\rho_0 \qquad R_{hm} = \frac{1-f/4}{1-f}\,R_{h_0} \tag{5.22}$$

where ρ_0 and R_{h_0} refer to the resistivity and Hall coefficient of the same but homogeneous specimen. For nonconducting cylindrical inclusions parallel to the magnetic field and perpendicular to the applied electric field Juretschke *et al.* (ref. 26) have shown that

$$\rho_m = \frac{1+f}{1-f}\,\rho_0 \qquad R_{hm} \simeq R_{h_0} \tag{5.23}$$

For nonconducting cylindrical inclusions parallel to the applied electric field and perpendicular to the magnetic field

$$\rho_m = \frac{\rho_0}{1-f} \qquad R_{hm} = \frac{R_{h_0}}{1-f} \tag{5.24}$$

For nonconducting cylinders oriented at random with respect to the applied electric and magnetic fields

$$\rho_m = \frac{1 + f/3}{(1 - f)(1 - f/3)} \rho_0 \qquad R_{hm} = \frac{1 - 2f/3}{(1 - f)(1 - f/3)} R_{ho} \tag{5.25}$$

The effects of finite conductivity inclusions σ_1 embedded in a matrix of conductivity σ have been calculated by Herring (ref. 27) under the assumption that the mobility of the charge carriers in the inclusions and the matrix is the same. For cylindrical inclusions parallel to the magnetic field

$$\begin{aligned} R_{hm}/R_{ho} &= 1 + 4f/9 \qquad &\text{for } \sigma_1/\sigma = 1.2 \\ R_{hm}/R_{ho} &= 1 - 8f/9 \qquad &\text{for } \sigma_1/\sigma = 2.0 \\ R_{hm}/R_{ho} &= 1 - 4f \qquad &\text{for } \sigma_1/\sigma = \infty \end{aligned} \tag{5.26}$$

For cylindrical inclusions perpendicular to the magnetic field Herring obtained, for any values of σ_1/σ, the equation

$$R_{hm}/R_{ho} = 1 + f[1 - (\sigma_1/\sigma)] \tag{5.27}$$

Similar relations applicable to spherical inclusion of conductivity σ_0 embedded in a medium of conductivity σ were developed by Voronkov *et al.* (ref. 28)

$$\frac{\sigma_m}{\sigma_0} = 1 + \frac{3f(\sigma_1 - \sigma)}{\sigma_1 + 2\sigma} \tag{5.28}$$

and the corresponding measured mobilities are

$$\mu_m/\mu_0 = (1 + 3f)^{-1} \qquad \text{for } \sigma_1/\sigma = \infty \tag{5.29}$$

$$\mu_m/\mu_0 = (1 - 9f/4)(1 - 3f/2)^{-1} \qquad \text{for } \sigma_1/\sigma = 0$$

The latter agrees only qualitatively with $\mu_m = R_{hm}/\rho_m$ from Eq. (5.22).

Volger (ref. 29) proposed a geometric model for an inhomogeneous material which consists of conducting grains separated by more resistive intergranular material, such as shown in Fig. 5.4. An ordered two-dimensional array of cube-shaped structures with length per side ℓ, and resistivity ρ_1 are separated from each other by a distance ℓ_2 containing a medium (such as an oxide) with a higher resistivity ρ_2. For $\ell_2 \ll \ell_1$ and $\rho_2 \gg \rho_1$ the mean resistivity is

$$\langle\rho\rangle = \rho_1 + (\ell_2/\ell_1)\rho_1 \tag{5.30}$$

He estimated the mean Hall coefficient $\langle R_h \rangle$ to be not much different from R_{h1} while $\langle\rho\rangle$ can be much greater than ρ_1. Bube (ref. 30) and Blount *et al.* (ref. 31) have attempted to extend the implications of this resistive network model to include Hall effect, photo-Hall effect, thermoelectric and photothermoelectric

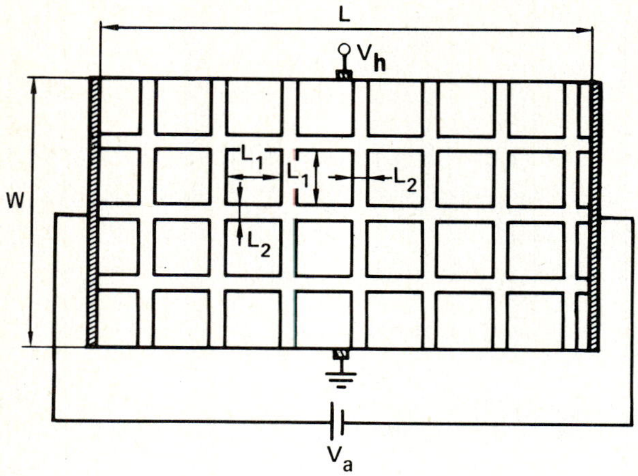

Fig. 5.4. Schematic representation of a material containing a low-resistivity two-dimensional array of ordered cubic inclusions of length ℓ_1 and resistivity ρ_1 in a high-resistivity matrix of length ℓ_2 and resistivity ρ_2. [after J. Volger, Phys. Rev., 79(1950)1023]

measurements made on inhomogeneous layers by assuming that the interbarrier electric field, in terms of Fig. 5.5(a), $E_3=V_a/L$ where V_a is the external voltage applied to the specimen and L is length, as shown in Fig. 5.4. Helesviki and Salo (ref. 32) have pointed out, however, that (in refs. 30 and 31) no account had been taken of the requirement for the equality of the tangential components of the electric fields E_1, in region 1, and E_3, in region 3 of Fig. 5.5(a) at the boundary between them; with $E_2 \gg E_1$ the field E_3 is much smaller than the mean field over the specimen, V_a/L. For the equivalent Hall voltage circuit shown in Fig. 5.5(b) Helesviki and Salo (ref. 32) derived the expressions

$$v_{1h} = R_{1h}\ell_1\gamma, \qquad v_{2h} = R_{2h}\ell_1\beta\gamma, \qquad v_{3h} = R_{2h}\ell_2\gamma(\rho_1/\rho_2) \tag{5.31}$$

where v_{1h}, v_{2h} and v_{3h} are the Hall voltages generated by a weak magnetic field in the corresponding regions of Fig. 5.5(a), with $\beta=\ell_2/\ell_1$ and $\gamma=B(V_a/L)(1+\beta)(\rho_1+\beta\rho_2)^{-1}$. The expressions for v_{1h} and v_{2h} are the same as those derived by Blount *et al.*(ref. 31) but their expression for v_{3h}, which did not consider the equality of the tangential components of the electric field at their boundary, is different from Eq. (5.31). The mean Hall coefficient is

$$\langle R_h \rangle \simeq R_h[1 + (\ell_2/\ell_1)(\mu_2/\mu_1)] \tag{5.32}$$

and the ratio of the measured carrier concentration $\langle n \rangle$ to the low resistivity region n_1 is $\langle n \rangle/n_1 \simeq (1+\beta\mu_2/\mu_1)^{-1}$; thus Hall measurements yield, to first order, the carrier concentration of the low resistance region.

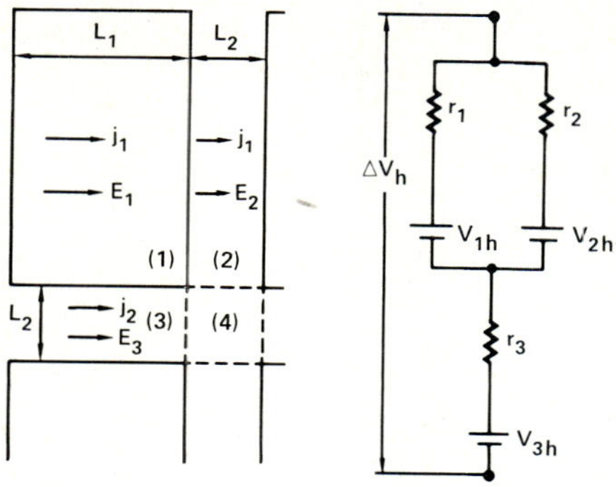

Fig. 5.5. (a) Current densities and electric fields in regions corresponding to Fig. 5.4; (b) Equivalent Hall voltage circuit for the corresponding regions of ordered array. [after J. Helesviki and T. Salo, J. Appl. Phys., 43(1972)740]

The inhomogeneities described thus far make the measured mobility of a specimen appear to be smaller and, occasionally, much smaller than the actual mobility of the same homogeneous specimen. This is, in part, a consequence of the averaging of the Hall field between the regions with different conductivities and, in part, may be due to an increase in charge carrier scattering at the boundaries between the inclusions and the matrix. Anomalously high mobilities have also been observed in materials such as silicon if an accumulation layer is induced on a high resistance n-type specimen by an HF treatment (ref. 33). Anomalously high mobilities have also been reported for PbSnTe (ref. 34) and GaAs (ref. 35). Wolfe *et al.* (refs. 35-37) have described a model for such anomalously high mobilities; it is based on a cylindrically symmetric structure, such as shown in Fig. 5.6 containing a coaxial conductive inclusion with an abrupt boundary. If the conductivity of the central inclusion is σ and that of the matrix is σ_0, and if $\sigma \gg \sigma_0$, then the inclusion and its boundary can be considered to be an equipotential. A solution of Laplace's equation in cylindrical coordinates applied to the structure shown in Fig. 5.6 leads to

$$\rho_m(\alpha) = \frac{1}{\sigma_0 \ell n 2} \sum_{n=1}^{\infty} \left[2 \cdot \frac{1 - \alpha^{2n}}{1 + \alpha^{2n}} - \frac{1 - \alpha^{4n}}{1 + \alpha^{4n}} \cdot \frac{(-1)^{n+1}}{n} \right] \tag{5.33}$$

where the ratio of the radii in Fig. 5.6, $\alpha = b/a$ with a current, i, applied between the contacts oriented on $\phi = \Pi/2$ and $\phi = 0$ and with the potential difference measured between the contacts at $\phi = \Pi$ and $\phi = 3\Pi/2$.

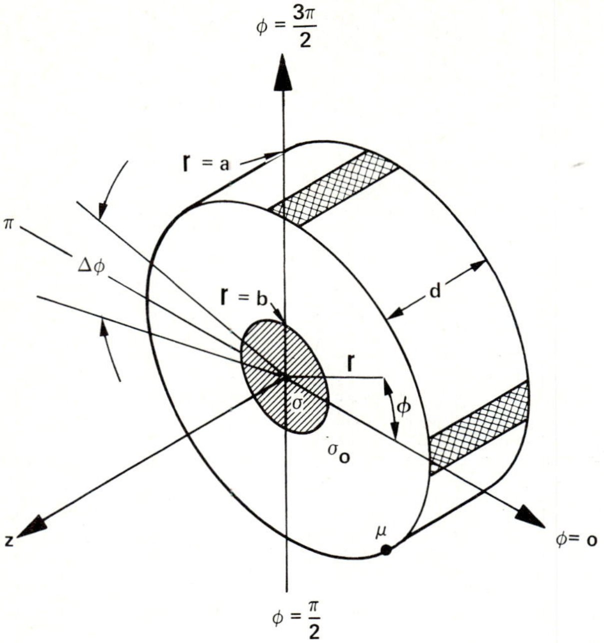

Fig. 5.6. Cylindrical spec...en of thickness d, radius a, conductivity σ_0 and mobility μ_o containing a coaxial conductive inclusion of radius b, conductivity σ and mobility μ with electrodes at 90° positions used for the measurement of the resistivity ρ_m and Hall coefficient R_{hm} of this inhomogeneous structure.

If the specimen is homogeneous then, evidently, $\alpha=0$ and the summation in Eq. (5.32) reduces to $\ell n2$; hence $\rho_m=1/\sigma_0$. If σ is not much larger than σ_0 then ρ_m depends on the ratio σ/σ_0 as well as on α. In particular, if the resistivity of the homogeneous specimen is ρ_0 and that of the inhomogeneous specimen is ρ_m then ρ_m/ρ_0 decreases with increasing α and α/α_0. However, even for small values of α there is an appreciable dependence of ρ_m/ρ_0 on σ/σ_0 for $1<\sigma/\sigma_0<10$. Thus relatively small changes in σ/σ_0 can produce significant changes in ρ_m/ρ_0 even if the fractional volume of the inhomogeneity is small.

The Hall coefficient of the inhomogeneous specimen in Fig. 5.6 can be determined by applying a current i between the electrodes located at $\pi/2$ and $3\pi/2$ and measuring the Hall voltage between the other two contacts leads; the Hall coefficient is

$$R_{hm}(\alpha,\beta) = \frac{4\mu}{\Pi\sigma}\sum_{n=1}^{\infty}\frac{(1+\beta^2)\cdot\left[\left(1-\alpha^{4n-2}\right)\cdot\left(1+\alpha^{4n-2}\right)^{-1}\right]^2\cdot\left(-1\right)^{n+1}}{1+\beta^2\left[\left(1-\alpha^{4n-2}\right)\cdot\left(1+\alpha^{4n-2}\right)^{-1}\right]^2\cdot(2n-1)} \tag{5.34}$$

where $\beta=\mu\beta$.

For $\alpha=0$ the summation of Eq. (5.33) reduces to $\pi/4$ and $R_{hm}=\mu/\sigma_0$. For $\mu B<<1$, the measured Hall coefficient R_{hm} decreases with increasing α. The measured resistivity ρ_m also decreases with increasing α in accordance with Eq. (5.32). However, the decrease of ρ_m with α can be considerably greater than that of R_{hm}. Therefore the apparent mobility calculated from the measured parameters ρ_m and R_{hm}, $\mu_m=R_{hm}/\rho_m$, can be as much as one order of magnitude larger than μ. For a given value of α, as B is increased, R_{hm} increases with B, and, since ρ_m is measured in B=0, μ_m also increases. For $(\mu B)^2>>1$ the measured Hall coefficient approaches μ/σ_0, the Hall coefficient of the homogeneous portion of the specimen. Physically, this implies that in the high field limit the current streamlines are distorted out of the inhomogeneous central portion of the specimen and current flow takes place only in the homogeneous region. The experimental measurements made by Wolfe *et al.*(ref. 35) on thin epitaxial GaAs layers deposited onto semi-insulating GaAs substrates, with metallic-alloyed gallium inclusions simulating the inhomogeneity, indicate good agreement between theory and experiment. For example, for $\alpha=0.5$, μ_m/μ is nearly independent of μB for μB values between 0.01 and 1.0, and increases by a factor of 6 as μB increases from 1 to 10. These results suggest that the electron mobility of bulk and epilayer semiconductors used as a material quality index should be treated with caution unless additional information is made available on their homogeneity. Anomalously high electron mobilities, in excess of the theoretical lattice scattering mobility, have been measured on some GaAs epilayers by Wolfe *et al.*(ref. 35). They found that these layers contained a random dispersion of conductive precipitates $\sim 10^9$ cm^{-3} in concentration. They concluded that anomalously high electron mobilities are likely to be measured in epilayers containing conductive inclusions if (a) the apparent mobilities calculated from resistivity and Hall measurements exceed theoretical expectations; (b) the measurements from different portions of the same wafer are not reproducible; (c) the Hall coefficients have an anomalous magnetic field dependence, and (d) the degree of calculated compensation appears to be unusually low.

The magnetic field dependence of the resistivity of a medium containing conductivity striations produced primarily by impurity fluctuations and, therefore, by periodic variations in carrier concentration has been calculated by Herring (ref. 27). For a current flowing in the x-direction and a magnetic field B oriented along the z-axis, the orientation of the current streamlines deflected by the transverse magnetic field vary between the striations due to their different Hall angles. The resistivity tensor, for InSb, as shown by Weiss (ref. 39), can be expressed as

$$|\rho| = \frac{1}{e\mu\langle n\rangle}\left\{\begin{vmatrix} 1 & \mu B_z & -\mu B_y \\ -\mu B_z & 1 & \mu B_x \\ \mu B_y & -\mu B_x & 1 \end{vmatrix} + \begin{vmatrix} A & 0 & 0 \\ 0 & 0 & 0 \\ 0 & 0 & 0 \end{vmatrix}\right\}$$

$$A = (\langle n\rangle\langle 1/n\rangle - 1)\cdot\left[1 + \frac{\mu^2(B^2 - B_x{}^2)}{1 + (\mu B_x)^2}\right] \tag{5.35}$$

where <n> and <1/n> are the spatial averages of the electron density <n> and its reciprocal <1/n> and the tensor in Eq. (5.35) consists of the resistivity tensor of the homogeneous material corrected by n/<n> and an additional tensor which represents the strong anisotropy of the magnetoresistance. The dependence of the magnetoresistance on the angle ψ between the current i, flowing perpendicular to the striations and the magnetic field B (ref. 40) is

$$\frac{\Delta\rho}{\rho_0} = \left[1 - \frac{1}{\langle n\rangle\langle 1/n\rangle}\right]\frac{(\mu B)^2 \sin^2\psi}{1 + (\mu B)^2 \cos^2\psi} \tag{5.36}$$

If the magnetic field B has no component in the direction of current flow then $\Delta\rho/\rho_0$ is proportional to B^2. Weiss found good agreement between experimentally measured and calculated magnetoresistance for conductivity striations in InSb in terms of Eq. (5.36).

Vinetskii *et al.* (ref. 41) have investigated theoretically and experimentally the effects of γ-radiation on the electron mobility of Czochralski-grown Ge containing a periodic one-dimensional fluctuation of Sb(donor) impurity concentration introduced during crystal growth. They found that γ-radiation produces a substantial increase in the anisotropy of the electron mobility with a much smaller change in the direction normal to the impurity gradient than in a direction parallel to it; for small radiation doses the temperature dependence of μ_{11} is invariant with radiation and is a slowly varying function of temperature as the radiation dose is increased.

5.3 Polycrystalline and polyphase materials

A polycrystalline material which is macroscopically isotropic has electrical and galvanomagnetic properties which are a function of the specific conductivity of the same single crystal material. If the dimensions of an individual crystallite are small with respect to that of the polycrystalline specimen and if the distribution of the crystallites is considered to be statistically homogeneous and isotropic and there is no phase difference nor surface space charge effects at the crystallite boundaries then estimates of the conductivity of the polycrystalline material can be made in terms of (ref. 42) the average conductivity

$$\langle\sigma\rangle = (1/3)(\sigma_1 + \sigma_2 + \sigma_3) \tag{5.37}$$

where σ_1, σ_2, and σ_3 are the conductivity components along the principal directions of the single crystal. Alternatively, an estimate of the average resistivity (ref. 43) can be expressed as

$$<\rho> = (1/3)(\rho_1 + \rho_2 + \rho_3) \tag{5.38}$$

Equations (5.37) and (5.38) are, respectively, upper and lower bounds on the effective macroscopic conductivity of a polycrystalline material (ref. 44). Attempts have been made to narrow these bounds by imposing various constraints on the arrangement of the crystallites which would predict the conductivity more accurately (refs. 45-48). However, Schulgasser (ref. 49) has shown that no universal relation exists between the conductivity of a single crystal and a polycrystalline material and experimental evidence was adduced to support this conclusion and that Eqs. (5.37) and (5.38) are the best, generally valid bounds on the conductivity of a polycrystalline material.

The existence of intergrain barriers in polycrystalline materials such as the lead salts has been investigated by Petritz (ref. 50). He proposed that the resistivity of an intergrain barrier resistivity, ρ_b, in series with the resistivity of a grain, ρ_c, represents the total resistivity, $\rho=\rho_b+\rho_c$, of a polycrystalline material, that the current-voltage relation has an exponential dependence which is a function of the barrier height, ψ, referred to the valence band edge. The macroscopic conductivity $\sigma=e\mu_p p \exp(-e\psi/kT)$ where μ_p is the hole mobility and p their density. If the conductivity is expressed in terms of $\sigma=ep\mu_p^*$, where μ_p^* is an effective mobility, then μ_p^* has a strong temperature dependence and also depends on the number of intergrain barriers in series as well as on the height of the barrier with respect to temperature. The Hall mobility is presumed to be a direct measure of μ_p^* and includes the effects of barriers. Snowden and Portis (ref. 51) used a similar model to describe experimental measurements made on polycrystalline PbS as a function of frequency of the applied electric field in the microwave range. They concluded that the DC electrical conductivity is limited by barriers but the DC Hall mobility is not barrier limited. Kuznicki (ref. 52) has considered intergrain barriers much greater than kT/e such as those which appear in CdSe thin films. He showed analytically and experimentally that for such constraints the effective current density, mobility and resistivity are exponential functions of the longitudinal applied electric fields.

A model for the electrical resistivity of polycrystalline metallic films was developed by Mayadas *et al.* (refs. 53,54). The model presumes that three types of electron scattering mechanisms need to be considered in such films: an isotropic background scattering due to phonons and point defects, scattering due to the external boundary surfaces of a film and grain boundary scattering, represented in terms of a series of partially reflecting, randomly spaced, planar grain boundaries. The intrinsic or bulk resistivity was calculated by solving the Boltzmann transport equation for both grain boundary and background scattering. If σ_0 is the conductivity of a film in the absence of grain boundaries, and σ_g is the conductivity due specifically to grain boundary scattering, then provided that the interplanar spacing between the grain boundaries, d, is comparable to the measured average grain diameter, D,

$$\sigma_g/\sigma_0 \equiv \rho_0/\rho_g = 3[1/3 - (1/2)\alpha + \alpha^2 - \alpha^3 \ell n(1 + 1/\alpha)] \qquad (5.39)$$

where $\alpha=(\ell_0/d)\ R(1-R)^{-1}$. Here ℓ_0 is the background scattering mean free path and R is an electron reflection coefficient. In the limit, Eq. (5.38) becomes

$$\rho_g/\rho_0 \simeq 1 + 1.5\alpha \text{ for } \alpha \ll 1 \qquad \rho_g/\rho_0 \simeq (4/3)\alpha \text{ for } \alpha \gg 1 \qquad (5.40)$$

The total resistivity is obtained by imposing boundary conditions (ref. 55) which depend on the film thickness relative to the electron mean free path (Fuchs size-effect) and the effective mobility is thus a function of background scattering, grain boundary scattering and surface scattering. An evaluation of experimental measurements suggest that (a) the grain boundary reflection coefficient of aluminum, $R \approx 0.15$; it is slightly larger in copper; (b) the observed thickness dependence of the resistivity of thin films is due to a combination grain boundary scattering and Fuchs size-effect; (c) the usual observation that single crystal films have lower resistivities than polycrystalline films is attributed to grain boundary scattering rather than to variations in the specularity of surfaces. Ramey and McLennon (ref. 56) have investigated grain boundary and surface scattering of vacuum-deposited germanium films and found good agreement for the thickness and temperature dependence of their mobility using the dislocation scattering model of Dexter and Seitz (ref. 57) for grain boundary scattering, an empirical parameter to account for surface scattering, lattice scattering by phonons and an impurity scattering processes with the latter interpreted in terms of the parameters of the bulk crystals.

High mobility III-V intermetallic semiconducting compounds in the form of polycrystalline layers are particularly sensitive to both grain boundary and surface scattering (ref. 58) which, in turn, are a function of the size of the crystallites and layer thickness. The Hall mobility of pyrolytically-deposited polycrystalline silicon has been shown by Kamins (ref. 59) to be an anomalous function of impurity concentration. The mobility for both n- and p-doped specimens has a peak for a free carrier concentration of $\sim 10^{18}/cm^3$ decreasing for both higher and lower impurity concentrations. The mobility decrease with decreasing carrier concentration was attributed to space-charge regions surrounding grain boundaries in the polycrystalline material, a situation which is consistent with the increase in mobility of the layers with increasing grain size and thickness.

The conductivity of two phase mixtures has been calculated in terms of the bulk conductivities of the individual phases by Maxwell (ref. 60) and Raleigh (ref. 61). The effect of an interfacial space-charge layer at the boundaries between the phases has been neglected. If metallic particles are imbedded in an ideal insulator matrix then the DC conductivity is zero. However, if account is taken of the space-charge layer between the metal-dielectric phases then the conductivity may be finite and can be calculated for some boundary conditions (ref. 62). Experimental evidence shows that the resistivity of metal or carbon particles imbedded in an insulating matrix decreases drastically (ref. 63) as the concentration of these microscopic inclusions exceeds a critical value. At this concentration essentially infinite chains of particles begin to appear and it is this fraction which contributes most significantly to the conduction process. The chains of connected metallic particles act like conducting filaments between a specimen's terminals. Bruggeman (ref. 64) and Landauer (ref. 65) developed an effective medium theory for two-phase systems in which a volume fraction C has a conductivity σ_0 and the remainder a conductivity σ_1 such that $\sigma_1 \ll \sigma_0$. In consequence, $\sigma = f\sigma_0$ where $f = a + (a^2 + 0.5x)^{1/2}$; $a = 0.5(1.5c - 0.5)(1-x) + 0.25x$ where $x = \sigma_1/\sigma_0$.

Nedoluha (ref. 66) developed a generalized theoretical treatment for the electrical conductivity of anisotropic inhomogeneous media in which the conductivity differences are small or a second phase of low density spherical inclusions is present in a matrix. Malliares and Turner (ref. 67) have investigated the influence of particle size on the resistivity of compacted mixtures of polymeric and metallic powders. They found that for polyethylene particles of mean size, R_p, and nickel particles of size, R_m, compacted at room temperature at a pressure of 10^3kg/cm^2 the metallic particles do not penetrate the polymer and that this results in a segregated distribution for high ratios of R_p/R_m with a resistivity of the order of 10^{16}ohm-cm. R_p/R_m ratios from 1 to 16 were investigated; at a critical ratio of metallic to polymer ratio, which is a function of R_p/R_m, the resistivity drops by as much as 20 orders of magnitude. The dependence of the electrical resistivity on composition was interpreted in terms of a model according to which nickel particles can adhere to the larger polymer particles forming a metallic monolayer coating; it can be represented as an accommodation of metallic particles on three mutually perpendicular sets of lattice planes. The critical composition for the sharp drop in electrical resistivity was attributed to the first non-zero probability for the formation of polymer-like metal particle chains.

5.4 Anisotropic materials

The measurements of the resistivity and of the Hall coefficient of materials, described in Chapters 2 and 3, have been concerned primarily with ρ and R_h treated as scalar quantities. However, if the material under consideration has a crystalline structure without any elements of symmetry then ρ is a tensor with six components and R_h is a tensor with nine components. Van der Pauw (ref. 68) has shown that the components of the resistivity tensor can be determined by using six plane-parallel specimens of arbitrary shape but oriented so that low index planes are perpendicular to the lines with direction cosines (100), (010), (001), (011), (101) and (110). For determining the Hall coefficient tensor components three specimens with (100), (010) and (001) orientations are adequate provided that each is measured for three different orientations of the applied magnetic field. Such techniques are hardly applicable to thin films or epitaxial layers, which are usually grown along preferred crystallographic orientations that promote the growth of homogeneous single-crystal layers. However, for an anisotropic material characterized by the tensor components ρ_x, ρ_y and ρ_z and provided that ρ_z is the component perpendicular to its principal surface, then $\langle\rho\rangle=(\rho_x\rho_y)^{1/2}$ yields the geometric mean of these tensor components from a van der Pauw-type measurement such as described in Chapter 1 (Section 1.5) for a circular symmetry or square specimen. This has been demonstrated by Hornstra and van der Pauw (ref. 69) and by Price (ref. 70); three measurements of different specimens each oriented with the x-, y- or z-axes normal to the specimen plane provide the necessary information for determining the respective tensor components. The components of the resistivity tensor in the plane of the specimen can be determined from a rectangular specimen cut in such a manner that its edges are parallel to the principal axes of the resistivity tensor, as shown by Price (ref. 70). Line contacts must be made to the corners of the rectangle whose sides have the dimensions a and b, respectively. From the solution of the potential distribution in an anisotropic rectangle he derived the relation

$$\left(\frac{\rho_x}{\rho_y}\right)^{1/2} = -\frac{b}{\alpha\pi}\,\ell n\left[\tanh\frac{\pi d R_{BC,DA}}{16(\rho_x\rho_y)^{1/2}}\right] \tag{5.41}$$

where the geometric mean $(\rho_x\rho_y)^{1/2}$ is obtained from the relation

$$\exp\left[-\frac{\pi d R_{AB,CD}}{(\rho_x\rho_y)^{1/2}}\right] + \exp\left[-\frac{\pi d R_{BC,DA}}{(\rho_x\rho_y)^{1/2}}\right] = 1 \tag{5.42}$$

and in-plane components of the resistivity tensor are therefore

$$\rho_x = (\rho_x\rho_y)^{1/2}\cdot(\rho_x/\rho_y)^{1/2} \qquad \rho_y = (\rho_x\rho_y)^{1/2}/(\rho_x/\rho_y)^{1/2} \tag{5.42}$$

An alternative method based on the colinear four-point probe method, described in Chapter 1 (Section 1.2), has been described by Airapetyants and Bressler (ref. 71) for measuring the resistivity of large anisotropic crystals by performing three different measurements along directions corresponding to the three principal directions of the resistivity tensor. Such measurements can be extended to the more general case of an arbitrarily-chosen coordinate system by the procedures described by van der Pauw (ref. 68).

Montgomery (ref. 72) has described a method for measuring the electrical resistivity of anisotropic crystals which provides some advantages for measuring the resistivity of small strongly temperature-dependent specimens in that it requires only a single specimen with fixed electrodes for measuring two components of the resistivity tensor. The specimen is cut in the form of a rectangular prism one of whose faces has its edges in the direction of the two resistivity components to be measured. Small electrodes are attached to the corners of this face. The ratio of the voltage V, measured between two adjacent electrodes, to the current i, between the other two electrodes, is determined and then the current and voltage contacts are commutated. Calculation of the tensor components involves two steps: (a) use of a simple theorem for mapping an anisotropic solid of arbitrary shape to an equivalent solid, based on the work of van der Pauw (ref. 68); (b) relating the bulk resistivity of the isotropic rectangular prism to the voltage-current measurements and to its dimensions. The geometrical factor which allows the calculation of the two tensor components have been tabulated (refs. 72, 73). If a third component is required then a different face of the same specimen or a different specimen with a different crystallographic orientation are suitable for measurements. Montgomery (ref. 73) made such measurements on specimens of V_2O_3 doped with a few percent Cr. This is a hexagonal crystalline material containing twinned regions with their c-axis in parallel. A thin square plate cut normal to the c-axis exhibited no anisotropy of the resistivity whereas a plate cut with one edge parallel to the c-axis showed evidence of substantial anisotropy. This method of measurement has particular advantages for the observation of resistivity changes in anisotropic crystalline materials which undergo crystallographic phase transitions. Logan *et al.*(ref. 73) have derived asymptotic expression and have tabulated the geometrical factor required for calculations of the tensor components of the resistivity (ref. 72) in terms of the specimen dimensions.

The measurement of the electrical resistivity of anisotropic metals such as cadmium or tin by an eddy-current method has been described by Lawson (ref. 74). He suggests the use of spherical specimens with the crystal c-axis (in the sphere) first parallel and then perpendicular to a uniform magnetic field produced by a solenoid. The switching on or off of the solenoid causes eddy currents to flow in concentric circles, principally in a plane perpendicular to the magnetic field, and a balanced search coil is used to determine the decay time constants of the eddy currents. These decay times were of the order of 20μs for tin and up to several seconds for pure cadmium at 1.3°K. The decay time τ_1 is inversely proportional to the resistivity and to the angle Ψ between the crystal sphere c-axis and the magnetic field. For $\Psi=0$ the currents flow in a plane which has an essentially isotropic resistivity and the decay time constant $\tau_1(0)$ is directly related to the essentially isotropic resistivity ρ_1. If the crystal c-axis is perpendicular to the magnetic fields the situation is more complicated since the eddy currents flow in a plane which has an anisotropic resistivity. To a first approximation $\tau_1(\Pi/2)$ corresponds to the average resistivity, $(1/2)(\rho_1+\rho_{11})$, and

$$\tau_1(0)/\tau_1(\Psi) = 1 + (1/2)(\alpha - 1)\sin^2\Psi \qquad (5.43)$$

where α is an anisotropy ratio ρ_{11}/ρ_1. The range of decay time measured experimentally is between 20µsec, for tin at room temperature, to several seconds for pure cadmium at 1.3°K. For materials with anisotropies $\alpha<1.5$ Lawson found good agreement between calibrated resistivity measurements made by the eddy-current decay method on single crystals of Zn and Cd and conventional four-point probe measurements. He found fair agreement with the measurements of the temperature dependence of the anisotropy determined by Alderson and Hurd (ref. 75) using more precise methods of measurement between 4°K and 300°K.

REFERENCES

1 A. F. Witt and H. C. Gatos, J. Electrochem. Soc., 113(1966)809.
2 A. Mueller and M. Wilhelm, Z. Naturforsch., 19a(1964)254.
3 A. F. Witt and H. C. Gatos, J. Electrochem. Soc., 115(1968)70.
4 F. Hasegawa and T. Saito, Japan J. Appl. Phys., 7(1968)1125.
5 F. Hasegawa and T. Saito, Japan, J. Appl. Phys., 7(1968)1540.
6 J. S. Harris, Y. Nannichi, G. L. Pearson and G. F. Day, J. Appl. Phys., 40 (1969)4575.
7 W. Y. Lum and H. H. Wieder, Appl. Phys. Lett., 31(1977)213.
8 C. M. Wolfe and G. E. Stillman in Semiconductors and Semimetals, ed., R. K. Willardson and A. C. Beer, Academic Press, New York, Vol 10(1975)175.
9 R. L. Petritz, Phys. Rev., 110(1958)1254.
10 J. N. Zemel and R. L. Petritz, Phys. Rev., 110(1958)1263.
11 M. G. Buehler, Stanford Electron. Lab's Tech. Rept., 5101-1(1966)45.
12 R. Baron, G. A. Shifrin, O. J. Marsh and J. W. Mayer, J. Sppl. Phys., 40 (1969)3702.
13 J. W. Mayer, O. J. Marsh, G. A. Shifrin and R. Baron, Can. J. Phys., 45(1967) 4073.
14 J. A. Davies, G. C. Ball, F. Brown and B. Domeij, Can. J. Phys., 42(1964)1070.
15 N. G. E. Johansson, J. W. Mayer and O. J. Marsh, Solid State Electron., 13 (1970)317.
16 W. S. Johnson, Solid-State Electron., 13(1970)951.
17 D. M. Caughey and R. E. Thomas, Proc. IEEE, 55(1967)2192.
18 H. Mueller, F. H. Eisen and J. W. Mayer, J. Electrochem. Soc., 122(1975)651.
19 V. K. Subashiev and A. S. Poltinnikov, Sov. Phys.-Solid State, 2(1960)1059.
20 O. N. Tufte, J. Electrochem. Soc., 109(1962)235.
21 A. F. Kravchenko, B. V. Morozov and E. M. Skok, Sov. Phys. Semicond., 6(1972) 257.
22 N. I. Pavlov, Sov. Phys. Semicond., 4(1971)1644.
23 I. Hlasnik, Solid-State Electron., 8(1965)461.
24 H. J. Juretschke and R. Landauer, J. Appl. Phys., 27(1956)838.
25 E. Goldin and H. J. Juretschke, Trans. Metall. Soc. AIME, 212(1958)357.
26 H. J. Juretschke, R. Landauer and J. A. Swanson, J. Appl. Phys., 27(1956)838.
27 C. Herring, J. Appl. Phys., 31(1960)1939.

28 V. V. Voronkov, G. I. Voronkova and M. I. Iglitsyn, Sov. Phys. Semicond., 4 (1971)1949.
29 J. Volger, Phys. Rev., 79(1950)1023.
30 R. H. Bube, Appl. Phys. Lett., 13(1968)136.
31 G. H. Blount, R. H. Bube and A. L. Robinson, J. Appl. Phys., 41(1970)2190.
32 J. Helesviki and T. Salo, J. Appl. Phys., 43(1972)740.
33 D. Colman and D. L. Kendall, J. Appl. Phys., 40(1969)4662.
34 G. A. Antcliffe and J. S. Wrobel, Mater. Res. Bull., 5(1970)747.
35 C. M. Wolfe, G. E. Stillman, D. L. Spears, D. E. Hill and F. V. Williams, J. Appl. Phys., 44(1973)732.
36 C. M. Wolfe and G. E. Stillman, Appl. Phys. Lett., 18(1971)205.
37 C. M. Wolfe, G. E. Stillman and J. A. Rossi, J. Electrochem. Soc., 119(1972) 250.
38 C. M. Wolfe and G. E. Stillman in Semiconductors and Semimetals, R. K. Willardson and A. B. Beer, ed., Academic Press, New York, 10(1975)175.
39 H. Weiss, J. Appl. Phys. (Suppl.) 32(1961)2064.
40 H. Weiss in Semiconductors and Semimetals, R. K. Willardson and A. C. Beer, ed., Academic Press, New York, 1(1966)315.
41 V. L. Vinetskii, N. V. Kukhtarev and A. K. Semenyuk, Sov. Phys. Semicond., 6(1972)879.
42 W. Voigt, Lehrbuch der Krystallphysik, Teubner Verlag, Leipzig (1928)410.
43 A. Reuss, Z. Angew. Math. Mech., 9(1929)49.
44 J. Molyneux, J. Math. Phys., 11(1970)1172.
45 Z. Hashin and S. Shtrikman, Phys. Rev., 130(1963)129.
46 V. Bolotin and V. Moskalinko. J. Appl. Mech. Tech. Phys., 8(1967)3.
47 E. Kröner, J. Mech. Phys. Solids, 15(1967)319.
48 J. K. Alstad, R. V. Colvin and S. Legvold, Phys. Rev., 123(1961)418.
49 K. Schulgasser, J. Appl. Phys., 47(1976)1880.
50 R. L. Petritz, Phys. Rev., 104(1958)1508.
51 D. P. Snowden and A. M. Postis, Phys. Rev., 120(1960)1983.
52 Z. Kuznicki, Solid-State Electron., 19(1976)894.
53 A. F. Mayadas, M. Shatzkes and J. F. Janak, Appl. Phys. Lett., 11(1969)345.
54 A. F. Mayadas and M. Shatzkes, Phys. Rev., B1(1970)1382.
55 See, for example, E. H. Sondheimer, Adv. Phys., 1(1952)1.
56 R. L. Ramey and W. D. McLennan, J. Appl. Phys., 38(1967)3491.
57 D. L. Dexter and F. Seitz, Phys. Rev., 86(1952)964.
58 H. H. Wieder, Intermetallic Semiconducting Films, Pergammon Press, Oxford, 10(1970)122.
59 T. I. Kamins, J. Appl. Phys., 42(1971)4357.
60 J. C. Maxwell, A. Treatise on Electricity and Magnetism, 2nd ed. Clarendon Press, Oxford, 1(1881)435.
61 Lord Raleigh, Phil. Mag.(5), 34(1892)481.
62 C. Wagner, J. Phys. Chem. Solids, 33(1972)1051.
63 S. M. Aharoni, J. Appl. Phys., 43(1972)2463.
64 D. A. G. Bruggeman, Ann. Phys.(Leipzig)), 24(1935)636.
65 R. Landauer, J. Appl. Phys., 23(1952)779.
66 A. Nedoluha, Z. Phys., 148(1957)248.
67 A. Malliares and D. T. Turner, J. Appl. Phys., 42(1971)614.
68 L. J. van der Pauw, Philips Res. Repts., 16(1961)187.
69 J. Hornstra and L. J. van der Pauw, J. Electron. Control, 7(1959)169.
70 W. L. V. Price, J. Phys. D, 5(1972)1127.
71 S. V. Airapetyants and M. S. Bressler, Sov. Phys. Solid State, 1(1959)134.
72 H. C. Montgomery, J. Appl. Phys., 42(1971)2971.
73 B. F. Logan, S. O. Rice and R. F. Wick, J. Appl. Phys., 42(1971)2975.
74 N. S. Lawson, J. Phys. E., Sci. Instrum., 10(1977)786.
75 J. E. A. Alderson and D. M. Hurd, Phys. Rev., B12(1975)501.

CHAPTER 6

SURFACE SPACE CHARGE-CONTROLLED MEASUREMENTS

6.0 Introduction

Two-terminal capacitance vs. voltage (C-V) and conductance vs. voltage (G-V) measurements made on reverse biased Schottky barrier diodes, p-n junctions or metal-insulator-semiconductor (MIS) structures have been used for many years for determining the fundamental properties of semiconductor surfaces and of their interfaces with metal or dielectric materials. Inherent surface properties, p-n junctions or interfacial contacts cause a redistribution of charge carriers in the surfaces of a semiconductor producing a space charge layer with many of the attributes of a capacitor. The effective width, δ, of the space charge layer is a function of the impurity concentration of the semiconductor, of the band bending of its energy bands and of the bandgap of the semiconductor (ref. 1). A large impurity concentration reduces the width of the space charge layer; it makes the corresponding capacitance, C_{sc}, large in comparison with that of the same pure material. The space charge capacitance is, also a function of the "gate" voltage, V_g, the potential applied to the metal electrode of a Schottky barrier or the metal electrode of a MIS structure.

Figure 6.1.(a) shows the potential profile of a metal and n-type semiconductor before interfacial contact is made between them; ϕ_m is the thermionic work function of the metal and represents the work required to remove from it an electron, whose energy corresponds to its Fermi level, and move it to free space; χ is the corresponding work function of the semiconductor. The difference $\phi_m-(\phi_n+\chi)$ is defined as the contact potential difference. Figure 6.1.(b) shows the potential profile of the metal-semiconductor junction, at equilibrium, after interfacial contact has been established between them. The Fermi level of the semiconductor has been lowered by an amount equal to the difference between the two work functions, a negative charge has been induced on the metal surface and an equal and opposite positive charge is induced in the semiconductor and is distributed in its surface space charge region. Free carriers are depleted from a layer of width δ. Since ϕ_m and χ are of the order of several electron volts, the electric field across this depletion layer is $\sim(\phi-\chi)/e\delta$; it can reach very high values for δ of the order of a fraction of a μm. The conduction and valence bands are bent relative to the Fermi level, as shown in Fig. 6.1.(b) where eV_D is defined as a diffusion potential measured between the conduction band edge of the surface and deep within the bulk of the semiconductor. The maximum barrier height

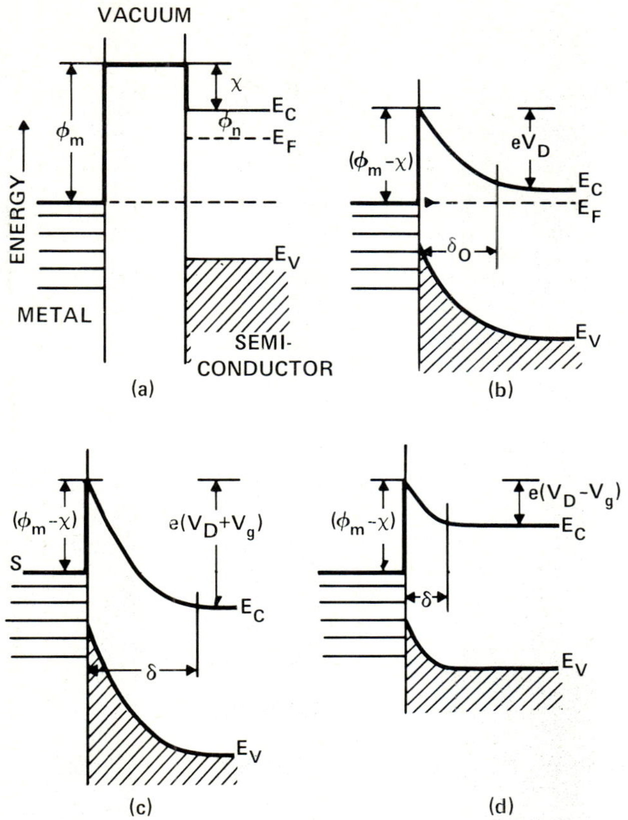

Fig. 6.1. Potential profile and energy band diagrams of Schottky barrier: (a) before metal-semiconductor interfacial contact; (b) at equilibrium after interfacial contact; (c) with reverse gate bias applied; (d) with forward gate bias applied.

is $\phi_{Bmax}=\phi_m-\chi$ for an n-type semiconductor and $e\phi_{Bmax}=E_g-e(\phi_m-\chi)$ for a p-type semiconductor; consequently $E_g/e=\phi_{Bn}+\phi_{Bp}$ (ref. 2). If a gate voltage V_g is applied to such a Schottky barrier diode then the band edges are either raised or lowered with respect to the Fermi level, as shown in Figs. 6.1.(c) and (d), depending on the polarity of V_g. The height of the barrier increases when V_g is negative (for the n-type semiconductor in Fig. 6.1.(c); this represents a reverse bias. The height of the barrier is lowered for positive V_g (i.e., forward bias) with the onset of a non-equilibrium state produced by current flow across the diode. If $V_g=V_D$ then no barrier is present; this corresponds to flatband.

A first order interpretation is feasible of reverse-biased Schottky barriers in terms of a gate voltage-dependent space charge capacitance per unit area,

$C_{sc}=\kappa_s\varepsilon_o/\delta$, where κ_s is the dielectric constant of the semiconductor and ε_o is the free space permittivity. An analytical representation of C_{sc} under equilibrium conditions can be made by a one-dimensional solution of Poisson's equation applied to the space charge layer, subject to the boundary conditions that for the coordinate x normal to the space charge layer (for an n-type semiconductor), the charge density, $\rho \approx eN_d$, for $x<\delta$ and that for $x>\delta$ the space charge density and the electric field of the space charge, $d\phi/dx$, are identically zero. The results obtained for a Schottky barrier are similar to those of an asymmetrical p-n junction in which the junction space charge layer extends much further into the lower doped side in accordance with the analytical derivations of Shockley (ref. 3) in his development of the depletion layer theory of p-n junctions.

The space charge per unit area of such an n-type semiconductor is

$$Q_{sc} = eN_d\delta = \left[2e\varepsilon_o\kappa_s(V_{bi} - V_g - kT/e)\right]^{1/2} \tag{6.1}$$

where V_{bi} is a "built in" potential which is a function of the bandgap and the impurity concentration of the semiconductor, k is Boltzmann's constant and T, the absolute temperature. The space charge capacitance per unit area is

$$C_{sc} = \frac{\partial Q_{sc}}{\partial V_g} = \frac{\kappa_s\varepsilon_o}{\delta} = \left[\frac{e\kappa_s\varepsilon_o N_d}{2(V_{bi} - V_g - kT/e)}\right]^{1/2} \tag{6.2}$$

From Eq. (6.2) the impurity concentration is

$$N_d = \left(\frac{2}{e\kappa_s\varepsilon_o}\right)\cdot\left[- \frac{dV_g}{d(1/C_{sc}^2)}\right] \tag{6.3}$$

Equation (6.3) states that if N_d is constant through the depletion region then a plot of $1/C_{sc}^2$ vs. V_g is a straight line whose slope is inversely proportional to N_d.

Figure 6.2 shows, schematically, the features of an MIS diode and its energy band diagram for $V_g=0$. The metal-semiconductor work function difference ϕ_{ms} for either an n- or a p-type semiconductor is

$$\phi_{ms} = \phi_m - (\chi + E_g/2 - \psi_B) = 0 \tag{6.4}$$

where ψ_B is the bulk potential, i.e., the potential difference between the actual Fermi level E_F and the intrinsic Fermi level E_i, as shown in Fig. 6.2. If an ideal MIS diode is biased by means of a gate voltage then the energy bands are bent relative to E_F and its surface may be either in accumulation, depletion or inversion, as shown on the corresponding potential diagrams of Fig. 6.3. If V_g is negative then holes are attracted to the metal-insulator interface of a p-type semiconductor; its surface becomes accumulated, as shown in Fig. 6.3.(a). For a small positive V_g holes are repelled from the interface producing a surface

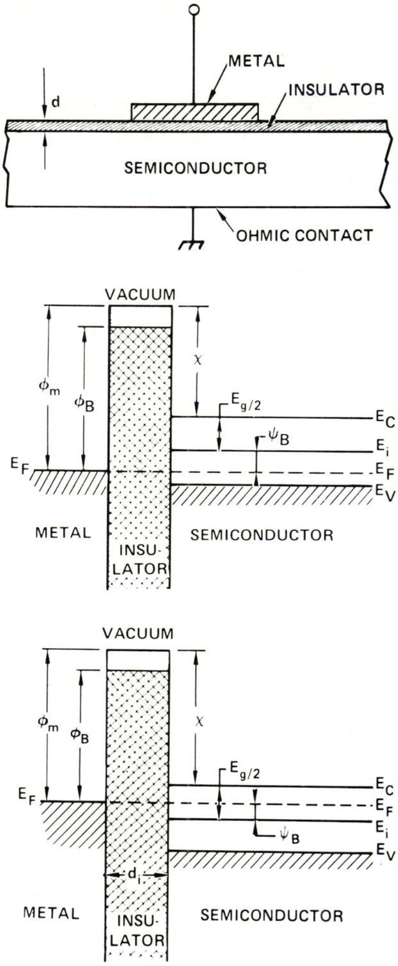

Fig. 6.2. (a) Metal-insulator-semiconductor (MIS) structure; (b) energy band diagram of an ideal MIS structure with p-type semiconductor for $V_g=0$; ϕ_B=potential barrier between metal and insulator, ϕ_m=metal work function, χ=semiconductor work function, E_g=bandgap, E_F=Fermi level, E_i=intrinsic Fermi level; (c) corresponding energy band diagram of an n-type semiconductor.

space charge and depletion region, as shown in Fig. 6.3.(b). A further increase of positive V_g causes an increase of the depletion width δ (ref. 4); however, as the bands are bent further, the conduction band will come close to the Fermi level and the density of electrons (which had been attracted to the semiconductor-insulator interface by the positive gate voltage) increases sharply. This is the inversion process shown in Fig. 6.3.(c) and the negative surface charge is

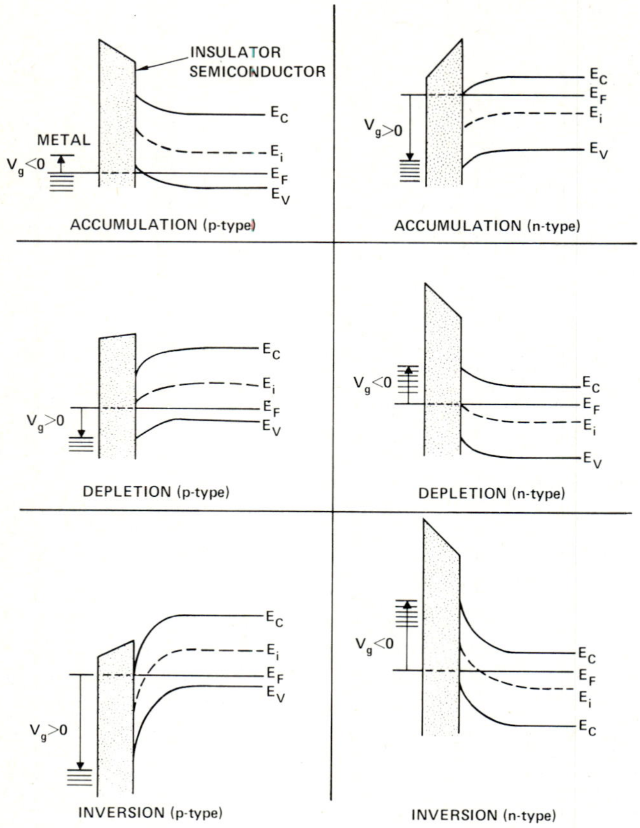

Fig. 6.3. Energy band diagrams of ideal n- and p-type MIS structures as a function of the applied gate voltage, V_g.

localized in a thin inversion layer. Once an inversion layer is formed the depletion depth reaches a maximum value, δ_m and the charge per unit area induced in the semiconductor is $Q_s=Q_n-eN_a\delta_m$, where Q_n is the surface charge density of electrons and N_a is the bulk acceptor density. It is evident that for inversion, electrons must be generated and transported to the interface and holes must be removed from the depletion region; therefore, the characteristic properties of MIS structures depend on the generation-recombination rate of charge carriers and the rates at which they are transported to the interface. If the gate voltage is a function of time then $V_g(t)$ should be slow enough so that charge rearrangement in the space charge region is feasible. In this case, an analysis based on the assumption of equilibrium conditions is applicable. Such a process may be achieved by making C-V measurements under quasi-static conditions.

6.1 Equilibrium and quasi-static properties

If the insulating layer of a MIS structure is an ideal dielectric material then the equilibrium space charge capacitance, C_{sc}, of an n-type semiconductor, in terms of its extrinsic Debye length,

$$L_D = \sqrt{\frac{2kT\kappa_s\varepsilon_o}{e^2 n_{no}}} \tag{6.5}$$

and the normalized surface potential $v_s = e\psi_s/kT$ (where the surface potential ψ_s corresponds to the band bending of the intrinsic Fermi level in Fig. 6.3 relative to its bulk value), is

$$C_{sc} = \frac{\partial Q_{sc}}{\partial \psi_s} = \left(\frac{\kappa_s\varepsilon_o}{L_D}\right)\left\{\frac{(\exp v_s - 1) + (p_{no}/n_{no})[1 - \exp(-v_s)]}{F(v_s, p_{no}/n_{no})}\right\} \tag{6.6}$$

where p_{no} and n_{no} are, respectively, the equilibrium densities of the bulk holes and electrons and the function F, obtained from a solution of the one-dimensional Poisson's equation, is

$$F(v_s, p_{no}/n_{no}) = \left\{[\exp(v_s) - v_s - 1] + \frac{p_{no}}{n_{no}}\left[\exp(-v_s) + v_s - 1\right]\right\}^{1/2} \tag{6.7}$$

and similar relations have been developed for p-type semiconductors (refs. 4,5)

The relation between the gate voltage, surface potential and surface potential-dependent space charge density is

$$V_g = \psi_s + (\phi_{ms}/e) - Q_{sc}/C_i \tag{6.8}$$

where ϕ_{ms} is the work function difference between the metal and semiconductor and the capacitance per unit area of the dielectric is

$$C_i = \kappa_i\varepsilon_o/d_i \tag{6.9}$$

with κ_i the dielectric constant and d_i the thickness of the dielectric layer. The actual capacitance which might be measured on an ideal MIS structure is the series circuit of C_{sc} and C_i,

$$C_m = (C_i C_{sc})\cdot(C_i + C_{sc})^{-1} \tag{6.10}$$

Figure 6.4 shows the dependence of C_m on V_g calculated by neglecting ϕ_{ms}, using Eqs. (6.5) through (6.10) for arbitrary values of ψ_s and $n_o=10^{15}/cm^3$ at T=300°K

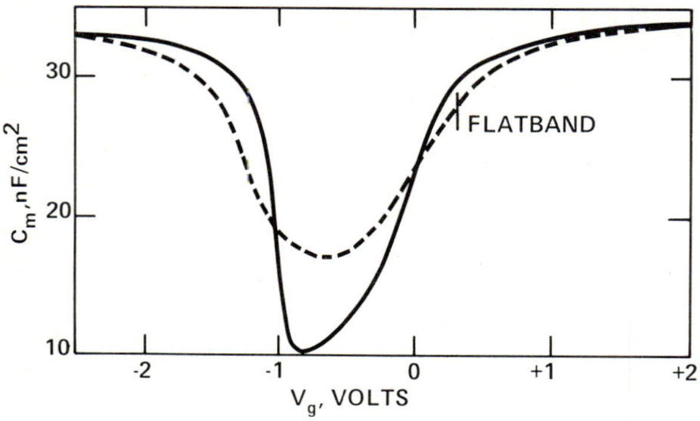

Fig. 6.4. Calculated gate voltage dependence of capacitance of an n-type silicon MOS structure; continuous curve without interface states, dashed curve calculated for the interface state density described in text.

determined on an n-type silicon MIS structure having a 0.1μm thick SiO_x dielectric layer (ref. 6). For positive values of V_g the surface of the semiconductor is accumulated; in accumulation, the semiconductor behaves essentially as a resistor in series with C_i. In strong inversion, holes appear in the inversion layer shielding the space charge region whose width remains constant beyond the minimum C_{sc} (which corresponds to the maximum depletion width); thus in strong inversion C_m approaches C_i. The analysis is more complicated if interface states are present in the bandgap. The charge of such interface states is determined by their energy position relative to the Fermi level. Interface states can have a donor-like or an acceptor-like character. The former are either positive or neutral; the latter are either negative or neutral. In any case, an interface state is in its more positive state of charge when it is above the Fermi level. The probability $f(E_t)$ that an interface state is in its more positive state is given by a Fermi-Dirac distribution

$$1 - f(E_t) = [1 + g \exp(E_F - E_t)/kT]^{-1} \tag{6.11}$$

where E_t is the energy of the interface state (charge trap) and g is a spin degeneracy factor. An interface state can add one elementary charge per state. Their sum produces a charge Q_{ss} and a corresponding capacitance C_{ss} which is a well-defined function of the surface potential and thus of V_g; a peak in C_{ss} occurs for the value of V_g for which the Fermi level crosses a specific interface state. The interface state charge density, integrated over the bandgap,

$$Q_{ss}(\psi_s) = -e\int N_{ss}(E_t)f(E_t)dE + e\int N_{sd}(E_t)dE \qquad (6.12)$$

where the energy-dependent density of interface states,

$$N_{ss}(E_t) = N_{sd}(E) + N_{sa}(E) \qquad (6.13)$$

with $N_{sd}(E)$ the donor-like and $N_{sa}(E)$ the acceptor-like states. To illustrate the effect of surface states consider a surface state density defined by

$$N_{ss}(E) = a\cosh[b(E - E_{is})] \qquad (6.14)$$

where $a=10^{11}eV^{-1}cm^{-2}$, $b=5.45\ eV^{-1}$ and the surface intrinsic Fermi level, $E_{is}=E_i=e\psi_s$; let the capacitance C_{ss} be calculated from

$$C_{ss} = -\frac{dQ_{ss}}{d\psi_s} = -e^2\int N_{ss}(E)\frac{df}{dE}dE \qquad (6.15)$$

with -df/dE considered to be a delta function at the Fermi energy. The effective MIS capacitance, in the presence of surface states, is

$$C_m^{-1} = C_i^{-1} + (C_{sc} + C_{ss})^{-1} \qquad (6.16)$$

and the relation between the surface potential and gate voltage is

$$V_g = \psi_s + (\phi_{ms}/e) - Q_{sc}(\psi_s)/C_i - Q_{ss}(\psi_s)/C_i \qquad (6.17)$$

The corresponding dependence of C_m on V_g, calculated by means of Eqs. (6.16) and (6.17), neglecting ϕ_{ms}, is shown in Fig. 6.4. The interface state density calculated from Eq. (6.14) is of the order of $10^{11}\ eV^{-1}cm^{-2}$ at midgap and $\sim 10^{12}\ eV^{-1}cm^{-2}$ at the band edges. These interface states produce a distortion of the C_m vs. V_g curve compared to that calculated for $N_{ss}=0$, a shallower minimum in C_m and a positive shift of flatband ($\psi_s=0$) from $V_g=0$ along the V_g axis. Other distortions of the C-V curves, a shift along the V_g axis and hysteresis with respect to the polarity of V_g may be observed experimentally. These may be due to a non-ideal insulator or to interface states. Fixed charge storage, ion or proton transport and ionic or electronic space charges with long time constants contribute to artifacts attributed to the oxide. Capture and emission of charge carriers from interface states are associated with time delays which can be represented as variable RC time constants. Equilibrium measurements must be performed at a frequency which is lower than the largest reciprocal time constant and at a rate which can accommodate equilibrium minority carrier generation and recombination processes. Such a frequency may have to be smaller than 1 Hz for many semiconductors and measurements at such low frequencies are difficult to make (ref. 7). Castagne (ref. 8), Kerr (ref. 9) and Kuhn (ref. 10) have described a slow ramp or quasi-static method for equilibrium C_m vs. V_g measurements. The circuit used for this purpose (ref. 10), shown in Fig. 6.5 is an analog differentiator. If a linear ramp voltage with a time dependence $V_g = V_1 + \alpha t$ is applied to the circuit

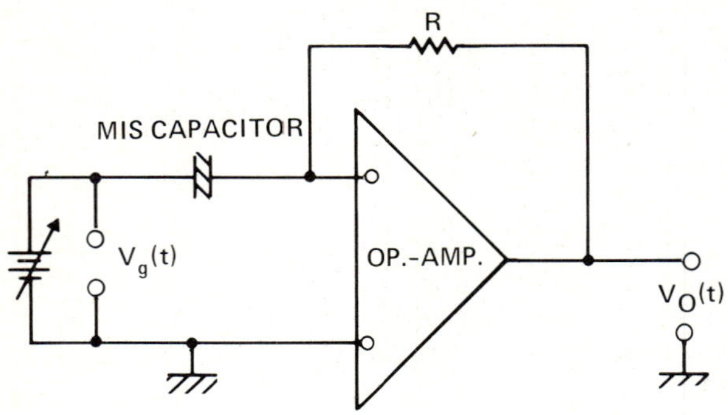

Fig. 6.5. Analog differentiator for quasi-static C-V measurements.

then the output voltage is $V_o(t) = \mp \alpha R C_m(t)$. A recorded display of $V_o(t)$ vs. $V_g(t)$ yields a scaled C_m vs. V_g curve with time as an independent parameter. Typical voltage sweep rates of the ramp are in the range between 5 and 5×10^2 mV/sec. and displacement current densities are usually less than 10^{-8} A/cm^2. Berglund (ref. 11) has shown that the surface potential may be determined to within an additive constant directly from equilibrium C-V curves by integrating them from a gate voltage which corresponds to strong accumulation toward inversion so that

$$\psi_s(V_g) = \int_{V_g(acc.)}^{V_g} \left[1 - \frac{C_m(V_g)}{C_i} \right] dV_g + \Delta \qquad (6.18)$$

where Δ is a constant of integration.

An ideal $C_{sc}(\psi_s)$ can be generated from the known doping density and insulator capacitance. The surface potential change can then be compared with that determined experimentally by means of Eq. (6.18) in order to assure the absence of gross non-uniformities. A plot of the ideal C_m/C_i vs. ψ_s can then be compared against a similar plot obtained experimentally. The two curves should coincide near strong accumulation and strong inversion and the additive constant Δ can thus be evaluated. The surface state density can then be calculated by the manipulation of the data and by the use (ref. 10) of

$$N_{ss}(\psi_s) = \frac{1}{e}\left\{C_m(V_g)\left[1 - \frac{C_m(V_g)}{C_i}\right]^{-1} - C_{sc}(\psi_s)\right\} \tag{6.19}$$

Integration of the C_m vs. V_g quasi-static curves may be made numerically by means of a digital computer or by electronic integration (ref. 12). The additive constant Δ may be evaluated, alternatively, from a calculation of the flatband voltage. The surface potential plot may be shifted linearly in order to align the surface potential corresponding to flatband with ψ_s=0. Distortion in shape of the C_m-V_g curves may arise from two sources: (a) series resistance due to poor contacts made to the MIS structure (this may be a particularly troublesome problem in epitaxial layer or thin film MIS measurements); (b) non-equilibrium effects associated with fast ramp sweep rates (such problems may be encountered in materials in which minority charge carriers have a long lifetime, at low temperatures and for low surface state densities).

Apparatus used for quasi-static C-V measurements described by Lopez (ref. 13) is shown in Fig. 6.6. The ramp generator is designed around an operational amplifier with an input impedance of the order of 10^{12} ohms. The sweep rates may be

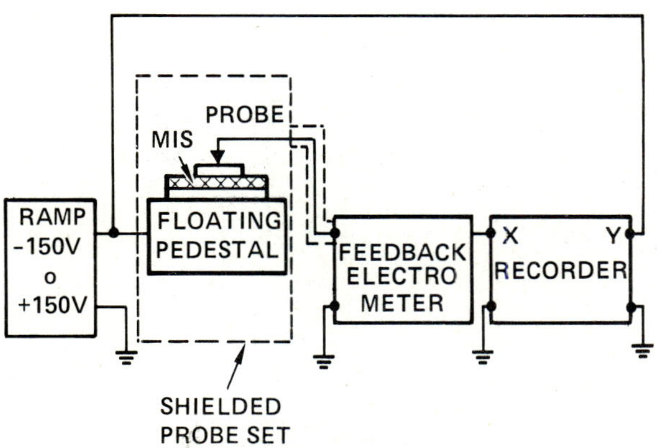

Fig. 6.6. Circuit diagram for quasi-static measurements. The connection between electrometer and probe should be short and rigid. [after A. D. Lopez, Rev. Sci. Instr., 44(1973)200]

varied from a few mV/sec. to ∿4 V/sec. A feedback electrometer is used to measure the displacement current. An X-Y recorder is used with an active filter module giving a response time of ∿0.25 sec. With the probe removed the apparatus measures the displacement current due to all the stray capacitances in the circuit. Therefore, it is necessary to subtract this contribution from the total displacement current measured with the MIS specimen in place in order to obtain the true quasi-static curve. A convenient way of obtaining the zero current reference is to sweep the ramp with the probe removed from the specimen. The magnitude of the displacement curve, i(v) will depend on the MIS capacitance C(v) and sweep rate dv/dt through $i(v)=C(v)\cdot dv/dt$. Thus knowing the sweep rate and displacement current the corresponding value of C can be calculated. The apparatus was used for quasi-static measurements made on silicon metal-oxide (MOS) structures with a resolution of $\sim 10^{10}$ states/eV·cm^2 near midgap.

A quasi-static technique for precise measurements of charge, low frequency capacitance, surface potential and interface states has been described by Ziegler and Klausmann (ref. 14). The circuit is shown, schematically, in Fig. 6.7. The fixed capacitor C_o is chosen to have a value comparable to that of the insulator

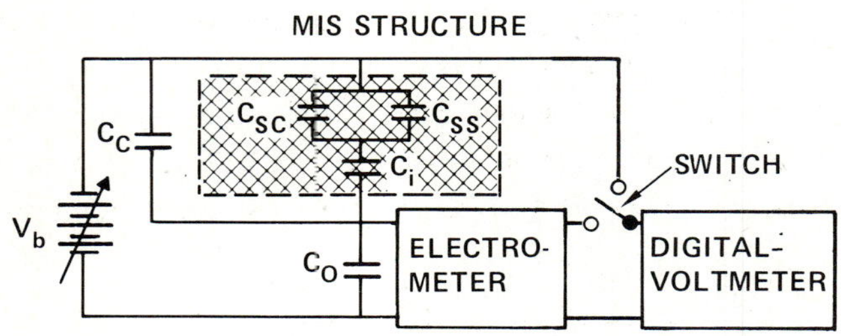

Fig. 6.7. Circuit for quasi-static measurements of MIS capacitance and surface potential. [after K. Ziegler and E. Klausmann, Appl. Phys. Let., 26(1975)400]

capacitance, C_i of the specimen. C_c is the stray capacitance of the specimen holder and connections. A potential V_b is applied to the series circuit of C_o and C_c in parallel with the MIS capacitor and V_b is measured by means of a digital voltmeter. After steady-state is reached the voltage drop V_o across C_o is determined from the output of an operational amplifier with a high input impedance used as a voltage follower. The instantaneous voltage across the MIS capacitor, $V=-(V_b-V_o)$, and the charge which flows onto the gate electrode is

$$Q_m = - C_oV_o - C_cV \tag{6.20}$$

The charge Q_m can be measured as a function of V by varying V_b incrementally or continuously. The quasi-static capacitance is

$$C_m = dQ_m/dV \tag{6.21}$$

The surface potential car be determined from

$$\psi_s = V - Q_m/C_i + \psi_{so} = \phi_s + \psi_{so} \tag{6.22}$$

where ϕ_s is the increment in surface potential produced by the voltage V and can be obtained by means of Eq. (6.20) as

$$\phi_s = (1 + C_c/C_i)V + (C_o/C_i)V_o \tag{6.23}$$

The value of the surface potential for V=0 is ψ_{so} and the flatband voltage is the value of V for which $\psi_{so}=Q_m/C_i-V$. The maximum error of the capacitance measurements (at the extreme values of the C-V curve) was determined to be 0.2%. In accumulation and inversion the error is less than 0.05% and that of ψ_{so} is smaller than ±0.002 V. Near bandgap N_{ss} values of a few times 10^9 states/eV-cm^2 could be resolved.

6.2 High frequency and non-equilibrium C-V measurements

If the gate voltage applied to an MIS structure has a sinusoidal time dependence then the space charge and interface charges may no longer be in equilibrium with V_g. At sufficiently high frequencies neither the interface charges nor the minority carriers are able to follow $V_g(t)$. The space charge capacitance becomes practically constant when the semiconductor surface is inverted with a capacitance which corresponds to the maximum depletion width δ_m as shown in Fig. 6.8. Analytically, this saturated space charge capacitance per unit area may be calculated (refs. 15,16), for an n-type MIS structure, from

$$C_{ssc} = \frac{1}{2}\left[\frac{N_d e \kappa_s \varepsilon_o}{(kT/e)\cdot \ell n(N_d/n_i)}\right]^{1/2}. \tag{6.24}$$

where n_i is the intrinsic electron density at the temperature T. A sufficiently high frequency must be chosen for such measurements in order to prevent any interface states with sufficiently small time constants from responding to $V_g(t)$. Although 1 MHz is usually adequate for Si MOS structures this may not be the case for other semiconductors; in the case of GaAs MIS structures a frequency of 0.15 GHz was needed to obtain high frequency C-V behavior (ref. 17).

The circuit and apparatus shown in Fig. 6.9 is intended for high frequency C-V measurements. A high frequency AC excitation (<50 mV peak-to-peak) is obtained from an oscillator which is an integral part of the high frequency self-balancing bridge shown in Fig. 6.9. The AC excitation is superposed on the slowly rising or falling ramp voltage applied to the MIS specimen. Typical high frequency C-V data such as shown in Fig. 6.7 can be obtained experimentally; however, the problems mentioned earlier, leakage currents through the dielectric layers, ionic transport, stored charges and space charges in the insulator introduce hysteresis and distort the C-V curves.

The gate voltage-dependent high frequency capacitance, C_h, of an MIS structure used in conjunction with quasi-static measurements made on the same specimen (ref. 14) can be used with the aid of Eq. (6.23) to determine the impurity

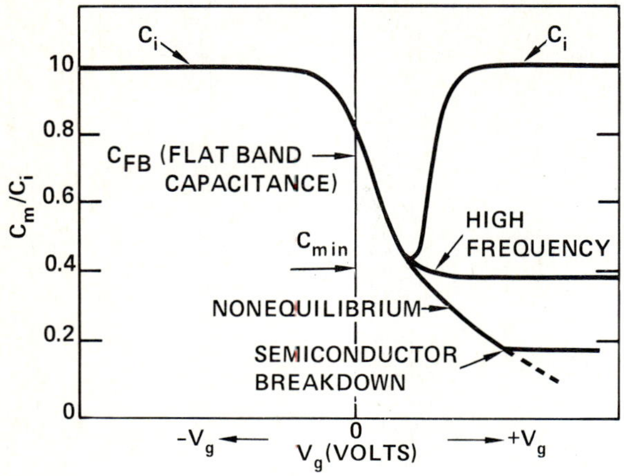

Fig. 6.8. Metal-insulator- p-type semiconductor normalized capacitance vs. gate voltage curves showing qualitatively the low frequency equilibrium, high frequency and non-equilibrium C_m-V_g curves.

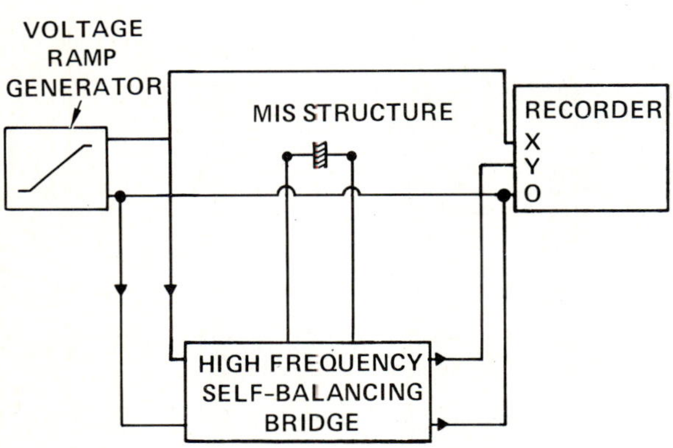

Fig. 6.9. Schematic of apparatus used for high frequency measurements; a slow ramp voltage is applied to the MIS specimen together with a low level high frequency AC signal. The high frequency bridge circuit is of the self-balancing type.

concentration, N_d, of an n-type MIS specimen and of ψ_{so}, the surface potential which corresponds to V=0, from

$$\left(\frac{C_i}{C_h} - 1\right)^2 = -\frac{2(C_i)^2}{e\varepsilon_o\kappa_s N_d}\left(\phi_s + \psi_{so} + \frac{kT}{e}\right) \qquad (6.25)$$

In depletion, a plot of $(C_i/C_h-1)^2$ vs. ϕ_s is a straight line whose slope yields N_d and whose intercept is ψ_{so}. The additive constant Δ, from Eq. (6.18), can then be calculated from

$$\Delta = -\psi_{so} - kT/e \tag{6.26}$$

The experimentally measured high frequency MIS capacitance C_{mh} is affected by a series resistance R_B which is particularly troublesome in epitaxial layer MIS structures. For a radian frequency ω the measured high frequency capacitance is

$$C_{hm} = C_h\,(1 + \omega^2 R_B^2 C_h^2)^{-1} \tag{6.27}$$

If the gate voltage is swept rapidly so that minority carriers are not collected near the semiconductor surface of an MIS structure, even in the bias range which corresponds to inversion, then the C-V curve corresponds to the non-equilibrium deep depletion case shown in Fig. 6.8. The normalized capacitance for a p-type MIS structure in deep depletion, calculated by Grove *et al.*(ref. 4) is

$$\frac{C_d}{C_i} = \left[1 + \frac{2\kappa_i^2\varepsilon_o/d_i}{e(N_d - N_a)\kappa_s}\,(V_g - \phi_{ms} + Q_{ss}/C_i)\right]^{-1/2} \tag{6.28}$$

Experimental measurements confirm the validity of Eq. (6.28); the capacitance keeps decreasing with increasing V_g because of the unlimited increase in the depletion region width which occurs if minority carriers cannot collect near the surface, up to the gate voltage which corresponds to the breakdown of the dielectric layer.

6.3 Complex admittance measurements

The density and energy distribution of interface states can be derived from an experimentally-measured frequency dispersion of the complex admittance of Schottky barrier diodes and MIS structures. Nicollian and Goetzberger (ref. 18) have described such methods of measurement with specific reference to Si MOS devices. A detailed review of the relevant theory and measurement techniques has been described by Goetzberger *et al.*(ref. 19). Frequency-dependent room-temperature measurements of both the real and imaginary portions of the complex admittance and their extension to low temperatures (ref. 20) provide the density N_{ss} of interface states over a larger portion of the bandgap and of the charge carrier capture cross-sections σ_c with greater resolution and smaller systematic errors than those based on capacitance measurements alone. Conductance measurements provide more information because losses in MIS capacitors vanish at high and low frequencies whereas the change in capacitance with frequency is much smaller.

The total small signal complex admittance of an MIS device can be represented as a parallel combination of a capacitance and conductance in terms of an equivalent circuit such as shown in Fig. 6.10. The small signal interface state capacitance $C_p(V_g,\omega)$ and conductance $G_p(V_g,\omega)$ are to be determined from measurements of the equivalent circuit shown in Fig. 6.10.(b) in which $C(V_g,\omega)$ and $G(V_g,\omega)$

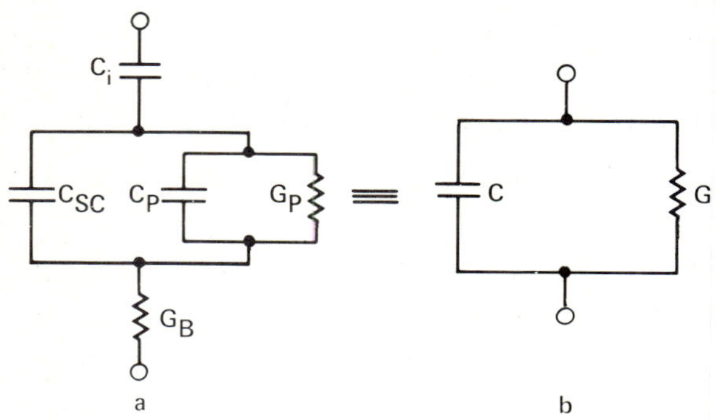

Fig. 6.10. An equivalent circuit model for MIS capacitor; (a) G_B is series conductance of semiconductor body, C_p is capacitance and G_p is conductance due to interface states; (b) equivalent capacitance and conductance measured experimentally.

contain the respective capacitances and conductances of Fig. 6.10.(a) and a circuital transformation is required of the component elements of Fig. 6.10.(a) into 6.10.(b) (ref. 19). As the gate voltage is swept so that the semiconductor surface moves from accumulation to inversion C_p rises from zero to a maximum value $C_{p_{max}}$ and G_p/ω passes through a bell-shaped resonance whose maximum is $(G_p/\omega)_{max}$. A typical frequency dependence of the capacitance and conductance associated with interface states, for an MIS device biased in depletion or weak accumulation, is shown in Fig. 6.11. The magnitude and width of the G_p/ω peak allows the calculation of N_{ss} and the radian frequency ω_{max} is used to calculate the capture cross-section σ_c. Figure 6.12 shows an experimentally-determined (ref. 6) conductance vs. gate voltage curve measured at room temperature on a Si MOS structure with a constant frequency of 20 kHz. It shows the characteristic peak due to surface states; the portion of the $G(V_g)$ curve for V_g>-1V is attributed to the bulk conductance $G_B(V_g)$. Its gate voltage dependence is due to that of the space charge capacitance in Eq. (6.27). An accurate measurement of the shape of the G_p/ω peak requires a frequency range at least one decade above to one decade below ω_{max}; the value of ω_{max} is exponentially-dependent on V_g and T, therefore the frequency range of available equipment imposes severe limitations on measurements of the dispersion of the complex admittance.

Admittance measurements involve the excitation of the unknown admittance with a known sinusoidal potential having an amplitude of the order of kT/e (~10 mV rms

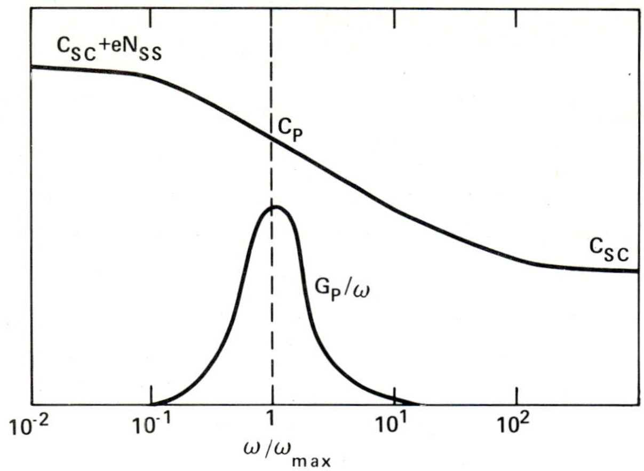

Fig. 6.11. Normalized frequency dependence of conductance, G_p/ω and capacitance of interface states in depletion or weak accumulation.

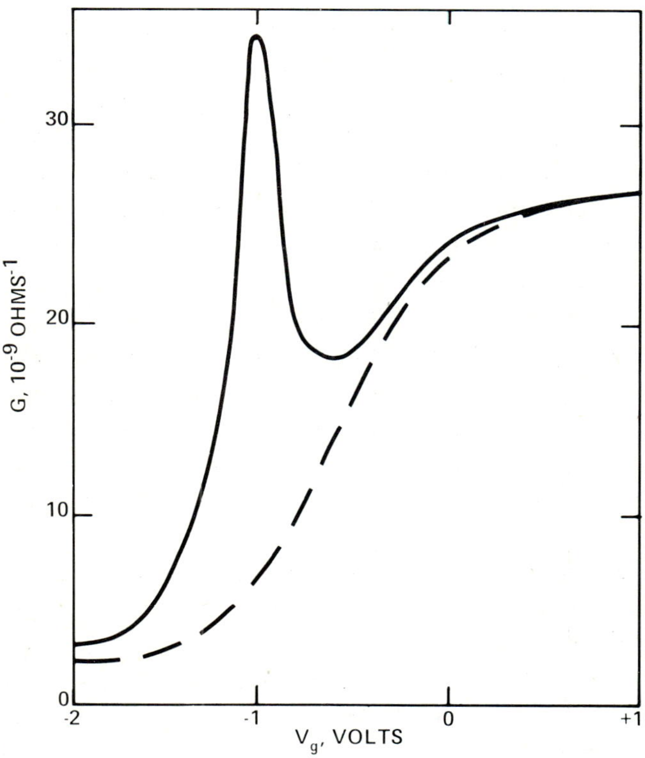

Fig. 6.12. Gate voltage dependence of conductance of Si MOS capacitor measured at room temperature with frequency fixed at 20 kHz (continuous curve); anomalous rise in $G(V_g)$ for V_g>-1 V is due to bulk conductance, whose gate voltage dependence [Eq. 6.27] is shown by dashed line.

at room temperature) while a fixed gate voltage is simultaneously applied to it. The AC current is

$$i_{ac} = V_{ac}[G(V_g,\omega) + j\omega C(V_g,\omega)] \tag{6.29}$$

The amplitude and phase then yield the real and imaginary components of the admittance. If a bridge is used then an equal and opposite voltage, $-V_{ac}$, is applied to standard calibrated capacitors and resistors and the current through them is added to i_{ac}. A detector is used to determine the bridge null which is obtained only if the standard conductance and capacitance correspond, respectively, to the unknown components. In commercially available bridges the 180° out-of-phase potentials are usually produced by tightly coupled secondary windings on a transformer. Such bridges allow accurate three-terminal capacitance measurements, even in the presence of large connecting cable capacitances, in the frequency range between 10^2 and 10^6 Hz. For higher frequencies, where only two-terminal measurements are feasible the impedance of the connecting cable and specimen holder limit the accuracy and resolution of measurement. The overall accuracy may depend on the phase relation between V_{ac} and $-V_{ac}$, the smallest detectable unbalance signal and the adjustments or fine-tuning feasible of the standard admittance elements of the bridge. The in-phase and quadrature components of the bridge output signal can be determined by means of phase-sensitive detectors (refs. 21-25). In general, the operation of such detectors depends on combining the signal and noise voltage vectorially with a reference signal having a specific frequency and phase and rectifying the combination so that over a period which is long compared to the frequency of the reference voltage the noise components are cancelled. The resultant rectified output voltage has an amplitude which is directly proportional to the desired phase component of the imput signal. A self-balancing bridge circuit employing phase-sensitive detection has been described by Schewchun and Waxman (ref. 26). It can be used to measure five orders of magnitude in admittance and conductances as low as $10^{-7}(\text{ohm-cm})^{-1}$ with a detection sensivity of 3.2 nV in a signal-to-noise ratio of $6.5\text{x}10^{-3}$ between 10 Hz and 10^5 Hz.

The real and imaginary components of the unknown admittance may also be measured by "lock-in" techniques such as illustrated in Fig. 6.13. The AC current flowing through the unknown admittance is applied to a current sensitive preamplifier and then to the lock-in amplifier (ref. 27). The latter is a phase-sensitive voltmeter and demodulator able to detect small signals in the presence of noise. When provided with an appropriate reference signal it will measure the in-phase or quadrature component of the AC current. Its DC output voltage is proportional to $i_{ac}\cos\phi$ where ϕ is the phase angle with respect to some reference waveform. The phase is not known, as a rule. It is calibrated by substituting a standard capacitor whose conductance is negligible for the unknown admittance. Having

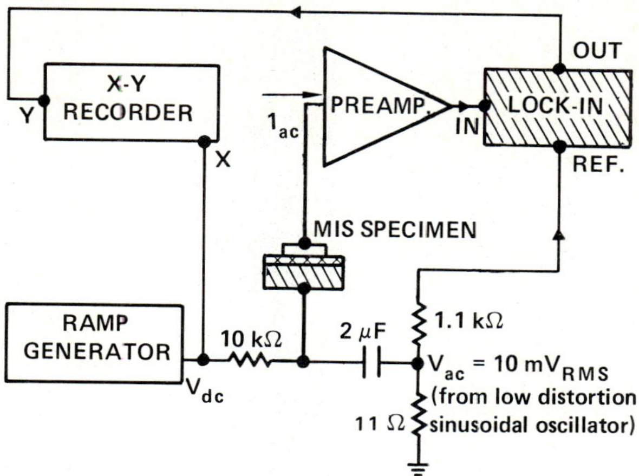

Fig. 6.13. Schematic circuit arrangement for measuring components of admittance of an MIS structure by means of "lock-in" amplifier and related procedures.

calibrated the phase, the unknown specimen may be replaced and $C(V_g,\omega)$ is measured. Then the phase is shifted by $\pm\Pi/2$ and $G(V_g,\omega)$ is measured. Figure 6.14 shows a comparison between G vs. V_g data obtained by means of a bridge and

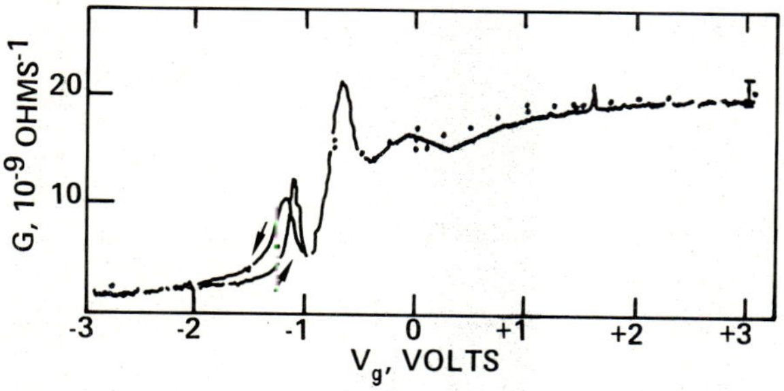

Fig. 6.14. A comparison of bridge measurements (single data points, 1-1/2 hours/ 20 points) and lock-in measurements (solid lines, 100 sec./2 sweeps) on the same sample ("5-dot" Si MOS wafer) for V_{ac}=10mV_{rms} and f=5 kHz.

measurements made by means of a lock-incircuit, which is shown in Fig. 6.13 (ref. 6). The agreement between the two methods is good. It appears, however, that for measurements made at a single frequency, bridge measurement techniques provide higher precision and a greater accuracy of measurement. On the other hand, the lock-in technique generates a considerably larger amount of data for a given time over a larger frequency range.

An apparatus for automatic admittance measurements made on MOS structures which operates in the frequency swept mode between 10 Hz and 10^5 Hz has been described by Boudry (ref. 28). It can be used to measure admittances with a total uncertainty in $G/\omega C$ of the order of 10^{-4} over the full frequency range. The accessible interface state energy range, can be extended by varying the temperature (since this varies the Fermi level of the semiconductor), as well as V_g and ω and by repeating the measurements at successive combinations of V_g and T curves of N_{ss} and σ may be constructed from the data.

Relatively few measurements have been made, thus far, on MIS structures compared to those made on Si MOS structures. To a large extent this is due to the lack of a compatible low loss dielectrics, such as provided by the Si-SiO_x technology, used in conjunction with other semiconductors such as the III-V compounds.

6.4 Free carrier concentration profiles

The impurity profile under a Schottky barrier can be determined from the reverse-biased gate voltage dependence of the capacitance, in accordance with Eq. (6.3), as described by Schottky (ref. 29). Unless the shape of the doping profile is known in advance (i.e., whether it is constant or graded in a linear manner) a differentiation of the raw $1/C^2$ vs. V_g data becomes necessary. If the impurity concentration varies rapidly then a large number of experimental data points are needed in order to minimize averaging over the calculated impurity distribution, in order to determine the effective impurity profile under the gate.

$$N_d(z) = -\left(\frac{C^3}{e\kappa_s\varepsilon_o}\right)\cdot\left(\frac{dC}{dV}\right)^{-1} \qquad (6.30)$$

Equation (6.30), in conjunction with the traditional expression for the electrical capacitance of a Schottky barrier or asymmetrical p-n junction, $C=\kappa_s\varepsilon_o/\delta$, is used, as a rule, in differential capacitance measurements of the impurity distribution. Kennedy *et al.* (ref. 30) have demonstrated that Eq. (6.30) is only applicable to conditions of charge neutrality and that some inherent errors can arise if this technique of measurement is applied to materials containing only a small impurity density. Their conclusion is that Eq. (6.30) measures the majority free carrier concentration, $n=N_d-N_a$ rather than the impurity doping level. The analysis assumes a sharp edge in the distribution n(z) of the free carriers. The semiconductor is assumed to be divided into two distinct regions: a layer that is entirely depleted of charge carriers and an interior region of perfect charge neutrality. This is equivalent to assuming a zero value for the Debye length in the vicinity of the edge of the space charge region. If the doping profile has large gradients and if a spatial resolution of the order of a few Debye lengths is attempted in the vicinity of these gradients then large errors are introduced in the calculation of the free carrier profile deduced from C-V measurements, as described by Johnson and Panousis (ref. 31). They have shown that a step in the free carrier concentration profile cannot be resolved to less than several Debye lengths and that a ramp cannot be distinguished from a step in n(z) unless its width is appreciably greater than a Debye length. A different source of error in the calculation of the free carrier profile determined from C-V measurements is the series resistance of the semiconductor. The measured capacitance, in terms of Eq. (6.27), is quite different from the actual capacitance C_h; large errors can be introduced in the evaluation of n(z) as described by Wiley (ref. 32), particularly for thin films or epitaxial layers. Wiley demonstrated by computer simulation and experimental verification that the existtence of such errors can be inferred from RF phase angle measurements obtained during the C-V profiling and that this information can be used to correct the distorted profiles. A useful rule of thumb is that the in-phase (conductance) signal not be greater than 18% of the out-of-phase (capacitance) current. A source of systematic error in such measurements is a determination of the effective area of the capacitor. It should be borne in mind that C represents the capacitance per unit area in Eq. (6.30) while actual measurements of C include the area, A, of the device. Thus n(z) is proportional to A^{-2}.

A method used for determining the free carrier profile of MIS capacitors has been described by van Gelder and Nicollian (ref. 33). It is based on the measurement of the slope of $1/C_h^2$ vs. V where C_h is the high frequency capacitance per unit area and V is the potential across the MIS capacitor

$$n_o(\delta) = \left(\frac{2}{e\kappa_s\varepsilon_o}\right)\cdot\left[\frac{d(1/C_h^2)}{dV_g}\right]^{-1} \tag{6.31}$$

The frequency of measurement is presumed to have been chosen high enough so that interface states do not respond and the range of measurements is restricted to the depletion regime.

Brews (ref. 34) has shown that if interface states are present producing the usually observed "stretch-out" of C-V curves (ref. 35) then the free carrier profile can be expressed by

$$n(\delta) = \left(\frac{2}{e\kappa_s\varepsilon_o}\right)\left(\frac{1 - C_\ell/C_i}{1 - C_h/C_i}\right)\left[\frac{d(1/C_h)^2}{dV}\right]^{-1} \tag{6.32}$$

where C_ℓ is the quasi-static low frequency capacitance per unit area. That is, the corrected profile of Eq. (6.31) expressed in terms of Eq. (6.32) is

$$n(\delta) = \left[\frac{1 - C_\ell/C_i}{1 - C_h/C_i}\right]\cdot n_o(\delta) \tag{6.33}$$

Equation (6.32) has a further advantage over Eq. (6.31) whether or not interface states are present; it applies in inversion as well as depletion. The range of distance over which the profile can be measured is roughly from 3 Debye lengths to the maximum depletion widths obtainable to ~5% in accuracy provided that C_ℓ is a true quasi-static curve. If it is not then a distortion of the measured profile will result due to an incorrect assessment of interface state "stretch-out." Departures from the quasi-static C-V curve are most likely in the weak to strong inversion range. Also in the region where $(1-C_\ell/C_i)$ is rapidly varying an accurate value of the derivative in Eq. (6.32) is needed in order to minimize errors in the calculation of the profile. Brews (ref. 34) obtained experimental verification of these relations by measuring the doping profile of a Si MOS capacitor before and after a number of bias, temperature, and voltage stress cycles.

The frequency dependence of differential capacitance measurements and errors introduced by various free carrier profile measurement techniques used on MIS capacitors has been described by Zohta (ref. 36). He showed analytically and experimentally the manner in which interface states and majority carriers influence the frequency dependence of the differential capacitance.

A method of profiling which is based on the evaluation of the harmonic content of reverse-biased diodes was described by Meyer and Guldbrandsen (ref. 37) and subsequently refined by Copeland (ref. 38). The diode is driven by a constant RF current of frequency f and the potential drop across the diode is measured. The component of frequency f is proportional to δ and the second harmonic (2f) component is proportional to $1/n(\delta)$.

A technique of measurement developed by Miller (ref. 39) may be described as a constant field method of feedback profiling. If Eq. (6.30) is expressed in terms of finite increments, rather than differentials, and if it is related to the incremental change $\Delta\delta$ of the depletion width produced by an incremental change in capacitance ΔC such that $\Delta\delta=\kappa_s\varepsilon_o/\Delta C$, then

$$n\,\Delta\delta = (\kappa_s\varepsilon_o/e)\cdot\Delta V_g/\delta \qquad (6.34)$$

If V_g is constrained so that the product $n\cdot\Delta\delta$ is a constant then $\Delta\delta$ is proportional to $1/n$; $\Delta V_g/\delta$ may be considered as an electric field increment. The diode is part of a feedback loop in which it is driven by an RF voltage so chosen as to keep the amplitude of the capacitive component of the RF current equal to a fixed reference current source. If the admittance of the diode is purely capacitive then V_{RF} is directly proportional to δ. Therefore n can be obtained directly from Eq. (6.34). This method of measurement provides some specific advantages which include the mounting of the specimen at the end of coaxial cables of essentially arbitrary lengths, an advantage if carrier concentration profiles must be determined on a specimen maintained in a cryogenic environment.

Hillibrand and Gold (ref. 40) have used reverse-biased C-V measurements to determine the impurity distribution of diffused junction diodes. The one-dimensional dependence on position normal to the plane of the junction was approximated by

$$N(x) = N_B[\exp(- bx) - 1] \qquad (6.35)$$

where N(x) is the net impurity concentration at a distance x from the metallurgical junction, N_B is the impurity concentration of the bulk semiconductor before diffusion, and b is a constant which can be determined in terms of the location of either edge of the depletion layer by imposing the requirement of net space charge neutrality. They found that the accuracy of impurity level determination is better than 20% being limited primarily by the error in the junction area measurement. The shape of the impurity profile can be determined to within 10% from data obtained from C-V measurements using a least squares analysis processed by means of a computer. An automatic apparatus for measuring doping concentration profiles of p-n junctions has been described by Reutlinger *et al.* (ref. 41) based on Eq. (6.30). The test set is capable of measuring 100 C-V points at the rate of 15 points/min. with a total measurement time per diode of $\sim$4 min. Ambrozy (ref. 42) has described a rapid and simple dC/dV measurement intended for the evaluation of doping or impurity profiles of hyperabrupt p-n junction diodes without requiring numerical differentiation of C-V curves. Knowing the source of nonlinearities relatively high $\Delta C/\Delta V$ increments may be used with good accuracy. The impurity profile is calculated from

$$N(\delta)\ \delta^2 = \left(\frac{\kappa_s \varepsilon_o}{2e}\right) \left| \frac{V_{RF}}{V_{out}} \Delta V_o \right| \qquad (6.36)$$

where ΔV_o is a low frequency trapesoidal voltage which modulates the DC reverse bias $V_b (\Delta V_o << V_b)$ applied to the diode, V_{RF} is a bridge AC excitation signal much smaller than ΔV_o and usually of the order of a few tens of mV and V_{out} is the bridge output voltage.

An ingeneous method for profiling the impurity concentration of a silicon four-terminal enhancement mode-field effect transistor (MOSFET) from about three Debye lengths from the oxide-semiconductor interface to a maximum depth limited by avalanche breakdown in silicon, has been described by Buehler (ref. 43). Connections are made to the gate (G), source (S), drain (D), and body (B). The MOSFET is operated in the linear current region and the depletion region is translated through the body of the transistor by using the back-gate bias effect (ref. 44). As the source-body voltage is changed the source-drain current is held constant by adjusting the gate-source voltage. The profile depth is proportional to dV_{SB}/dV_{GS} and the impurity concentration is inversely proportional to d^2V_{SB}/dV_{GS}^2 with only the thickness and dielectric constant of the oxide and the dielectric constant of the semiconductor needed for the computation of the impurity profile.

If the space charge region of a thin film driven by means of an applied gate voltage penetrates to its opposite surface where a fixed depletion layer is present between the film and its substrate then, as shown by Lehovec (ref. 45), a calculation of the free carrier concentration profile derived from such measurements is invalid. The analysis considered a Schottky barrier contact made to a conducting epitaxial layer of GaAs on a semi-insulating GaAs substrate; a depletion layer is assumed to be present between the latter due to a contact potential difference between them. It is also considered to apply to thin films deposited on insulating substrates such as silicon on sapphire and to interfaces between an n-conducting channel and a p-substrate or vice versa. Artifacts in the form of erroneous charge carrier distribution profiles are derived from the calculation of $n(\delta)$ if the space charge region extends through the entire layer leading to a near "pinch-off" of the conducting channel.

6.5 Gate-controlled conductivity and Hall measurements

It is highly advantageous to supplement free carrier profile measurements made by means of C-V techniques with more direct conductivity and Hall measurements. Such measurements are relatively simple although their interpretation can lead to severe difficulties if the gradient in impurity concentration is large or the layers on which such measurements are made are of the order of a few Debye lengths. A typical specimen used for such measurements is shown in Fig. 6.15. A six-arm bridge specimen, similar to that shown in Fig. 2.3, has a dielectric layer deposited on the surface of the semiconductor. Conventional conductivity and Hall measurements, such as described in Chapters 1 and 2, are made as a function of

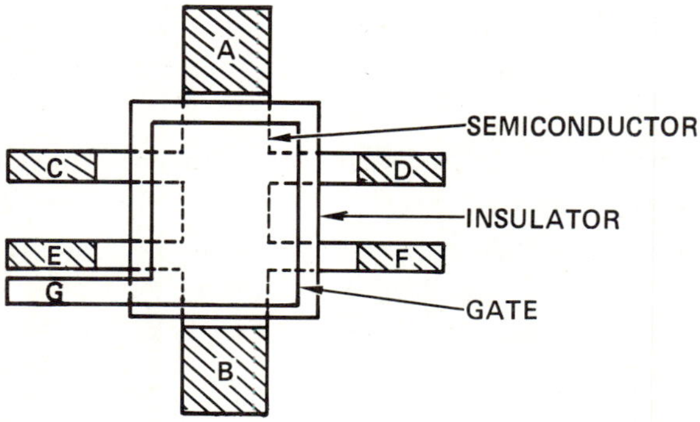

Fig. 6.15. A top view of the specimen configuration used for gate-controlled conductivity and Hall measurements; A and B are current contacts, C, D, E, F are contacts for conductivity and Hall measurements, G is the contact to the gate.

gate bias applied between the gate electrode and terminals A or B of the specimen. Assuming the effect of the gate voltage is to reduce the effective thickness of the conducting channel then an analysis similar to that of Section 5.1 can be used to determine the position-dependent carrier concentration and mobility, provided that the thickness of the depleted region is known or can be determined from subsidiary C-V measurements.

Ipri (refs. 46,47) has described such measurements made on $\sim$1 μm thick n- and p-doped silicon on sapphire MOS structures. He found that the calculated mobility and free carrier concentrations decrease monotonically with distance from the Si-SiO_2 interface. He postulated a model which requires the presence of a nonconducting region in the silicon layer within $\sim$0.2 μm of the silicon-sapphire interface, a constant impurity profile and linearly decreasing mobility with depth. Ipri and Zemel (ref. 48) have extended such gated-conductivity and Hall measurements made on silicon on sapphire MOS structures to include their temperature dependence. They found an energy level at $\sim$0.1 eV below the conduction band, in addition to the intentionally-introduced arsenic impurity level. The former has been associated with a SiO_4^+ complex present in oxygen-rich silicon and it appears that the density of this additional impurity increases with depth into the film and might also contribute to the experimentally-observed mobility gradient.

They found evidence for a temperature-independent neutral impurity scattering mechanism and suggested that stress may be responsible for the experimentally-observed reduction in electron and hole mobilities at elevated temperatures where phonon scattering is dominant. Elliot and Anderson (ref. 49) have made gated measurements on 0.6 to 2.0μm thick n-doped silicon on sapphire layers on which a composite insulating layer was deposited consisting of $\sim$0.04μm thick silicon oxide and 0.08μm silicon nitride. They found that in the vicinity of the sapphire interface the room temperature mobility is limited by charged defect scattering while in accumulation layers near the oxide interface phonon scattering is the dominant mobility limiting mechanism and a substantial portion of the oxide interface scattering is specular. Ham (ref. 50) pointed out that surface transport may be a dominant mechanism in thin layers; if the maximum depletion depth is comparable to the layer thickness then the conducting portion of the layer may be reduced significantly or even eliminated entirely. To illustrate the effects of the surface transport he compared measurements made on silicon on sapphire layers exposed to the room environment with identical specimens processed in a six-arm Hall bar MOS configuration subjecting these to a controlled surface potential. He determined the bulk properties (with the term "bulk" meaning unaffected by surface charges or surface currents). With the gate biased at flatband (this condition determined from C-V measurements) the channel under the gate is conducting only bulk current. He found that the Hall voltage increases with gate bias from heavy accumulation into depletion; at the onset of inversion the depletion depth ceases to increase and the v_h vs. V_g curve flattens. For the n-type material conductivity and Hall measurements made on specimens having "free" surfaces agree with those measured on the gated specimens at flatband provided that they are not allowed to remain for extended periods in the room environment. The p-type material, on the other hand, appears to be depleted when the surface is not protected. While the lightly-doped electron mobility decreases with increasing depletion depth more heavily-doped Si layer MOS structures exhibit little if any decrease in mobility primarily because the maximum depletion depth for heavily-doped material is small. Hynecek (ref. 51) also made such measurements on silicon on sapphire deep depletion MOS structures. He concluded that for n-type specimens ($N_d<10^{17}/cm^3$) an assumption of position-independent mobility is reasonable. In p-type specimens impurity diffusion from the sapphire substrate is presumed to produce a drop in mobility in agreement with the observation of Ipri (ref. 47). A different type of MIS structure in which the depletion layer is formed at the interface between the silicon layer and its sapphire substrate was described by Lehovec and Shi-Tron Lin (ref. 52). The silicon layer was processed in the form of an interdigitated test structure and the gate applied to the obverse side of the sapphire wafer. Such a structure enabled a determination of the sheet conductance of the silicon layers as a function of biasing field using a distributed network analysis. They found good agreement with theoretical calculations of the conductivity mobility at flatband and in the depletion regime. An electron mobility of 1.2×10^3 cm^2/V-sec. was found in the silicon layer beyond $\sim0.2\mu m$ from the silicon-sapphire interface. A detailed analytical treatment of the functional relation between the charge carrier mobility and electrostatic potential in n- or p-type thin films has been given by Swart and Campbell (ref. 53) for a symmetrical metal-insulator-semiconductor-insulator-metal structure. They developed analytic expressions relating donor density and potential-dependent carrier drift mobility to measureable Hall effect parameters and indicated a numerical procedure for solving the integral equations relating drift mobility to the mean Hall mobility.

The charge carrier transport coefficients of heteroepitaxially-grown pure single-crystal InAs layers deposited on semi-insulating GaAs substrates have an anomalous magnetic field and temperature dependence (refs. 54,55) attributed to the shunting effect of their accumulated surfaces. Appreciable errors can be introduced in the evaluation of electrical and galvanomagnetic measurements made on an epilayer of thickness $d \simeq 20\mu m$ if transport of electrons in its accumulated surface region is neglected. The gate voltage dependence of the conductivity $\sigma(V_g)$ and of the Hall coefficient $R_h(V_g)$ of specimens prepared such as the one shown in Fig. 6.16 were determined by means of van der Pauw-type measurements in

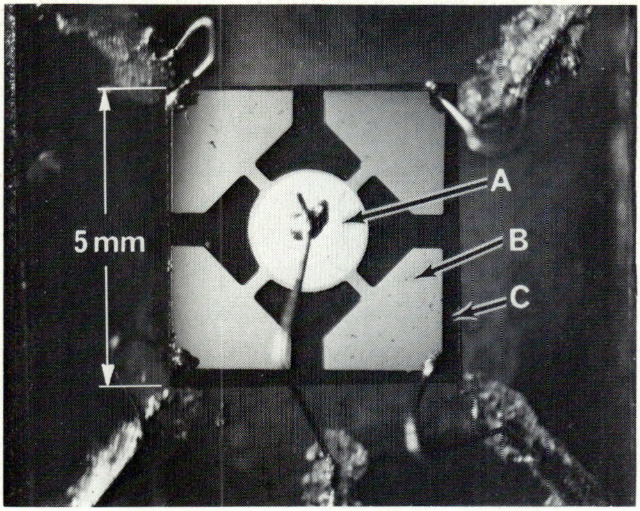

Fig. 6.16. A photolithographically-processed InAs layer for gate-voltage-controlled van der Pauw-type conductivity and Hall measurements: A, aluminum gate electrode over an SiO_2 layer 10^3 A thick; B, electrode pad; C, the darker area is the semi-insulating GaAs substrate.

a magnetic field B=0.04T and are shown in Fig. 6.17. Starting from V_g=0 the conductivity decreases and the Hall coefficient increases as V_g is made more

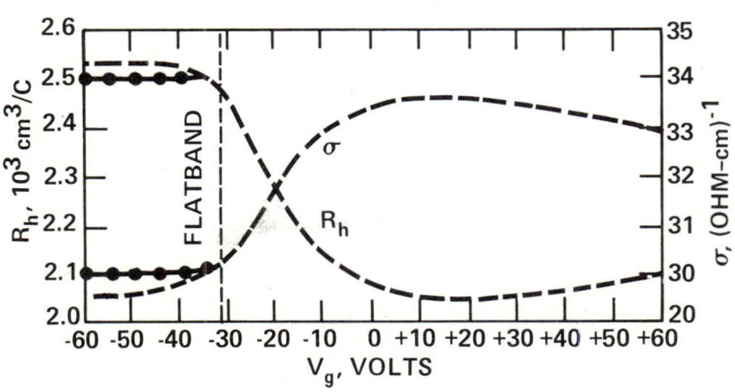

Fig. 6.17. Gate voltage dependence of Hall coefficient and of conductivity of InAs epitaxial layer measured at room temperature; square data points are data corrected for reduction of effective thickness of conducting channel due to depletion.

negative and driven towards flatband; in depletion both σ and R_h are nearly independent of V_g. In the depletion regime a correction factor must be introduced in $R_h(V_g)$ and $\sigma(V_g)$ to account for the reduction of the effective epilayer thickness by the depletion depth δ. If the Hall coefficients and conductivities measured in the depletion regime are R_{hm} and σ_m respectively then, to first order,

$$R_h = R_{hm}(1 - \delta/d) \qquad\qquad \sigma = \sigma_m(1 - \delta/d)^{-1} \tag{6.37}$$

where d is the thickness of the epilayer. These are represented by the square data points in Fig. 6.17. They determine the "bulk-like" properties of the electron density, $n_b=2.5\times10^{15}/cm^3$ and mobility $\mu_b=7.5\times10^4 cm^2/V\text{-sec}$. A model of charge carrier transport in the epitaxial layer in the accumulation regime may be constructed to first order in terms of an arbitrary number of interfacially connected homogeneous layers of thickness d_i. Let the applied electric field E_x produce in each layer a corresponding current density J_{xi}, while the applied transverse magnetic induction B induces in each layer a corresponding Hall field E_{yi} and Hall current J_{yi}. Let the average circulating Hall current J_y in the entire lamina be zero; then the constitutive ohm-Hall equations for the lamina can be expressed as

$$J_x = \frac{1}{d}\sum_i J_{xi}d_i \qquad\qquad J_y = \frac{1}{d}\sum_i J_{yi}d_i = 0 \tag{6.38}$$

where $d=\sum_i d_i$ and the corresponding J_{xi} of each layer are

$$J_{xi} = \alpha_i(E_x + \beta_i E_y) \qquad\qquad J_{yi} = \alpha_i(E_y - \beta_i E_x) \tag{6.39}$$

where $\alpha_i=\sigma_i(1+\beta_i^2)^{-1}$ and $\beta_i=\mu_i B$. The average tensor components of the conductivity may be expressed in terms of the parameters α_i and $\gamma_i=\alpha_i\beta_i$ as

$$\alpha = \frac{1}{d}\sum_i \alpha_i d_i \qquad\qquad \gamma = \frac{1}{d}\sum_i \gamma_i d_i \tag{6.40}$$

For a two-layer model (refs. 56,57) characterized by the "bulk-like" parameters $\alpha_b,\beta_b,\gamma_b$ and d_b and "surface-like" parameters $\alpha_s,\beta_s,\gamma_s$ and d_s the constitutive ohm-Hall equations are

$$\begin{aligned} J_x d &= (\alpha_b d_b + \alpha_s d_s)E_x + (\gamma_b d_b + \gamma_s d_s)E_y \\ 0 &= (\alpha_b d_b + \alpha_s d_s)E_y - (\gamma_b d_b + \gamma_s d_s)E_x \end{aligned} \tag{6.41}$$

The effective conductivity $\sigma(B)=J_x/E_x$ is then,

$$\sigma(B) = \frac{(\alpha_b d_b + \alpha_s d_s)^2 + (\gamma_b d_b + \gamma_s d_s)^2}{d(\alpha_b + \alpha_s d_s)} \tag{6.42}$$

The magnetic-induction-dependent Hall coefficient $R_h(B)=E_y/J_xB$ is

$$R_h(B) = \frac{(d/B)(\gamma_b d_b + \gamma_s d_s)}{(\alpha_b d_b + \alpha_s d_s)^2 + (\gamma_b d_b + \gamma_s d_s)^2} \qquad (6.43)$$

Assuming the accumulation layer to be degenerate (ref. 54) the two-layer model was found to be in good agreement with measurements made on epilayers approximately 17μm thick in magnetic fields between 0.04 and 6.0 T and in the temperature range 20-400 K. However, if the contribution of surface electron transport to total carrier transport is not taken into account, then the 300 K electron mobility may be underestimated by about 22% and the electron density may be overestimated by about 71%; at 84 K the corresponding error in μ is about 11% and that in n is about 30%; these errors may be considerably larger in thin (d<10μm) epilayers. A different measurement technique used for determining the carrier concentration and mobility profiles by means of a Schottky barrier gate on a specimen with a geometry suitable for van der Pauw-type measurements, shown in Fig. 6.18, has been described by Tansley (ref. 58). A fixed reverse bias voltage V_b is applied to the gate; this determines the effective depth Δz below the

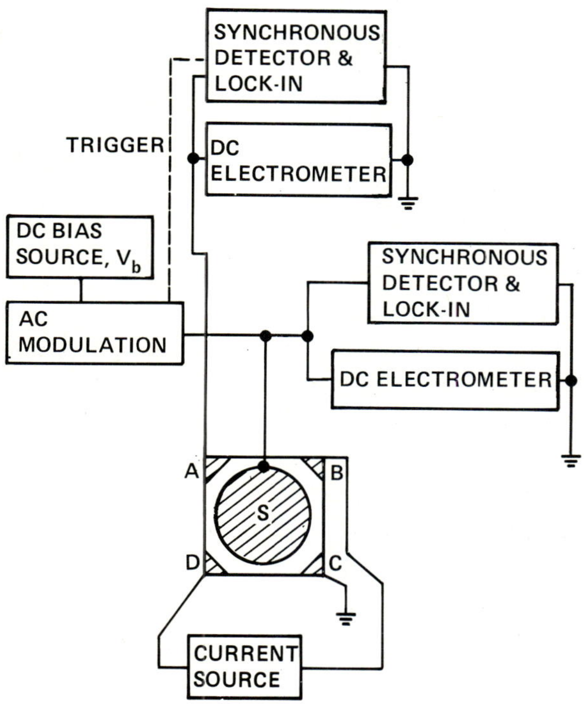

Fig. 6.18. Apparatus used for making gate-controlled van der Pauw-type measurements of carrier concentration and impurity profiles of semiconductors.

surface in which the specimen is depleted of charge carriers. The superimposed low level AC modulation on V_b provides the incremental signals which are demodulated and detected synchronously. The DC electrometer is intended to have a sufficiently slow response to be unaffected by the modulation signals detected by the synchronous detection system. Commutation between the appropriate electrodes provides the quantities $V_{AB,CD}$, $V_{BC,DA}$ and v_h and the modulation in the resistivity and Hall voltages provides $\Delta V_{AB,CA}$, $\Delta V_{BC,DA}$ and Δv_h. The effective thickness Δz is determined from capacitance vs. voltage measurements and the relation $\Delta z = \kappa_s \varepsilon_o$, where κ_s is the dielectric constant, ε_o the permittivity of free space, A the gate area and C the capacitance measured for a specific bias voltage. In terms of the relations described in Section 1.5 and Eq. (1.44) the incremental change in the sheet conductivity, $\Delta\sigma_s = \Delta(\sigma d)$, is

$$\Delta\sigma_s = \frac{2i\,\ell n2}{\Pi f}\left[\frac{\Delta V_{AB,CD} + \Delta V_{BC,DA}}{(V_{AB,CD} + V_{BC,DA})^2}\right] \tag{6.44}$$

The effective carrier concentration n(z) and mobility $\mu(z)$ can then be determined from

$$n(z) = \frac{iB}{e\delta}\frac{t^2}{(\Delta v_h - 2tv_h)} \qquad \mu(z) = \frac{2\ell n2}{\Pi fB}\left[\frac{\Delta v_h - 2tv_h}{\Delta V_{AB,CD} + \Delta V_{BC,DA}}\right] \tag{6.45}$$

where

$$t = \frac{\Delta V_{AB,CD} + \Delta V_{BC,CD}}{V_{AB,CD} + V_{BC,DA}} \tag{6.46}$$

The gated van der Pauw measurements suffer from the same limitation associated with C-V measurements: the electric field must be maintained below the critical breakdown field. This sets the limit for the ionized impurity density in Si to $\sim 2.5\times10^{12}\,cm^{-2}$, which corresponds to a maximum depletion depth of approximately 10μm for material with a resistivity of about 2Ω-cm but a depth of only 10 nm for a material with a resistivity of about $10^{-2}\,\Omega$-cm. The drift mobility profile can also be determined by measurements made on MIS structure, such as shown schematically in Fig. 6.19, described by Hsu and Scott (ref. 59). The gate of the FET shown in Fig. 6.19 is so biased that the semiconducting layer is partly depleted. Let the width of the depletion region be δ and the length and width of the gate be, respectively, L and W. The conductance between source and drain is

$$G = (W/L)\int_{\delta}^{d} Q(x)\mu(x)dx \tag{6.47}$$

and the gate voltage is given by

$$V_g = V_{FB} - \frac{1}{C_i}\int_\delta^0 Q(x)dx + \frac{1}{\kappa_s\varepsilon_o}\int_\delta^0\int_\delta^x Q(x')dx'dx \tag{6.48}$$

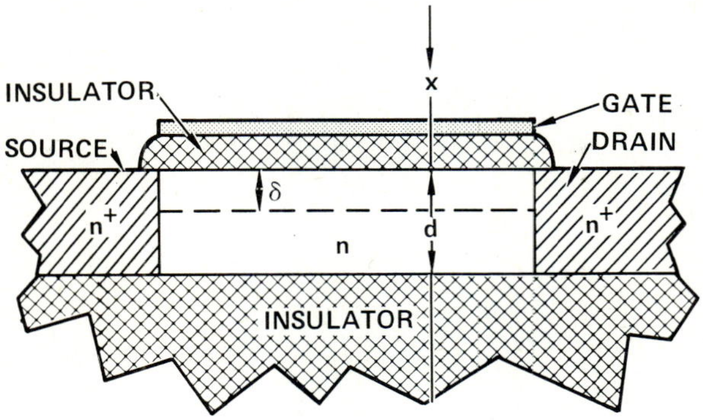

Fig. 6.19. Schematic cross-section of ideal insulated gate field effect transistor on insulating substrate.

where V_{FB} is the gate voltage at flatband and C_i is the capacitance per unit area of the dielectric. Differentiating Eqs. (6.47) and (6.48) with respect to δ leads to

$$\frac{\partial G}{\partial \delta} = -\frac{W}{L} Q(\delta)\mu(\delta)$$

$$\frac{\partial V_g}{\partial \delta} = \frac{Q(\delta)}{C_i} + \frac{\delta}{\kappa_s\varepsilon_o} Q(\delta) = \frac{Q(\delta)}{C_g(\delta)} \tag{6.49}$$

where $C_g(\delta)$ is the series capacitance of the space charge capacitance per unit area and the dielectric capacitance. Combining the two equations in (6.49) leads to

$$\frac{dG}{dV_g} = -\frac{W}{L}\mu(\delta)C_g(\delta) \tag{6.50}$$

If the DC voltage between the source and drain is V_D then in the limit, for $V_D \to 0$, $G=i_D/V_D$ and the depletion depth-dependent mobility is

$$\mu(\delta) = -\lim_{V_D\to 0}\left[\frac{g_m L}{W V_D\, C_g(\delta)}\right] \tag{6.51}$$

where g_m is the small signal transconductance in the limit $V_D \to 0$. Thus measurement of the gate capacitance and of the small signal transconductance as a function of gate bias provides the charge carrier drift mobility at a distance δ from the insulator-semiconductor interface. A similar deviatior was made by Pucel and Krumm (ref. 60); however, in the derivation of the transconductance they introduced the effective resistance of the undepleted channel under the gate, R_c, the parasitic resistance between gate and source, R_s, and the parasitic resistance between the gate and drain, R_d with $R_T = R_s + R_d + R_c$. The transconductance is then,

$$g_m = -\frac{eW}{L}\mu(\delta)n(\delta)V_D\left(\frac{R_c}{R_T}\right)^2\cdot\left(\frac{\partial\delta}{\partial V_g}\right) \tag{6.52}$$

the depletion depth is

$$\delta = \kappa_s\varepsilon_o WL\left(\frac{1 - g_m R_s}{C_g}\right) \tag{6.53}$$

and the depletion depth-dependent mobility is

$$\mu(\delta) = \frac{g_m}{C_g}\cdot\frac{L^2}{V_D}\cdot\left(1 - \frac{R_s + R_d}{R_T}\right)^{-2} \tag{6.54}$$

For GaAs they found a drift mobility which is about 15% smaller than the Hall mobility a result consistent with the estimated Hall factor for GaAs. Adequate care is required, however, in accounting for the effect of surface states in the measurement of depletion depth as discussed in the earlier sections of this Chapter.

REFERENCES

1 A. Many, Y. Goldstein and N. B. Grover, Semiconductor Surfaces, North Holland Publ. Co.(1965)136.
2 H. K. Henisch, Rectifying Semiconductor Contacts, Clarendon Press, Oxford (1957)182.
3 W. Shockley, Bell Syst. Tech. J., 28(1949)435.
4 A. S. Grove, B. E. Deal, E. H. Snow and C. T. Sah, Solid-State Electron., 8(1965)145.
5 S. M. Sze, Physics of Semiconductor Devices, Wiley-Interscience, New York (1969)431.
6 A. K. Nedoluha, L. J. Messick and C. R. Zeisse, Naval Electron. Lab. Ctr., TN-3033(1975)5.
7 N. S. Clayton, J. Phys. E, J. Sci. Instr., Sec. 2, 1(1968)662.
8 R. Castagnè, C. R. Acad. Sci. Paris, 267(1968)866.
9 D. R. Kerr, Intnatl. Conf. MIS Structures, Grenoble, J. Bovel ed., (1969)303.
10 M. Kuhn, Solid-State Electron., 13(1970)873.
11 C. N. Berglund, IEEE Trans. Electron. Dev., ED-13(1966)701.
12 D. Braunig and H. G. Wageman, IEEE Trans. Electron. Dev., ED-21(1974)241.
13 A. D. Lopez, Rev. Sci. Instr., 44(1973)200.
14 K. Ziegler and E. Klausmann, Appl. Phys. Lett., 26(1975)400.
15 K. Lehovec and A. Slobodskoy, Solid-State Electron., 7(1964)59.
16 K. Lehovec, Solid-State Electron., 11(1968)135.
17 L. Meiners, J. Vac. Sci. Technol., 15(1978) July/August(to be published).
18 E. H. Nicollian and A. Goetzberger, Bell Syst. Tech. J., 46(1966)1055.
19 A. Goetzberger, E. Klausmann and M. J. Schulz, Crit. Rev. Solid-State Sci., 6(1976)1.
20 H. Deuling, E. Klausmann and A. Goetzberger, Solid-State Electron., 15(1972) 559.

21 H. L. Cox, Jr., Rev. Sci. Instr., 24(1953)307.
22 R. Chidambaram and S. Krishnan, Electron. Eng. 32(1960)158.
23 W. R. Patterson and J. Shewchun, Rev. Sci. Instr., 35(1964)1704.
24 G. R. Branner, E. M. Friar and G. Medicus, Rev. Sci. Instr., 34(1963)231.
25 D. E. Thomas and J. M. Rowell, Rev. Sci. Instr., 36(1965)1301.
26 J. Shewchun and A. Waxman, Rev. Sci. Instr., 37(1964)1195.
27 Princeton Applied Research, Tech. Note 112, "Use of Lock-in Amplifiers in C-V and G-V Systems for Characterizing Semiconductor Devices," (1974)-6K-6/74 MG.
28 M. R. Boudry, J. Phys. E, Sci. Instr., 11(1978)237.
29 W. Schottky, Z. Physik, 18(1957)458.
30 D. P. Kennedy, P. C. Murley and W. Kleinfelder, IBM J. Res. Dev., 12(1968) 399.
31 W. C. Johnson and P. T. Panousis, IEEE Trans. Electron. Dev., ED-18(1971)138.
32 J. D. Wiley, IEEE Trans, Electron. Dev., ED-22(1975)265.
33. W. van Gelder and E. H. Nicollian, J. Electrochem. Soc., 118(1971)138.
34 J. R. Brews, J. Appl. Phys., 44(1973)3228.
35 L. M. Terman, Solid-State Electron., 5(1962)285.
36 Y. Zohta, Solid-State Electron., 17(1974)1299.
37 N. I. Meyer and T. Guldbrandsen, Proc. IEEE, 51(1963)1631.
38 J. A. Copeland, IEEE Trans. Electron. Dev., ED-16(1969)445.
39 G. L. Miller, IEEE Trans. Electron. Dev., ED-19(1971)1103.
40 J. Hilibrand and R. D. Gold, RCA Rev., 21(1960)245.
41 G. W. Reutlinger, S. J. Regas, D. J. Sidor and B. Schwartz, Solid-State Electron., 12(1969)31.
42 A. Ambrozy, Solid-State Electron., 13(1970)347.
43 M. G. Buehler, Appl. Phys. Lett., 31(1977)848.
44 R. H. Crawford, MOSFET in Circuit Design, McGraw Hill Book Co., New York (1967)40.
45 K. Lehovec, Appl. Phys. Lett., 26(1975)82.
46 A. C. Ipri, Appl. Phys. Lett., 20(1972)1.
47 A. C. Ipri, J. Appl. Phys., 43(1972)2770.
48 A. C. Ipri and J. N. Zemel, J. Appl. Phys., 44(1973)744.
49 A. B. M. Elliot and J. C. Anderson, Solid-State Electron., 15(1972)531.
50 W. E. Ham, Appl. Phys. Lett., 21(1972)440.
51 J. Hynecek, J. Appl. Phys., 45(1974)2806.
52 K. Lehovec and S. T. Liu, J. Appl. Phys., 47(1976)2088.
53 P. L. Swart and C. K. Campbell, Thin Sol. Films, 44(1977)93.
54 H. H. Wieder, Appl. Phys. Lett., 25(1974)206.
55 J. R. Sites and H. H. Wieder, CRC Crit. Rev. Solid-State Sci., 5(1975)385.
56 A. Nedoluha and K. M. Koch, Z. Phys., 132(1952)608.
57 J. N. Zemel and R. L. Petritz, Phys. Rev., 110(1958)1263.
58 T. L. Tansley, J. Phys. E, Sci. Instr., 8(1975)52.
59 S. T. Hsu and J. H. Scott, RCA Review, 36(1975)240.
60 R. A. Pucel and C. F. Krumm, Electron. Lett., 12(1976)240.

CHAPTER 7

SPATIAL DISTRIBUTION OF ELECTRICAL PROPERTIES

7.0 Introduction

The technological evolution of integrated electronic circuits, with performance, reliability and cost reduction as driving forces, demands smaller device dimensions and higher packing densities per chip or wafer than those which can be achieved with presently available materials and technological processes. Conventional methods used heretofore to characterize materials are inadequate for these purposes. On the one hand, they lack the resolution necessary for determining material parameters within dimensions of the order of μm, comparable to those of individual devices, on the other hand they are not well-suited for mapping the local fluctuations in these parameters over a wafer whose dimensions may be of the order of a few cm.

Resistivity gradients in a semiconductor wafer can affect adversely the performance and reliability of electronic devices fabricated from such wafers. It is, therefore, of critical importance to eliminate those wafers which contain unacceptably large resistivity gradients. A method in present use (ref. 1) for such acceptance tests involves six four-point probe resistivity measurements: two at the center of the wafer and four more at ∿3.2mm from the periphery of the wafer along orthogonal diameters. A disadvantage of this method is the lack of information about resistivity variations between the center of the wafer and its periphery and the relatively large sampling region of such probes. One of the more critical aspects of four-point probe measurements involves the potential damage to the wafer surface caused by the mechanical stresses between the probe points and the wafer. Such stresses can generate dislocations and related defects during subsequent processing steps to which a wafer is subjected. Such defects may interact with impurities producing resistivity fluctuations, charge carrier scattering centers or non-radiative recombination centers. Alternative techniques used for the evaluation of the electrical properties of semiconducting materials are, therefore, considered to be desirable and are described in this Chapter.

7.1 Optical and photovoltaic scanning methods

The radial resistivity profile of a wafer may be determined by means of a photovoltaic-scanning method such as described by Oroshnik and Many (refs. 2,3).

The technique is based on the bulk photovoltaic effect: resistivity fluctuations in a semiconductor constitute microscopic (or larger) n^+-n or p-n junctions and incident radiation can excite a photovoltaic response from them. Since the photovoltage depends on the existence of a resistivity gradient, it is a derivative quantity. Therefore, the integrated photovoltaic signal corresponds to the variation in resistivity, in accordance with the rudimentary explanation of this effect shown in Fig. 7.1. Suppose that the specimen is illuminated at point x_3 then excess electrons generated by the light drift toward the lower energy region, to the left of the figure, and holes drift to the right. The separation of charges produces a positive photovoltaic signal detected by the electrometer E. At point x_1 the drift of electrons and holes generated by the incident radiation is opposite to that observed at x_1; the photovoltaic signal is negative as shown in Fig. 7.1.(c). Zero photovoltage signals correspond to maxima and minima in the resistivity profile. If the photovoltage is integrated along the direction of the displacement of the beam of light which scans the surface of the semiconductor then the integrated signal corresponds qualitatively to the changes in resistivity. Quantitative relations which allow a determination of the one-dimensional inhomogeneity of a filamentary specimen were derived under the following boundary conditions: (a) the resistivity is constant through the cross-section but varies along the length (taken as the x-direction) of the filament; (b) the end contacts are ohmic; (c) the lifetime of the charge carriers is homogeneous and isotropic; (d) the illumination is restricted to a region which is small compared to a diffusion length, and (e) the excess carrier densities generated by the light are much smaller than the equilibrium carrier concentrations. Under these conditions the resistivity gradient $d\rho/dx$ is a function of the mobility ratio, $b=\mu_n/\mu_p$, of the excess conductance of the filament due to the illuminated strip ΔG and of the cross-sectional area A and length L of the filament as well as the type of dominant impurity type

$$\frac{d\rho}{dx} = -\frac{e}{2kT}\cdot(1 + b)\cdot\frac{A}{L^2}\cdot\frac{U}{\Delta G} \qquad \text{(n-type)}$$
$$\frac{d\rho}{dx} = \frac{e}{kT}\cdot\left(\frac{1 + b}{b}\right)\cdot\frac{A}{L^2}\cdot\frac{U}{\Delta G} \qquad \text{(p-type)} \qquad (7.1)$$

Measurements made on Ge filaments using these techniques were found by Oroshnik and Many (ref. 2) to provide an accurate representation of resistivity gradients between a few percent to a few tens of percent. The main source of error seems to be that inherent in the non-uniformity of the lifetime along the filaments. In those filaments in which the lifetime was uniform the error was found to be no greater than 3%. Similar measurements have been described by Baev and Valyashko (refs. 4,5). The technique was adapted by Blackburn *et al.*(ref. 6) to the measurement of radial resistivity gradients in silicon wafers with the additional constraint that the scanning light beam is never closer to any of the

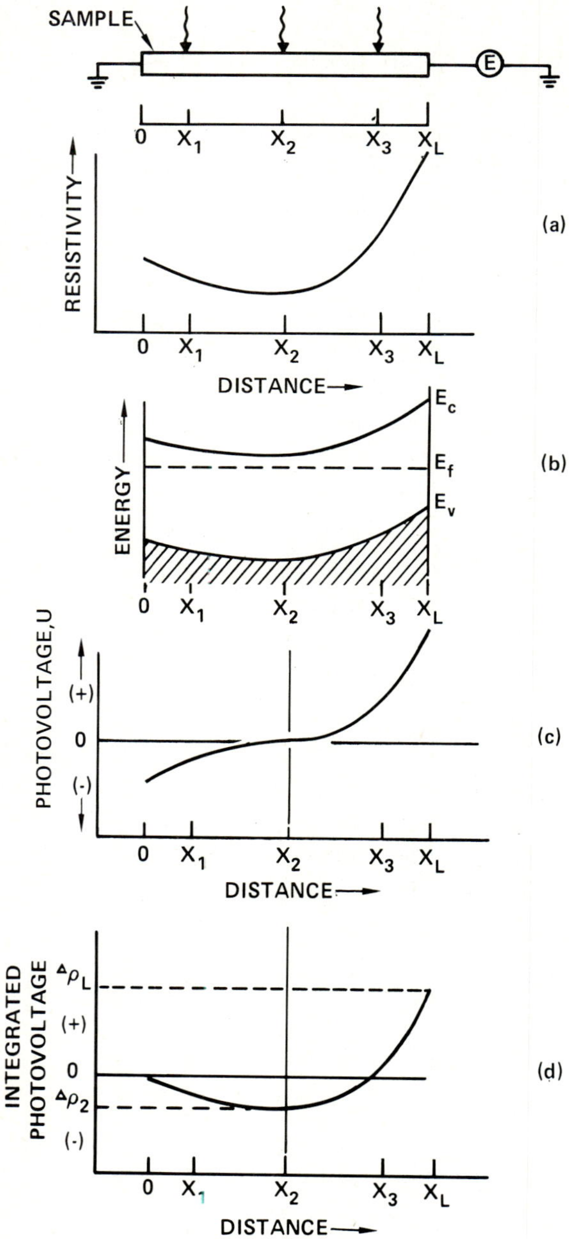

Fig. 7.1. Qualitative dependence of resistivity, localized energy band-structure and photovoltaic response of an inhomogeneous semiconductor wafer. [after J. Oroshnik and A. Many, Solid-State Electron., 1(1960)46]

specimen contacts than a distance which corresponds to several light spot diameters. This is due to the breakdown of geometrical correction factors as the light spot approaches the periphery of the wafer. The geometrical aspect of the measurement is shown in Fig. 7.2. If a circular wafer of radius b has a

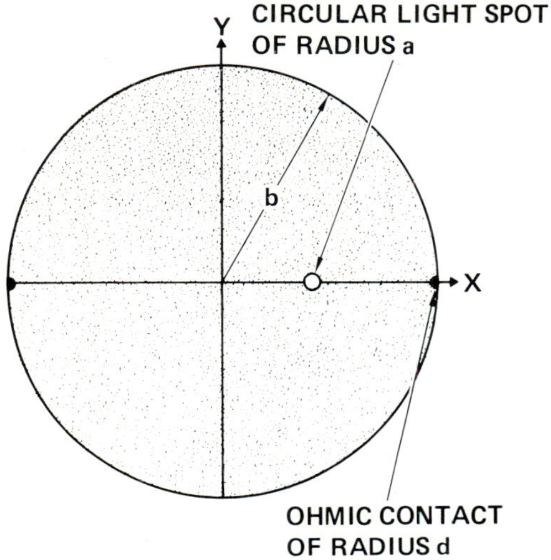

Fig. 7.2. Specimen geometry used for photovoltaic scanning along x of radial resistivity gradients.

light spot of radius a scanned along a diameter between two peripheral ohmic contacts of radius d then, in contrast with the earlier described case applicable to a filament, a current which shunts the illuminated region exists in the wafer when the photovoltage is generated in it. A conformal transformation can be used to derive the photovoltage V, measured between the contacts, in terms of the average dipole strength of the potential dipole produced by the light spot scanned along the coordinate x in Fig. 7.2. If R_0 is the resistance of the wafer measured between its contacts then the perturbation, ΔR, in R_0 produced by a change in resistivity $(\rho+\Delta\rho)$ in a spot of radius a can be calculated if $a \ll b$ and $d \ll b$. In that case, the resistivity gradient (ref. 6) is

$$\frac{d\rho}{dx} = -\left(\frac{3e}{4\pi kT}\right)\cdot\left[\frac{(1+\mu_M/\mu_n)}{1-(x/b)^2}\right]\cdot\left(\frac{\bar{\rho}^2}{b\delta}\right)\cdot\left(\frac{V(x)}{\Delta R(x)}\right) \tag{7.2}$$

Here μ_M and μ_n are, respectively, the majority carrier and minority carrier mobilities, $\bar{\rho}$ is the average resistivity and δ is the thickness cf the wafer; $V(x)$ and $\Delta R(x)$ are, respectively, the position-dependent photovoltaic signal and the corresponding change in resistance measured in the process of scanning the light beam along the diameter in accordance with the apparatus shown schematically in Fig. 7.3. Two pairs of knife-edge contacts are made to contact the wafer. With the switch in position 1 the photovoltaic response is measured as the light beam produced by a radiant source, whose energy is higher than the bandgap of the semiconductor, is traversed along the direction x in Fig. 7.2. The light

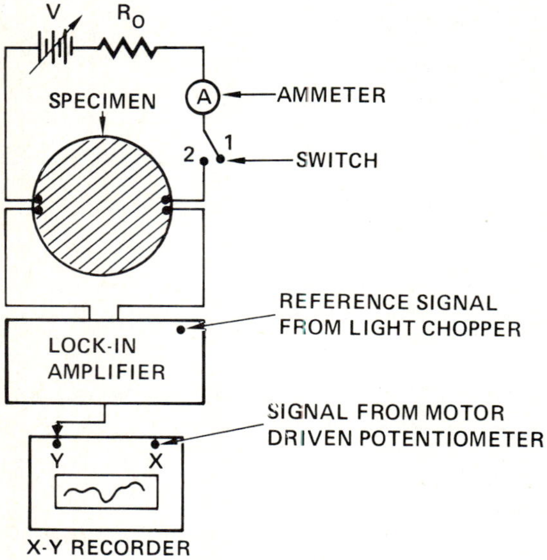

Fig. 7.3. Apparatus used to measure photovoltage and photo-induced changes in specimen resistance. [after D. L. Blackburn, H. A. Shaft and L. J. Swartzendruber, J. Electrochem. Soc., 119(1972)1773.

is modulated by a chopper at a rate such that the "on-time" of the light is adequate to achieve a steady-state charge carrier distribution. It is synchronized with the lock-in amplifier where it is detected and amplified. The photovoltaic signals are of the order of or less than 1 μV. With the switch in position 2 a predetermined current is applied to two of the contacts and the photo-induced changes in potential are monitored by the other two contacts in order to determine ΔR. The mean resistivity is determined from conventional van der Pauw-type measurements such as described in Chapter 1. A motor-driven potentiometer provides a potential proportional to the relative position of the light spot on the wafer and an X-Y recorder is used to plot the photovoltage or photoconductive response as a function of position along a wafer diameter. In practice values for $V(x)$ and $\Delta R(x)$ are determined for equal but arbitrary lengths for the entire diameter of the wafer and a summation is performed. The value obtained differs from $\rho(x)$ by the constant of integration ρ_o, the value of the resistivity at the point on the diameter where the summation begins. Figure 7.4 shows a comparison between resistivity gradients measured by using the

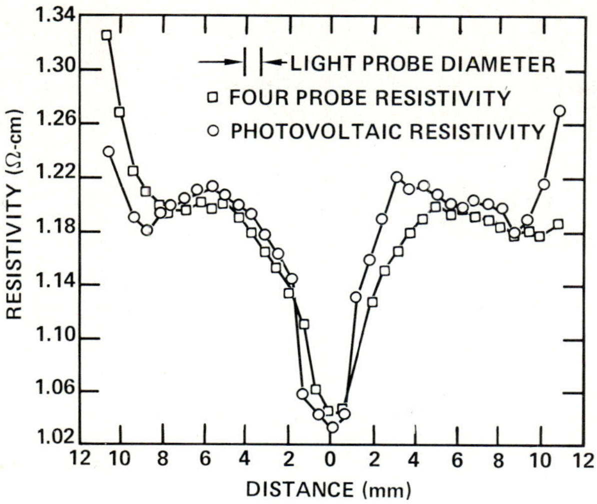

Fig. 7.4. Resistivity as a function of position determined at room temperature on a 1 ohm-cm silicon wafer.

photovoltaic technique and four-point probe measurements made on a 1 ohm-cm n-type silicon wafer. In general, the agreement between these measurements is good in the central portion of the wafer but it deteriorates somewhat beyond ∿0.5 the radius of the wafer. The reason might be attributed to the non-ohmic character of the knife-edge contacts which may generate a barrier photovoltage superposed on the bulk photovoltage. The spatial resolution of this photovoltaic technique appears to be superior to the four-probe method and is reproducible to within a few percent of the average specimen resistivity.

Relatively little has been done thus far towards an exploitation of essentially routine infrared transmission and reflection measurements used for a quantitative characterization of the carrier concentration and its distribution in semiconducting wafers. Black *et al.*(ref. 7) have proposed that such measurements made on GaAs wafers at room temperature be used to determine their free carrier concentrations. This can be done with an accuracy estimated to be ∿30% in the range $5\times10^{16}<n<10^{19}/cm^3$. Using such optical methods the region of a wafer that is intercepting an incident light beam can be varied by focusing or defocusing the optical beam. If infrared laser sources are used in the wavelength range of 5 to 15μm then a spatial resolution of the order of the wavelength can be used to map inhomogeneities by scanning the beam over the wafer. The carrier concentration can be determined from a measurement of the free carrier plasma resonance frequency $f_r=\omega_r/2\Pi$ which corresponds to the minimum in reflectivity (ref. 8),

$$\omega_r = \frac{ne^2}{m^* \varepsilon_o k_\ell} \tag{7.3}$$

If the effective mass, m* and the high frequency lattice dielectric constant k_ℓ are known quantities then the free carrier concentration can be determined from the measured plasma resonance frequency. Black *et al.*(ref. 7), using published data (ref. 9) have shown that for $n>10^{18}/cm^3$, f_r^2 of GaAs is proportional to the free electron concentration. For $n<10^{18}/cm^3$ infrared absorption measurements used to determine n are more sensitive than reflection measurements. The dependence of the transmission coefficient T on the absorption coefficient α, expressed in terms of the reflection coefficient R, is

$$T = \frac{(1 - R)^2 \exp(- \alpha\delta)}{1 - R^2 \exp(- 2\alpha\delta)} \tag{7.4}$$

where δ is the sample thickness; the dependence of the transmission coefficient of GaAs on the product $\alpha\delta$ is shown in Fig. 7.5; it may be calculated from Eq. (7.4) using R=0.28 for GaAs; this is a reasonable approximation because T is not a strong function of R, as shown in the insert of Fig. 7.5. Thus if T is measured experimentally and δ is known then α can be calculated for specimens with

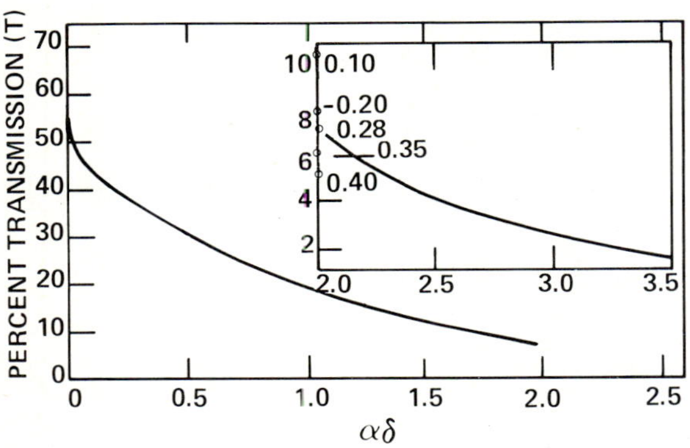

Fig. 7.5. Dependence of transmission coefficient of GaAs on absorptance, $\alpha\delta$ calculated for a reflectivity of 0.28; insert, at $\alpha\delta$=2.0 represents calculated values of T for chosen values of the reflectivity shown nearby.

specific carrier concentrations. The dependence of n on α (ref. 10), with the wavelength of the incident radiation as a running parameter, is shown in Fig. 7.6. Good agreement can be obtained between carrier concentration measurements made by means of Hall effect and plasma resonance measurements provided that $n \geq 3 \times 10^{17}/cm^3$. A comparison between values of n determined by means of Hall measurements made on the same specimens is shown in Fig. 7.7; they are essentially independent of the type of doping impurity. Using a beam condenser to

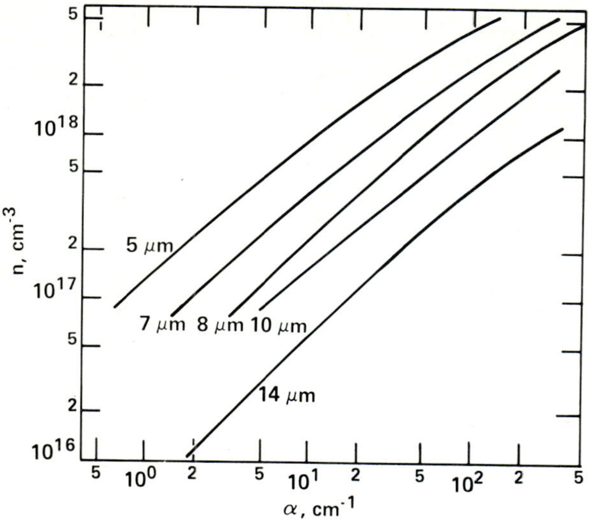

Fig. 7.6. Dependence of free carrier concentration on absorption coefficient with wavelength of incident radiation as a running parameter. [after W. G. Spitzer and J. M. Whelan, Phys. Rev., 114(1951)59]

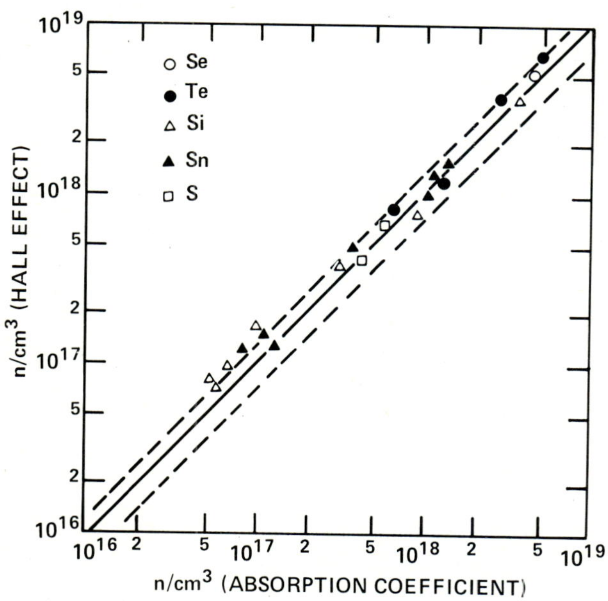

Fig. 7.7. Comparison of free carrier concentration determined from transmission measurements with those calculated from Hall measurements made on the same GaAs specimens doped with various n-type impurities. Dashed lines represent ±30% deviation from Hall data. [after J. F. Black, E. Lanning and S. Perkovitz, Infrared Physics, 10(1970)125]

focus the infrared radiation and relatively simple electromechanical X-Y scanners the homogeneity of a specimen can be determined and a map of the fluctuations in n can be made which is a function of the optical resolution of the system; the inhomogeneities do not have to have radial symmetry although thickness-dependent inhomogeneities cannot be evaluated directly.

Electroreflectance spectroscopy is a valuable tool for investigating the surface band structure of semiconductors (ref. 11). In order to make electroreflectance measurements a thin transparent semiconducting layer of SnO_2 or a thin transparent metallic Schottky barrier is deposited on the surface of the semiconducting specimen under investigation. By applying a reverse voltage to such a structure an electric field E is generated at the surface of the semiconductor. A beam of light whose energy is above the bandgap of the semiconductor is incident on its surface and the change in reflectance, ΔR, of the reflected beam is determined as a function of E. The relative change in reflectance at a fixed wavelength λ of the light depends on the electric field

$$(\Delta R/R)_{\lambda} = r(E) \tag{7.5}$$

where R is the reflectance at zero applied field. The electric field, in turn, depends on the voltage V_g applied to the Schottky barrier or heterojunction and, in terms of the relations discussed in Chapter 6, depends on the free carrier concentration. Thus $r(E)=r(V_g,n)$. The theoretical relation between the relative change in reflectance and these parameters has been given by Aspness (ref. 12); it is of the form

$$\Delta R/R = -(2e/\kappa_0\varepsilon_0)\cdot V_g n\cdot L(\lambda) \tag{7.6}$$

where V_g is the potential applied to the metal gate, κ_0 is the static dielectric constant and $L(\lambda)$ is the spectral line shape function which is entirely determined by the band structure.

In order to measure the free carrier concentration by electroreflectance methods the measurements of $\Delta R/R_0$ must be made on fully depleted space charge regions. An adequate test is an invariant spectral lineshape with increasing V_g and a linear dependence of the relative electroreflectance signal on V_g. Equation (7.6) is the basis for determining spatial variations of the free carrier concentration by means of reflectance measurements, as described by Sittig and Zimmermann (ref. 13). The apparatus is shown schematically in Fig. 7.8. Light from a xenon or deuterium lamp after traversing a monochromator is focused by means of a microscope objective lens onto the specimen surface. The image of the exit slit is reduced by about one tenth its actual size; by varying the dimensions of the exit slit the form of the light spot on the specimen can be controlled. The latter is mounted on a micromanipulator table which permits it to be positioned in the X-Y plane. Light reflected from the specimen surface passes

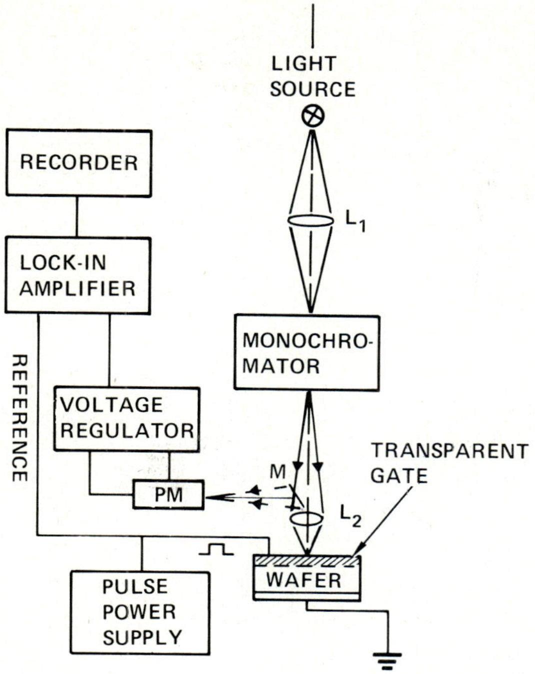

Fig. 7.8. Apparatus used for measuring inhomogeneous distribution of free carriers in a Schottky barrier or heterojunction by scanning electroreflectance; L_1 and L_2 are lenses, M is a front surface mirror, PM is a photomultiplier. [after R. Sittig and W. Zimmermann, Phys. Stat. Sol.(a), 12(1972)663]

again through the lens L_2 and is reflected by a mirror into a photomultiplier. The potential V_g is applied in the form of periodic rectangular pulses to the gate of the Schottky barrier or heterojunction and it is synchronized with the lock-in amplifier. For measuring the spatial variations of the electroreflectance signal a voltage V_g is chosen near the upper limit of the linear dependence of the electroreflectance signal on V_g.

For a spot width of ∿50μm about ten ΔR/R measurements are taken and integrated before the micromanipulator moves the specimen to another position. The time required for the full cycle is ∿1 sec. and a full 1 cm run of the specimen surface takes about 34 min. for a mean error of ±2% in the electroreflectance signal. Excellent results have been obtained with this technique, as shown by the data presented in Fig. 7.9, which compares the resolution which can be be obtained by means of electroreflectance measurements with four-point probe conductivity measurements made along a diameter of a silicon wafer. The specific profile was confirmed by subsequent electrochemical striation etching. Glosser (ref. 14) has suggested an additional application of such electroreflectance

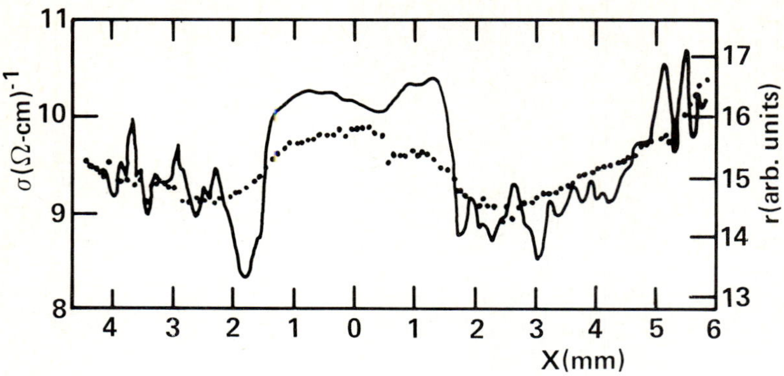

Fig. 7.9. Comparison between four-point probe conductivity measurement (•) and electroreflectance (-) along a diameter of an n-type silicon slice having a specified resistivity of 0.08 to 0.12Ω-cm. The light spot area was 50x100μm^2, the wavelength 374 nm. The scale of the electroreflectance was chosen to fit the magnitude of the conductivity measurement. The zero levels of both scales coincide. [after R. Sittig and W. Zimmerman, Phys. Stat. Sol.(a), 12(1972)633]

measurements made on MIS or electrolyte-semiconductor interfaces intended for a determination of the sign and magnitude of the surface potential of degenerate or nearly degenerate small bandgap semiconductors without requiring conventional C-V measurements.

7.2 Electron beam mapping of conductivity gradients

A phenomenon analogous to the photovoltaic effect is the electron-voltaic effect (refs. 15-18) produced in a junction or Schottky barrier subjected to irradiation by a high energy electron beam which generates electron-hole pairs in it; these are swept across the barrier by the electric field of the junction. In an inhomogeneous material there may be one more resistivity gradients made up of microscopic or macroscopic n^+-n or p^+-p junctions. If the e-beam traverses a semiconductor specimen with ohmic contacts on its periphery along a coordinate x containing fluctuations in resistivity then the incremental position-dependent potential measured between its contacts is a function of the variations in the resistivity $\rho(x)$. An electron microscope (ref. 19) can be used to reveal, qualitatively, the presence and distribution of such fluctuations. If a constant bias is applied to such a specimen producing in it an electric field E then e-beam irradiation produces a change in field and current density ΔE and ΔJ,

$$J + \Delta J = (\sigma + \Delta\sigma)(E - \Delta E)$$

$$\Delta J = E \cdot \Delta\sigma - \sigma \cdot \Delta E \qquad (7.7)$$

If the width of the incident e-beam is Δx then the incremental potential difference ΔV measured between the contacts of the specimen is

$$\Delta V = \int \Delta E dx \simeq (\Delta\sigma/\sigma) E \cdot \Delta x \tag{7.8}$$

Since only E is a variable in Eq. (7.8) ΔV is directly proportional to E; therefore

$$V = \frac{\sigma}{\Delta\sigma} \int \frac{\Delta V}{\Delta x} dx \tag{7.9}$$

Munakata (refs. 20,21) used such a relation to determine the potential distribution in an inhomogeneous Ge crystal with a mean resistivity of 0.8 ohm-cm and dimensions 0.5x0.9x5 mm. A 30 keV e-beam with a mean current of 2 nA modulated at a 1 kHz rate was used to irradiate and scan the specimen which had a potential difference of $\sim$50 mV applied to it. The beam diameter was less than 20μm, however, Δx is determined primarily by the diffusion length of the minority carriers generated by the incident beam rather than by the beam diameter. The output signal from the specimen after detection and synchronous demodulation was integrated and displayed as a function of the sweep-coordinate of the e-beam; good agreement with probe measurements were obtained by displacing the latter along the same coordinate traversed by the beam. A complete e-beam scan can be made in $\sim$10 sec.

An equivalent circuit representation for an e-beam-irradiated rectangular semiconducting wafer can be made (ref. 21) by considering a fraction of the specimen in the form of a narrow parallelipiped with a cross-sectional area S_o, length ℓ and resistance R in parallel with the resistance R_b of the rest of the wafer and in series with the load resistance r_o. The conductivity σ increases by $\Delta\sigma$ over a width Δx of this fractional volume along the direction scanned by the e-beam. Figure 7.10 shows the equivalent circuit with δV_o the increment of voltage produced by e-beam irradiation. The specimen has a biasing voltage V_s applied to it producing the electric field E.

$$\delta V_o = \int_0^{\ell} \delta E dx \simeq E(\Delta\sigma/\sigma) \cdot \Delta x = \frac{1}{\sigma^2} \left(\frac{V_s}{RS_o} \right) \Delta\sigma \cdot \Delta x \tag{7.10}$$

The incremental voltage δV detected across the output resistor r_o is

$$\delta V = \left(\frac{r_o R_t}{r_o + R_t}\right) \bullet \frac{\rho^2}{R^2 S_o} \bullet \Delta\sigma \bullet \Delta x \tag{7.11}$$

where the shunt resistance $R_t = R_b R \cdot (R_b + R)^{-1}$. If R, S_o, $\Delta\sigma$ and Δx are constant as the beam is displaced along the x-axis then the response of the equivalent circuit, $(\delta V)^{1/2}$, is proportional to the local resistivity. Good agreement has been

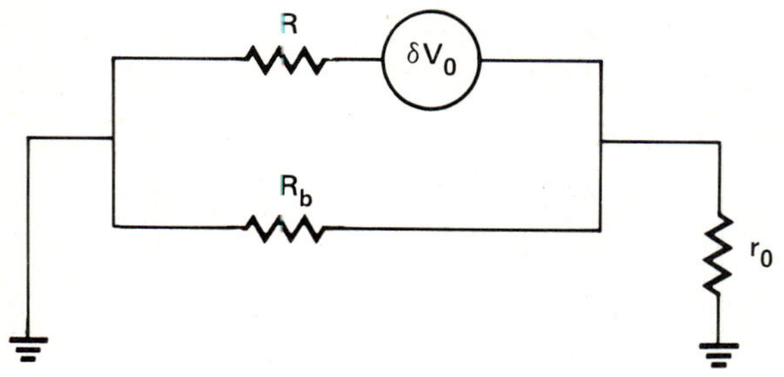

Fig. 7.10. Equivalent circuit of fractional volume element of resistance R_o length ℓ and rectangular cross-section S_o of a rectangular specimen of resistance R in series with a load resistor r_o. An increment of potential δV_o is produced in the fractional volume by an e-beam excitation of a surface element with side Δx. [after C. Munakata, J. Sci. Instr.(J. Phys. E) ser. 2, 1(1968)639]

obtained between such e-beam conductivity modulation measurements, electron-voltaic measurements and potential probe measurements made over the same path. However, if the resistivity varies along directions other than those of the scanning e-beam then δV no longer represents the resistivity fluctuations (ref. 21); instead it represents the local electric field distribution. As a rule, the resolution and sensitivity of the e-beam conductivity modulation technique are superior to those of electron-voltaic measurements. However, while the electron-voltaic method yields the resistivity gradients in and below the surface of a specimen, the e-beam conductivity modulation technique is strongly sensitive to surface topography and surface preparation.

Electron beam-induced conductivity modulation has been described analytically and experimentally by Gopinath (ref. 22). The analysis for a one-dimensional case is based on the model shown in Fig. 7.11. An e-beam incident on a fractional volume element approximately two diffusion lengths wide creates electron-hole pairs. The area over which these carriers diffuse or drift in the direction of the field (created by the bias voltage V_s) has a higher conductivity than the surrounding material. The net increase in conductivity makes an additional current fraction Δi to flow into the specimen derived from the constant voltage source V_s. This incremental current is

$$\Delta i \simeq [\Delta r(x_0)/R^2] \cdot V_s \tag{7.12}$$

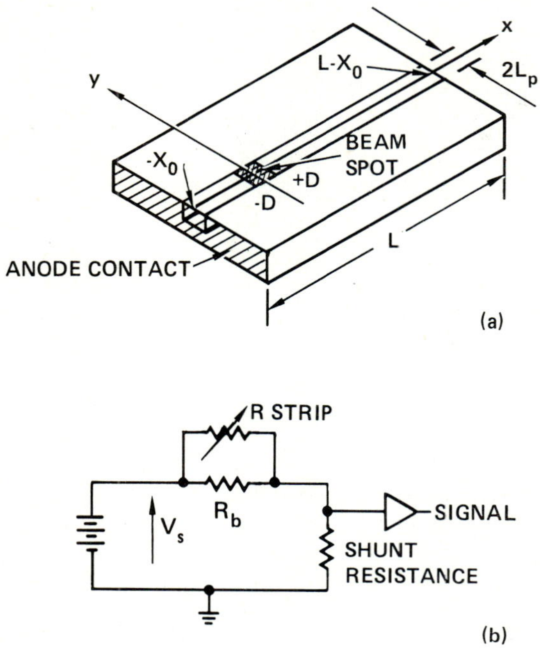

Fig. 7.11. (a) Model for irradiated strip in rectangular specimen with depth of strip a function of beam voltage and carrier diffusion length; (b) equivalent circuit with shunt resistance of strip $R \ll R_b$. [after A. Gopinath, J. Phys. D: Appl. Phys., 3(1970)467]

where R is the unperturbed resistance of the strip containing the area affected by the e-beam and $[R-\Delta r(x_0)]$ is the corresponding resistance of the strip with the beam spot at x_0. The continuity equation for the minority carriers, considered to be holes, for a constant electric field neglecting space charge effects and assuming the minority carriers to have a uniform distribution over the irradiated strip in Fig. 7.11 is

$$\frac{d^2\Delta p}{dx^2} - \frac{(E/E_c)}{L_p} \cdot \frac{d\Delta p}{dx} - \frac{\Delta p}{L_p^2} = - \frac{g\tau_p}{L_p^2} \tag{7.13}$$

where the diffusion length $L_p = (D_p \tau_p)^{1/2}$ with D_p the diffusion coefficient and τ_p the lifetime of the e-beam-generated holes of concentration Δp. The DC electric field produced by the biasing voltage is E, the diffusion field is $E_c = (kT/eL_p)$ and the generation rate of electron hole pairs is g. Equation (7.13) can be evaluated subject to the following boundary conditions:

(a) $\Delta p = 0$ at $x=-x_o$ and at $L-x_o$;

(b) Δp is continuous at $\pm D$, the spot boundaries; and

(c) $\Delta p/\Delta x$ has continuity of flow at $\pm D$.

Equation (7.13) was solved (ref. 22) subject to these boundary conditions and leads to

$$\Delta i = C \int_{-x_o}^{L-x_o} \frac{\Delta p(x)}{\sigma^2}\, dx \qquad (7.14)$$

where the constant of proportionality C is a function of the biasing voltage V_s. The e-beam induced current is thus a function of $1/\sigma^2$ and Δp. The latter is a function of L_p, E/E_c, τ_p and g, in terms of Eq. (7.13). If the electric field in the specimen is small and $L_p \ll L$ then the induced current is a function of $1/\sigma^2$ except near the electrodes of the specimen. Resistivity variations will appear in a one-dimensional scan, in accordance with Munakata (ref. 21), however, not only resistivity gradients but differences in generation rate will also show up in a scan of $\Delta i(x)$. If $L_p \gg L$ then Δp decreases linearly to zero from the beam spot toward the anode and cathode and localized variations in resistivity will not be made evident. If the biasing field is large, $E/E_c \gg 1$ and $L_p < L$ the minority carriers drift only toward the cathode and diffusion to the anode is negligible; any variation in resistivity or field will be indistinguishable in $\Delta i(x)$ unless the variation is large and affects the carrier generation rate. Between these extremes there is an intermediate range of electric fields which can be used to obtain a qualitative estimate of L_p and τ_p.

Munakata (ref. 23) has shown a way in which to improve the spatial resolution of e-beam-induced conductivity modulation used to reveal one-dimensional resistivity gradients in semiconductors by means of a pulsed electron beam. With a continuous electron beam if the electric field in the specimen is high enough to disturb the thermal diffusion of the carriers then the distribution should be characterized by the drift length $L_E = \mu_p \tau_p E$ (where μ_p is the minority carrier hole mobility) rather than by the diffusion length L_p; the resolving power decreases as the field intensity increases. In order to make the spatial resolution independent of E it is required that $L_E < L_p$ and by the use of the Einstein equation this inequality becomes

$$E < \left(\frac{kT}{e\mu_p}\right)^{1/2} \cdot \left(\frac{1}{\tau_p}\right)^{1/2} \qquad (7.15)$$

thus E can be allowed to be large if τ_p is made small. If the charge carriers are generated by a pulsed electron beam whose angular frequency is ω producing minority carrier holes with an AC lifetime $\tau_p(\omega)$ (ref. 24), given by,

$$\tau_p(\omega) = \tau_p(1 + j\omega\tau_p)^{-1} \tag{7.16}$$

where $j=(-1)^{1/2}$ then Eq. (7.16) introduced in Eq. (7.15) with $\omega\tau_p >> 1$ leads to the inequality

$$E < \left(\frac{kT\omega}{e\mu_p} \right)^{1/2} \tag{7.17}$$

Thus the maximum field intensity can be increased if the electron beam is pulsed at a high frequency. Munakata used a pulsed electron beam of 15 keV driven at a rate of 0.1 to 10^2 kHz with a diameter of less than 1 μm. The beam was displaced with a velocity of 0.13 mm/sec. along an n-type Ge wafer whose thickness, width and length were, respectively, 0.64, 1.37, and 8.15 mm. The hole mobility at x=2.57 mm was found to be 1780 cm^2/V-sec. and the corresponding lifetime, τ_p=12μsec. For a beam frequency of 0.1 kHz and V_s=±0.4V the measured conductivity modulation signal vs. x was found to be essentially the same as that obtained with a continuous beam; L_p and L_E were estimated to be 105μm and 230μm, respectively. When V_s was increased to ±3.6V a definite degradation of the experimentally observed spatial resolution was noted attributed to $L_E > L_p$; the former was estimated to be 945μm and the latter as 230μm. The resolution was restored when the beam frequency was raised to 100 kHz while the same value of V_S was maintained on the specimen. The e-beam conductivity modulation is a strong function of the surface recombination velocity (ref. 25) if the e-beam energy is low enough so that the penetration depth is shorter than the minority carrier diffusion length; it is essentially independent of the surface recombination velocity if the penetration depth is ∿20 times larger than the diffusion length. The absolute value of the signal derived from a pulsed e-beam excitation decreases monotonically with ω as the beam frequency exceeds the reciprocal of τ_p (ref. 26); the phase angle delay (referred to the beam frequency) increases from 0 to Π/2 as the frequency increases approaching $\omega\tau_p$.

A different high resolution method employing an electron beam acting on an extended Schottky barrier was described by Kamm (ref. 27) and by Kamm and Mueller (ref. 28). The focused electron beam penetrates the metal electrode M and generates electron-hole pairs in the depletion layer maintained by the reverse bias voltage. All the carriers generated in the depletion layer are considered to be collected by the diffusion field. If the biasing voltage is fixed then the depletion layer width is a function only of the free carrier concentration. Carriers generated outside the depletion layer in the region of the semiconductor in which charge neutrality prevails may reach the boundary of the depletion layer. In that case they contribute an increment of current which is a function of the diffusion length. If the diffusion length is small then nearly all the charge carriers generated in the neutral portion of the semiconductor recombine there; the current generated by the scanned e-beam is then determined only by the width of the depletion layer and, therefore, by the local fluctuations in free carrier concentration under the Schottky barrier. If the doping density is moderate to large then the depletion width is small and free carrier fluctuations caused by impurity concentration gradients do not influence the current generated by the e-beam; observed changes in signal current with position displacement of the e-beam over the Schottky barrier are caused by corresponding fluctuations in diffusion length. If the doping concentration is low then it is not possible to separate the contributions of free carrier concentration fluctuations from those of diffusion length variations. Therefore, the instrument response must be calibrated if it is to provide quantitative data.

One of the advantages of this method of evaluation of inhomogeneities is its high spatial resolution estimated to be of the order of 10μm. It requires, however, a large area high quality reverse-biased Schottky barrier. The evaluation can be made in terms of experimentally measured two-dimensional coordinate-dependent changes in the i-V characteristics of the Schottky barrier as the electron beam is scanned in the form of a raster pattern over its surface.

7.3 Probe mapping and iso-resistivity maps

Modern technological processes particularly those concerned with manufacturing of semiconductors and integrated circuits require that such measurements be made not only of the spatial average of macroscopic specimens but also of localized microscopic fluctuations in their resistivity. Such fluctuations may be present in virgin semiconducting wafers cut from a single crystal boule; they may be due to a radial and axial micro-segregation of impurities during growth or to defects generated by sawing, slicing or other forms of plastic deformation to which a wafer may be subjected before it is accepted for further processing. Changes in resistivity and its spatial distribution may also occur in a wafer during any one or more of the various stages of device or integrated circuit procedures. These process-induced changes can serve as diagnostic tools for monitoring and quality control of a fabrication process. Such measurements can be made by collinear four-point (CFP) probe techniques or by means of square four-point (SFP) probe techniques such as described in Chapter 1 used in conjunction with an automatic probe positioning apparatus in order to provide a map of the distribution of the sheet resistivity over the wafer. A circular wafer whose radius r is much greater than the interprobe spacing s of a CFP requires the proper geometrical correction factor k in order to calculate the position-dependence of the sheet resistivity, $\rho_s = k(\Delta V/i)$. If the probe is displaced from the center of the wafer then k is a function of the relative position of the probe and of its orientations with respect to the radius of the wafer (refs. 29-32). The topographic distribution of the sheet can be represented in various formats. One-dimensional profiles provide detailed information concerning point-to-point variation in doping; they are particularly useful for exhibiting rapid local fluctuations such as those which may occur in ion implanted layers (ref. 32). Two-dimensional maps may be made by dividing a wafer into a grid pattern with an arbitrary mesh of n test sites and determining the mean sheet resistivity $\bar{\rho}_s$ and the standard deviation σ_d given by

$$\bar{\rho}_s = \frac{1}{n}\sum_{i=1}^{n}(\rho_s)_i \qquad \sigma_d^2 = \left(\frac{1}{n-1}\right)\sum_{i=1}^{n}\left[(\rho_s)_i - \bar{\rho}_s\right]^2 \tag{7.18}$$

Figure 7.12 is an example (ref. 32) of such a map determined on a 7.62 cm diameter phosphorus-diffused p-type silicon wafer. Values of $(\rho_s)_i$ obtained from CFP measurements were used to compute deviations from the wafer mean resistivity according to the relation

$$\Delta i = [(\rho_s)_i/\bar{\rho}_s - 1] \times 100\% \tag{7.19}$$

A format such as that shown in Fig. 7.12(a) may be preferable in some instances since it presents deviations above and below the mean by(+) and (-) signs. It requires relatively little computation and can be implemented by means of a line printer. An alternative method of presentation of the data is in the form of two-dimensional histograms which associate measured values and interpolations with one of a number of intervals into which the range between $(\rho_s)_{max}$ and $(\rho_s)_{min}$ is divided. These intervals, each of which contains an equal number of values are identified by circles of various weight (thickness) and the symbols (-) and (+) in Fig. 7.12.(b) identify values above or below the median. A different representation of the data may be made in terms of contour lines which represent equal values of sheet resistivity. Figure 7.12.(c) shows such a contour map: the contour lines differ from each other by increments of 0.01 $\bar{\rho}_s$ and the symbols (+) and (-) represent experimental data greater or less than $\bar{\rho}_s$. The darker line in Fig. 7.121(c) represents the contour which corresponds to the mean sheet resistivity $\bar{\rho}_s$. Data interpolation and extrapolation has been used (refs. 33, 34) to provide for the continuity of the contour lines and their extension to the periphery of the wafer.

Crossley and Ham (ref. 35) have described a variety of test structures used to obtain parametric data for characterizing the processing steps used to fabricate silicon-on-sapphire integrated circuits. The circuit elements used were MOS capacitors which can provide information about the oxide thickness, charge, doping density, inversion voltage, minority carrier lifetime and maximum depletion depth before and after surface inversion. Enhancement mode and deep depletion MOS transistors, as test structures, can provide information on junction properties, the field effect mobility of majority carriers and related properties.

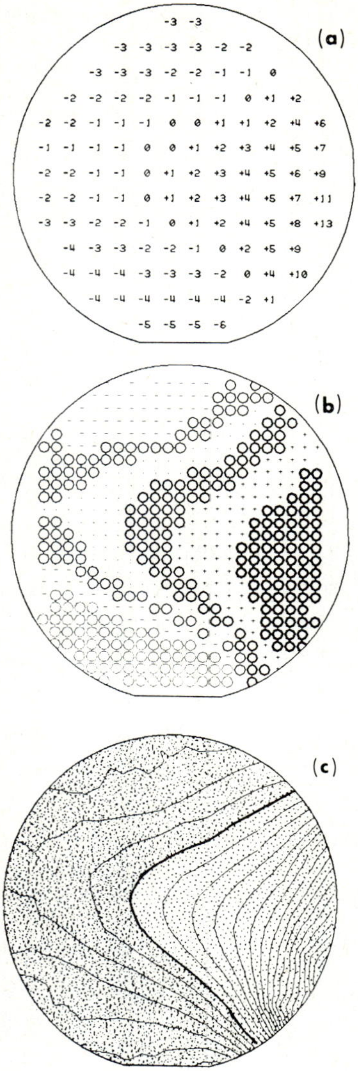

Fig. 7.12. Method of displaying spatial sheet resistivity variations obtained by means of collinear four-point probe measurements originally presented at the Spring 1976 Meeting of The Electrochemical Society, Inc., held in Washington, D.C.; (a) percent deviation from mean; (b) histogram; (c) iso-sheet resistivity. [after D. S. Perloff, F. W. Wahl and J. T. Kerr, Electron and Ion Beam Sci. & Tech., (1976)464]

Gated van der Pauw-type resistivity and Hall effect structures biased to flatband were used to make accurate resistivity measurements. Extensive data logging and computer-based analysis and correlation of the raw data was found necessary for a characterization of the capabilities and limitations of integrated circuit fabrication processes. The data were presented in the form of histograms, two-dimensional correlation plots or quasi-three-dimensional plots of parameter values or combinations of parameters as a function of wafer position.

A small planar four-probe test structure which is compatible with the fabrication processes used for bipolar transistors was developed by Buehler and Thurber (ref. 36). Analogous to a mechanical square four-point array, the planar four-probe array consists of a large area base diffusion broken at four points through which contact is made to the undiffused collector material. Including bonding pads such a structure occupies a square region 318μm on a side. A probe spacing of 57.2μm allows the resistivity of the silicon wafer to be measured with good spatial resolution. During fabrication a base is diffused into 0.013 to 12 ohm-cm n-type silicon except at the four-probe positions which are protected from the base diffusion by four base-oxide islands. These islands define four collector pipes which make contact with the collector material as shown in Fig. 7.13. These islands are squares 6.4μm on a side; they must be large enough to prevent lateral base diffusion from closing the collector pipes but, on the other hand they have to be small enough so that the collector pipes are point-contacts. This situation can be improved by increasing the probe spacing; if the probe spacing approaches the thickness of the wafer then assuming the substrate to be semi-infinite in extent the resistivity of the square array (ref. 37) is $\rho=10.726s(V/i)$. One of the important features of this test structure is its size, it is small enough to fit in an unused corner of a device chip. Its small size allows a dense packing of devices which, in turn, can be used for a detailed mapping of the bulk resistivity of the wafer. Resistivity maps made of entire wafers appeared to have radial symmetry with a lowest resistivity at the central core of the wafer, typical of Czochralski-grown crystals. Measurements made with mechanical four-point probes were found to be in agreement, to within ±3%, with similar resistivity mapping measurements made with the planar four-probe structures. The latter can also be used for measuring high resistivity materials; in these contact resistance problems make mechanical four-probe measurements difficult at best if they are at all feasible. Perloff *et al.* (ref. 34) have used CFP probes either on unprocessed wafers or in conjunction with photolithographically defined rectangular test patterns for the characterization of diffused or ion implanted layers. The test systems for the acquisition of sheet resistivity data consisted of an automatic X-Y prober, recording and sequencing devices operated by a minicomputer. Display of the data was made in the form of contour maps. An <111>-oriented n-type wafer was found to have a radial resistivity gradient. An n-type epitaxial layer grown on a p-type silicon substrate had a tear drop

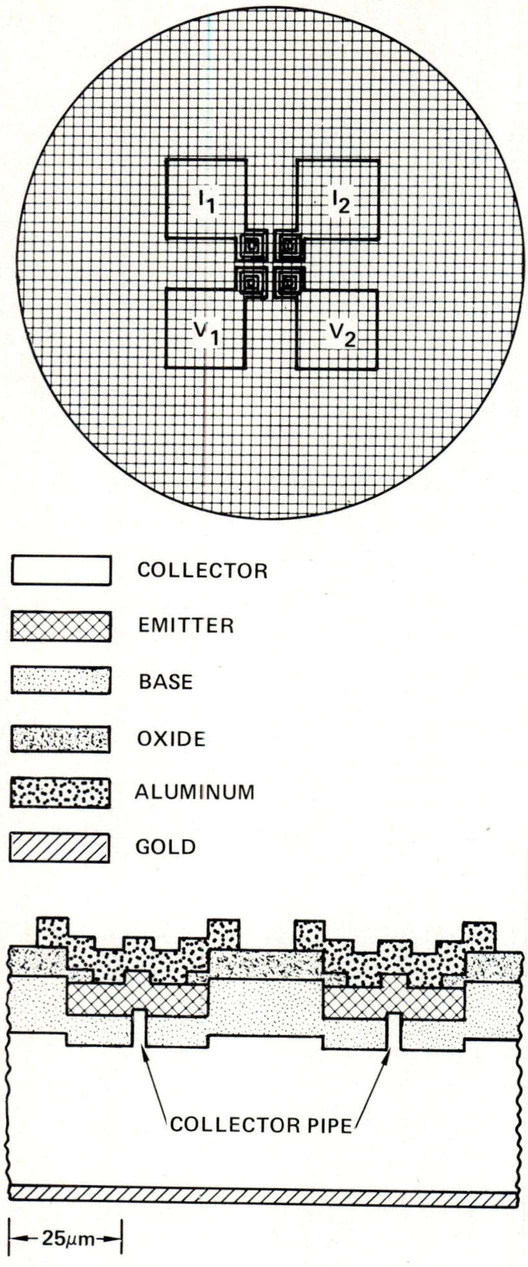

Fig. 7.13. (a) Planar square array four-probe test structure, I_1 and I_2 are current contacts, V_1 and V_2 are potential probes; (b) cross-sectional view. [after M. G. Buehler and R. Thurber, IEEE Trans. Electron. Dev. Ed-23(1976)968]

shaped resistivity contour pattern which reflected the spatial distribution of the reactant gases during the growth process.

The resistivity contour maps of silicon wafers which had been subjected to an open-tube flow gas-phase diffusion of impurities reflect the perturbations in flow of the gas through the reactor in contrast with gas-phase diffusion in a sealed quartz ampoule; such wafers were found to have a uniform resistivity over the entire wafer, to within $\sim$1%. Ion implantation offers a degree of doping uniformity which is superior to that of diffusion. However, geometric effects in electrostatic scanning and displacement velocity effects in mechanical scanning systems used for ion implantation limit the attainable uniformity (ref. 34). Such resistivity contour maps exhibit distinctive patterns which are a function either of a process step, equipment limitations or the manner in which such equipment is employed when handling a large number of wafers. They permit an assessment of parametric variations and once these parameters are established then periodic testing may be sufficient for verification and quality control. A method of quality control proposed by Buehler and Sawyer (ref. 38) relies on a process validation wafer (PVW) concept. A wafer composed entirely of a variety of test structures (ref. 36) undergoes integrated circuit processing steps on a periodic basis. This approach supplements and might provide an alternative to a test pattern on an integrated circuit chip and its use does not disrupt the flow line and fabrication cycle.

7.4 Conformal representation and method of finite differences

In a homogeneous and isotropic wafer the electric field and flux lines are functions of its shape, its electrodes and the applied electric and magnetic fields. Analytical and experimental techniques for mapping the distribution of equipotentials and streamlines in zero magnetic field, B=0 and as a function of various Hall angles are required in order to determine the perturbations introduced by localized macroscopic inclusions which may be present or those which may be introduced deliberately in an inhomogeneous material. Conformal representation is such a method. It can be used for transforming analytically a polygonal boundary into an infinite straight line or the perimeter of a circle. A differential equation of the polygonal structure is determined by inspection and in terms of the applicable boundary conditions it is integrated to yield the transformation equation. The Schwarz-Christoffel transformation (refs. 40,41) connects the real axis in one plane with the boundary of a polygon in another plane, as shown in Fig. 7.14, in such a way that the upper half of the first plane transforms into the interior of the polygon. The transformation from the real axis of the t-plane to the polygonal boundary in the z-plane is obtained by integrating the equation

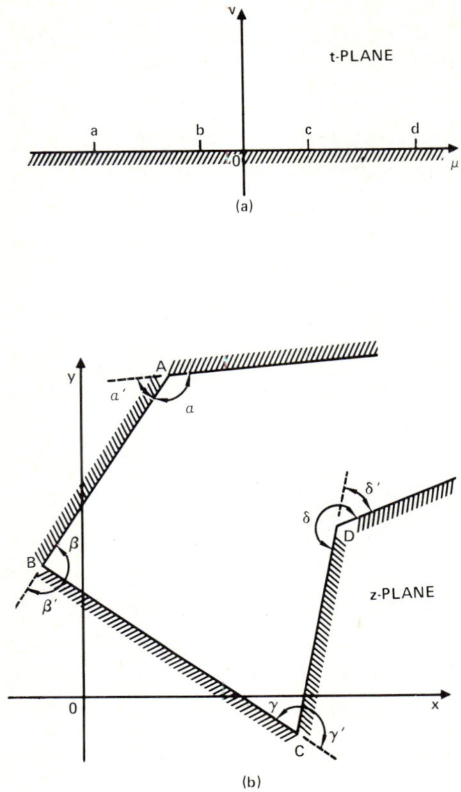

Fig. 7.14. Regions to be transformed from t-plane to z-plane by Schwarz-Christoffel transformation lie on unshaded side of boundary lines.

$$\frac{dz}{dt} = S(t-a)^{(\alpha/\Pi)-1}\ (t-b)^{(\beta/\Pi)-1}\ (t-c)^{(\gamma/\Pi)-1}\ (t-d)^{(\delta/\Pi)-1}.\ .\ .\ . \tag{7.20}$$

where S is a scale constant, a,b,c,d... are points on the real axis of the t-plane corresponding to the vertices of the polygon on the z-plane and α, β, γ, δ... are the interior angles of the polygon at A, B, C, D... respectively; the number of products in Eq. (7.20) equal to the number of vertices required to define the polygonal boundary. The Schwarz-Christoffel transformation introduces a different set of vertices in the z-plane in terms of the angles α', β', γ', δ',... with each primed variable equal to Π minus the corresponding unprimed one. These represent the change in direction of the z-plane boundary at the related vertex in Fig. 7.14.(b). Equation (7.20) then becomes

$$\frac{dz}{dt} = S(t-a)^{-(\alpha'/\Pi)}\ (t-b)^{-(\beta'/\Pi)}\ (t-c)^{-(\gamma'/\Pi)}\ (t-d)^{-(\gamma'/\Pi)}.\ .\ .\ . \tag{7.21}$$

and describes the dimension and relative angle of an increment dz in terms of the dimension and relative angle of the corresponding element dt. The argument of this equation defines, for continuously varying real values of t between $-\infty$ and $+\infty$, the inclination of dz increments tracing the boundary of the polygon. In this way, all the angles of the polygon are completely defined by those at vertices corresponding with finite points in the t-plane; if the point $t=\pm\infty$ corresponds with a vertex, then the angle at that vertex is fixed by the remaining angles since the sum of the interior angles of a polygon is $\pi(n-2)$, where n is the number of vertices. The number of factors in Eq. (7.20) is (n-1) when the point $t=\infty$ corresponds to a vertex, and is n when $t=\infty$ is a finite point on one side of the polygon. A judicious choice of corresponding points can often lead to a considerable simplification of the solution derived from Eq. (7.20). The modulus of Eq. (7.20) determines the length of an element dz in terms of the length of the corresponding element dt, and thus, in transforming the real axis of the t-plane into a polygon with given dimensions, it is necessary to choose the constants, a, b, c,... to give the desired vertex positions in the z-plane. For polygons which are infinite, the simplest solution is usually obtained by choosing the limits of the real axis to correspond with the limits of a pair of adjacent sides of the polygon which go to infinity. In cases for which this line passes through a vertex and the center of one side, the point is chosen to be at the vertex.

The literature is replete with papers concerned with a conformal representation of rectangular Hall plates. Lippmann and Kuhrt (ref. 42) have derived the geometry dependence and Hall angle dependence of the transverse magnetoresistance of an n-type semiconductor plate of length a and width b oriented along cartesian axes with electrodes at the coordinates (-b/2, 0), (b/2, 0) and (-b/2, a), (b/2, a) and a potential V applied between them. The boundary conditions were defined in terms of the potential function $\psi(x,y)$; in the interior of the rectangle $\Delta\psi(x,y)=0$ and at its boundaries $\psi(x,o)=V$, $\psi(x,a)=0$ and

$$-\left(\frac{\partial\psi}{\partial x}\Big/\frac{\partial\psi}{\partial y}\right) = 0 \tag{7.22}$$

Hauesler (refs. 41-45) has mapped the electric field distribution in rectangular Hall plates by means of a conformal representation taking into account the perturbation in the current streamlines in the immediate vicinity of both the current electrodes and Hall electrodes. Oberlander and Wilhelm (ref. 46) have used conformal transformation methods for mapping the electric fields and the conductivity of rectangular plates with symmetrically placed contacts. They evaluated the electric field perturbation at the contacts and developed appropriate geometrical correction factors for various length-to-width aspect ratios. A conformal representation of the electric field and current streamlines of a rectangular plate was described by De Mey (ref. 47) subject to the boundary conditions that the constitutive Ohm-Hall equation

$$\overline{E} = \rho\overline{J} + \mu\rho(\overline{J} \times \overline{B}) \tag{7.23}$$

is applicable, that the electric field is the gradient of the potential ψ in the Hall plate, $\overline{E}=-\nabla\psi$ and $\nabla\bullet\overline{J}=0$ which leads to Laplace's equation

$$\nabla^2\psi = 0 \tag{7.24}$$

At the metal contacts, $\overline{AB}$ and $\overline{CD}$ in Fig. 7.15, the electric field is normal to the contacts while at the other boundaries the current density is tangential

$$\begin{aligned} \overline{E}\bullet\overline{n} &= \mu\ \overline{B}\ \overline{E}\bullet\overline{k} \\ \overline{J}\bullet\overline{n} &= 0 \end{aligned} \tag{7.25}$$

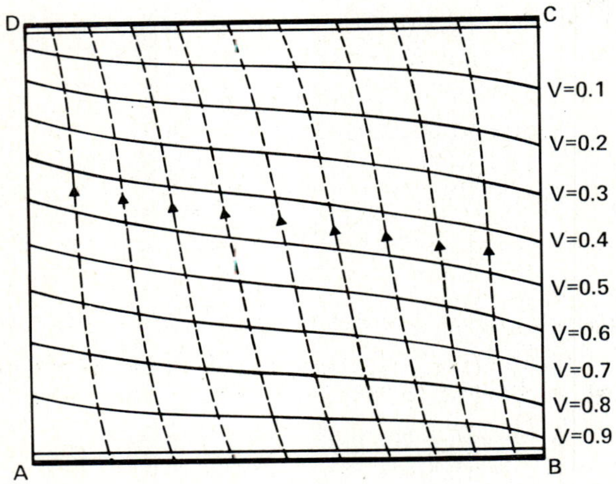

Fig. 7.15. Distribution of equipotentials (solid lines) and current streamlines (dashed lines) for a rectangular plate of length-to-width ratio of 0.8545; metallic electrodes are assumed to be present along lines $\overline{AB}$ and $\overline{CD}$. [after G. De Mey, Solid-State Electron., 16(1973)955]

where $\overline{n}$ is a unit vector perpendicular to the boundary and $\overline{k}$ is a unit vector tangent to the boundary oriented counterclockwise (depending on the direction of the magnetic field $\overline{B}$). Figure 7.15 represents the calculated distribution of the equipotentials and current streamlines for a Hall angle, θ=18° and a length-to-width ratio ℓ/w=1.709/2. A conformal transformation was also used by Jandl *et al.*(ref. 48) to map the electric field distribution in a six-arm bridge-shaped specimen such as shown in Fig. 1.12 in order to determine the reduction in Hall voltage produced by the lateral side arms. They found that the voltage measured between the ends of lateral arms with half the cross section of the main sample body and separated by eight sample widths, is only 1% less than the voltage between two point contacts of similar separation placed on the main sample body. To compare the measured effect of the lateral arms with experiment they built an electrolytic tank of the same shape as that of the six-arm bridge in Fig. 1.12 and found the side-arms, while deforming the potential quite appreciably in the region where they meet the main sample body, have little effect on the potential anywhere else in the sample.

A different form of mapping the field distribution in a Hall plate is that based on a resistance network analog (refs. 49-52). The conductivity of the plate is represented by a rectangular grid of equal resistances bounded by a short-circuit at the boundaries where the source current is introduced or removed. The Hall effect is represented by placing in each branch an emf which is proportional to the adjacent perpendicular currents. Newsome (refs. 53,54) described such a resistance network for mapping the fields in rectangular Hall plates based on a solution of Laplace's equation using the method of finite differences. The relation between the voltage at node point 0 in Fig. 7.16 and the surrounding nodes is the "five-point" formula

$$V_1 + V_2 + V_3 + V_4 = 4V_0 + h^2 \cdot \nabla^2 V + O(h^2) \tag{7.26}$$

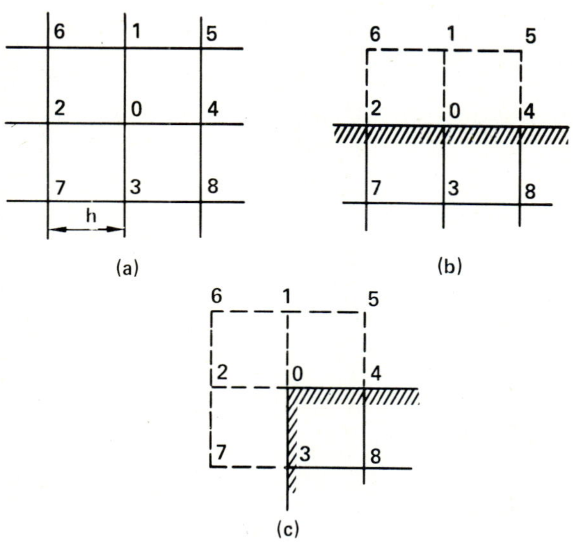

Fig. 7.16. Types of meshes and nodes used to solve potential distribution in a Hall plate by finite difference equations; (a) in the interior of the Hall plate; (b) at an edge; (c) at a corner.

This is derived by considering the Taylor series expansion about the node 0 (ref. 55). For interior regions where Laplace's equation is satisfied this reduces to the finite difference approximation

$$V_1 + V_2 + V_3 + V_4 - 4V_0 = 0 \tag{7.27}$$

A more powerful finite difference approximation with an error proportional to a higher power of the mesh interval is the "nine-point" formula (ref. 55) and simplifications in the computation may be introduced by symmetry considerations (ref. 56). Boundary nodes may differ from the interior nodes in that some of their surrounding nodes lie outside the specimen geometry, as shown in Fig. 7.16.(b) and (c) for edge and corner nodes. Using a Taylor series expansion (ref. 57) the "five-point" formula for an edge node in Fig. 7.16.(b),

$$V_2 + 2V_3 + V_4 = 4V_0 \simeq 0 \tag{7.28}$$

and for a corner node it is

$$2V_3 + 2V_4 - 4V_0 \simeq 0 \tag{7.29}$$

The set of finite difference equations obtained from the relation at each node may be solved by direct or iterative methods. A relaxation procedure may be used such that the residue at each node point in turn is reduced to zero and is continued until a solution of the desired accuracy is achieved. The residue at node 0 for a "five-point" formula can be expressed as

$$R = (1/4)(V_1 + V_2 + V_3 + V_4) - V_0 \tag{7.30}$$

The voltage V_0' at node 0 is

$$V_0' = V_0 + ZR \tag{7.31}$$

where Z is a relaxation factor. The rate of convergence is strongly dependent on the value assigned to Z. One pass through the mesh is an iteration cycle and the number of cycles are repeated until the solution is brought to the desired number of significant figures. White *et al.*(ref. 58) have used finite difference methods to calculate the Hall potential, geometrical correction factor, magnetoresistance, normalized current and normalized resistance of various Hall plate geometrical configurations. Provided that the Hall tangent is less than 0.5 a relatively coarse grid appears to be sufficient for a solution of the mesh equations by matrix inversion methods. De Mey (ref. 59) has solved the relevant potential equation in a Hall plate by an expansion method and then replaced the first order expression by a variational principle which allows the finite element mesh configuration to be applied to arbitrary geometrical structures. The distribution of equipotentials and flux lines in anisotropic Hall plates derived by method of images has been given by Price (ref. 60) with specific reference to van der Pauw-type measurements of the resistivity.

7.5 Spreading resistance techniques

Four-point probe techniques do not provide adequate spatial resolution for mapping the distribution of equipotentials in a specimen or the evaluation of resistivity gradients over dimensions of the order of μm. Of the various alternative probe measurement techniques one of the more interesting is the spreading resistance technique described by Mazur and Dickey (refs. 61,62) used extensively for the spatial characterization of the resistivity of Si wafers and epitaxial layers. Such measurements are made by placing one or two osmium-tipped probes in contact with a polished semiconductor surface. The probes may have a radius of curvature at the tip of $\sim 2.5\times 10^{-3}$ cm. They may be mounted on separate booms and loaded by calibrated weights of $\sim$50g. A biasing voltage of 10 to 15 mV is applied to a probe and the current through it is measured as a function of both forward and reverse bias. Adequate care is taken to insure that the nonlinearities in the i-V characteristics usually associated with metal-semiconductor contacts are absent; this is done by measuring the i-V characteristic well beyond the 10 to 15 mV used as a rule for spreading resistance measurements. If two probes are used then one is stationary and the other is displaced over the surface of the wafer to specific coordinates where the effective spreading resistance is measured in terms of the potential difference between the probes and the current passing through them.

A simple interpretation of spreading resistance can be made in terms of a spherical contact of radius a imbedded in an infinite medium of conductivity σ. If the potential of the spherical contact is V_o and the potential at a distance r is $V=(a/r)V_o$, then with the current density $J=-\sigma\cdot\nabla V$

$$J = \sigma V_o(r_o/a^2) \tag{7.32}$$

The total current flowing into or away from the spherical contact is $i=4\Pi a^2 J_o$ (with J_o the current density evaluated at r=a). The spreading resistance $R_S=V_o/i$ is, therefore,

$$R_s = \rho/(4\Pi a) \tag{7.33}$$

For a hemispherical contact imbedded in a semiconductor surface whose boundaries are essentially at infinity the spreading resistance is twice that of Eq. (7.33), $R_s=\rho/(2\Pi a)$. If instead of a hemisphere the contact is a circular disc of radius a then a solution of Laplace's equation leads to the spreading resistance

$$R_s = \rho/4a \tag{7.34}$$

Usually an extra voltage drop appears in the course of spreading resistance measurements which may be attributed (refs. 63,64) either to a barrier resistance or to surface asperities of the probe-semiconductor interface. This is usually taken into account by defining an effective spreading resistance,

$$R_{es} = k(\rho)/4a \tag{7.35}$$

where $k(\rho)$ is an empirical correction factor which is a function of the resistivity of the semiconducting wafer. A calibration of spreading resistance apparatus can be made in terms of a specimen whose resistivity is determined by other methods with a high degree of accuracy. Mazur and Dickey (ref. 61) found a good correlation between four-point probe measurements and spreading resistance measurements made on the same Si wafers. The spreading resistance was found to be proportional to the resistivity of Si specimens within the range of 10^{-3} to $\sim 10^3$ ohm-cm.

Spreading resistance measurements have been made on angle-lapped multilayer epitaxial structures, on p-n junctions and n^+-n junctions with a resolution of the order of μm. Specific requirements for performing precise quantitative spreading resistance measurements have been described by Mazur (ref. 62) in terms of the following conditions:

(a) the mechanical contacts have to be reproducible on a long term basis and be capable of being placed within ±1 μm of a given standard position,

(b) the spreading resistance technique is a comparison technique and unknown specimens have to be evaluated in terms of calibration curves produced on specimens of known resistivity and surface finish,

(c) the voltage across the contacts has to be small enough to avoid significant amounts of carrier injection, contact heating or related instabilities.

Cox and Strack (ref. 65) have used a semi-empirical method to express the spreading resistance of a semiconducting layer of thickness d as a function of a contact resistance, R_c, a spreading resistance term corrected for finite thickness, $R_S'=(\rho/4a)f(d/a)$, and a residual resistance R_o, of the form,

$$R_S = \frac{R_c}{\Pi a^2} + \frac{\rho}{4a}\cdot f\left(\frac{d}{a}\right) + R_o \tag{7.36}$$

By plotting the total resistance as a function of 1/a they were able to separate the three different components and from electrolytic tank experiments they concluded that the function f obeys the empirical relation

$$f(d/a) = (2/\Pi)\ \arctan(2d/a) \tag{7.37}$$

which has the proper values for the extremes of $d\to 0$ and $d\to\infty$.

Spreading resistance measurements for the evaluation of resistivity gradients can be made with an automatic probing system coupled to a computer for data acquisition and analysis. For accurate and reproducible measurements (ref. 62) the probe tips must be properly conditioned with an appropriate contour and chemical treatment. Mazur and Dickey (ref. 61) reported a reproducibility equal to or less than 15% for silicon specimens with resistivities in the range between

10^{-3} and $5x10^2$ ohm-cm. Krausse (ref. 66) obtained, within 30%, a proportional dependence between the average spreading resistance measured on silicon wafers and the resistivity measured by means of a four-point probe with a point spacing of 200μm. Gutai (ref. 67) has proposed the measurement of the relative change in the spreading resistance produced by a magnetic field applied either perpendicular or parallel to the interfacial contact. He calculated the spreading resistance in the weak magnetic field limit as a function of crystallographic orientation. From measurements made on n-type GaAs and Ge he showed that the local galvanomagnetic coefficients can be determined from such spreading magnetoresistance (SMR) measurements and that they are suitable for a simple determination of the Hall mobility because the SMR is proportional to the square of the Hall mobility. The errors caused by the finite specimen size are of the order of 1% provided that the specimen dimensions are a factor of 20 larger than the diameter of the spreading resistance contact. If a resistivity profile is required then the specimen must be angle lapped. In most cases the complete profile can only be obtained by spreading resistance measurements particularly for submicron-thick epitaxial layers such as required for many microwave devices. The utility of spreading resistance measurements to outline the interface between an epitaxial layer and its substrate makes it a powerful tool for investigating outdiffusion and autodoping, as shown by Morris (ref. 68). Outdiffusion is defined as the change in the resistivity profile resulting from solid-state diffusion processes. Autodoping is the change in resistivity profiles due to impurities which are evaporated from the substrate and reintroduced into the solid during epitaxial growth. The initial epitaxial profile can also be altered by heat treatment and annealing associated with oxidation and diffusion during device processing steps. Morris (ref. 68) suggests that the spreading resistance technique can provide an overall accuracy of ±10% for $n\text{-}n^+$ and $p\text{-}p^+$ junction profiles and ±20% for p-n junction profiles.

REFERENCES

1 Am. Soc. Test. Mat., ASTM F81-70 (1971) part 8.
2 J. Oroshnik and A. Many, J. Electrochem. Soc., 106(1959)3660.
3 J. Oroshnik and A. Many, Solid-State Electron., 1(1960)46.
4 I. A. Baev and E. G. Valyashko, Sov. Phys. Solid State, 6(1964)1357.
5 I. A. Baev and E. G. Valyashko, Sov. Phys. Solid State, 7(1966)2093.
6 D. L. Blackburn, H. A. Schafft and L. J. Swartzendruber, J. Electrochem. Soc., 119(1972)1774.
7 J. F. Black, E. Lanning and S. Perkovitz, Infrared Phys., 10(1970)125.
8 T. S. Moss, Optical Properties of Semiconductors, Acad. Press (1959)22.
9 K. Okada and T. Oku, Japan. J. Appl. Phys., 6(1967)276.
10 W. G. Spitzer and J. M. Whelan, Phys. Rev., 114(1959)59.
11 M. Cardona, Solid-State Phys., Suppl. 11(1969)165.
12 D. E. Aspness, Phys. Rev. Lett., 28(1972)913.
13 R. Sittig and W. Zimmermann, Phys. Stat. Sol.(a) 12(1972)663.
14 R. Glosser, Crit. Rev. Solid-State Sci., 5(1975)287.
15 W. Ehrenberg, Chi-Shi Lang and R. West, Proc. Phys. Soc., 64(1951)424.

16 P. Rappaport, Phys. Rev., 93(1954)246.
17 C. Munakata, Japan, J. Appl. Phys., 4(1965)697.
18 C. Munakata, Japan. J. Appl. Phys., 4(1965)815.
19 P. R. Thornton, K. A. Hughes, D. V. Sulway and R. C. Waite, Microelectron. and Reliab., 5(1966)291.
20 C. Munakata, Japan. J. Appl. Phys., 6(1967)548.
21 C. Munakata, J. Sci. Instr., J. Phys. E, ser 2, 1(1968)639.
22 A. Gopinath, J. Phys. D: Appl. Phys., 3(1970)467.
23 C. Munakata, J. Phys. D: Appl. Phys., 5(1972)1000.
24 S. M. Sze, Physics of Semiconductor Devices, Wiley Brothers, New York (1969) 108.
25 C. Munakata, Japan. J. Appl. Phys., 11(1972)869.
26 C. Munakata, Japan, J. Appl. Phys., 11(1972)1333.
27 J. D. Kamm, Solid-State Electron., 19(1976)921.
28 J. D. Kamm and R. Mueller, Solid-State Electron., 20(1977)105.
29 M. A. Logan, Bell Syst. Tech. J., 40(1961)885.
30 D. S. Perloff, J. Electrochem. Soc., 123(1976)1745.
31 L. J. Swartzendruber, Solid-State Electron., 7(1964)413.
32 D. S. Perloff, F. E. Wahl and J. T. Kerr, Proc. Electron and Ion Beam Science and Tech., 7th Internatl. Conf., Electrochem. Soc. (1976)464.
33 D. S. Perloff, F. E. Wahl and J. Conragan, J. Electrochem. Soc., 124(1977) 582.
34 D. S. Perloff, F. E. Wahl and J. D. Reimer, Solid State Tech., 20(1977)31.
35 P. A. Crossley and W. E. Ham, J. Electron. Mat., 2(1973)465.
36 M. G. Buehler and W. Thurber, IEEE Trans. Electron. Dev., ED-23(1976)968.
37 A. Uhlir, Jr., Bell Syst. Tech. J., 34(1955)105.
38 M. G. Buehler and D. E. Sawyer, Circuits Manuf., 17(1977)46.
39 W. R. Thurber and M. G. Buehler, Nat'l Bureau Stds. Special Publ. 400-36 "Microelectronic Test Pateern NBS-4" (1978).
40 H. A. Schwarz, J. Angew. Math., 70(1869)105.
41 E. B. Christoffel, Ann. Mat. Pura Appl., 1(1867)95.
42 H. J. Lippmann and F. Kuhrt, Z. Naturforsch., 13a(1958)462.
43 J. Hauesler, Z. Naturforsch., 17a(1962)506.
44 J. Hauesler, Solid-State Electron., 9(1966)417.
45 J. Hauesler, Archiv. Electrotech., 1(1968)11.
46 S. Oberlander and W. E. Wilhelm, Phys. Stat. Sol., 12(1965)569.
47 G. De Mey, Solid-State Electron., 16(1973)955.
48 S. Jandl, K. D. Usadel and G. Fischer, Rev. Sci. Instr., 45(1974)1479.
49 D. Midgely, Ind. Electron., 1(1963)383.
50 C. B. Burckhardt and M. J. O. Stratt, Z. Naturforsch., 18a(1963)44.
51 M. Jevons, Int. J. Eng'ng Ed., 4(1966)569.
52 M. Jevons, J. Sci. Instr., 44(1967)514.
53 J. P. Newsome, Proc. IEE, 110(1963)653.
54 J. P. Newsome, Solid-State Electron., 10(1967)183.
55 G. E. Forsythe and W. R. Wason, Finite-difference method for partial differential equations, John Wiley and Sons, New York (1960).
56 T. Mimizuka, Solid-State Electron., 14(1971)107.
57 M. A. Green and M. W. Gunn, Solid-State Electron., 14(1971)1167.
58 D. J. White, M. L. Knotek and M. H. Ritchie, J. Appl. Phys., 44(1973)1870.
59 G. De Mey, Radio and Electron. Eng., 45(1975)472.
60 W. L. V. Price, Solid-State Electron., 16(1973)753.
61 R. G. Mazur and D. H. Dickey, J. Electrochem. Soc., 113(1966)255.
62 Proc. Spreading resistance symposium, NBS Spec. Publ. 400-10, J. R. Ehrstein, ed. Gaithersburg, MD (1974).
63 W. A. Keenan, P. A. Schumann, R. P. Philips and A. H. Tong, Ohmic contacts to Semiconductors, B. Schwartz, ed., Electrochem. Soc., (1969)263.
64 P. J. Severin, Solid-State Electron., 14(1971)247.
65 R. H. Cox and H. Strack, Solid-State Electron., 10(1969)1213.
66 J. Krausse, IEEE Trans. Electron. Dev., ED-23(1976)974.
67 L. Gutai, Solid-State Electron., 16(1973)395.
68 B. L. Morris, J. Electrochem. Soc., 121(1974)423.

CHAPTER 8

ELECTRICAL MEASUREMENTS ON HIGH RESISTIVITY MATERIALS

8.0 Introduction

The measurement of the electrical properties of high resistivity semiconductors and insulators is complicated by factors which are usually considered to be of second order in low resistivity materials and often can be ignored. In high resistivity materials the accuracy of measurement is affected by contact resistance, by the presence of space charges which distort their internal electric fields and by the time and history dependent space charge-limited currents. The stray capacitance of leads and cables and the large specimen resistance produce long time constants for achieving steady-state conditions during the measurements. The large potentials, of the order of 10^2 to 10^3V, used for such measurements produce significant surface leakage currents and these must be suppressed if the true bulk resistivity is to be determined. The high specimen resistance requires high impedance auxiliary apparatus and shielding is mandatory if pick-up is to be avoided. It also leads to large noise contributions which must be eliminated or reduced . This Chapter outlines the nature of these problems and provides a description of various methods intended to circumvent them.

8.1 DC methods of measurement

A potentiometer-type of apparatus for measuring the resistivity of specimens with DC resistances in the range between 10^{-1} and 10^{12} ohms has been described by Fischer *et al.*(ref. 1); it is shown schematically in Fig. 8.1. A series of dry cell batteries supply the voltage V_B which can be varied between 0 and 100 V by means of a potentiometer. The switch S_2 provides for polarity reversal of V_B. The current applied to the specimen is measured by the electrometer E_1 in terms of the potential drop which appears across the shunt resistor R_0. Switch S_3 is a low leakage-resistance teflon-insulated switch; E_2 and E_3 are identical electrometers. For resistivity measurements S_3 connects either leads 1 and 2 or leads 3 and 4 to the electrometer null-indicators. The current is first applied to the specimen and then E_2 is nulled by means of potentiometer P_2. Thereafter, with S_4 in position a, the electrometer E_2 is balanced by means of P_1; this provides a direct measurement of the potential difference which appears between the potential probes.

For Hall measurements leads 1 and 4 or 2 and 3 are connected to the null indicators while the input current and the transverse magnetic field are applied to

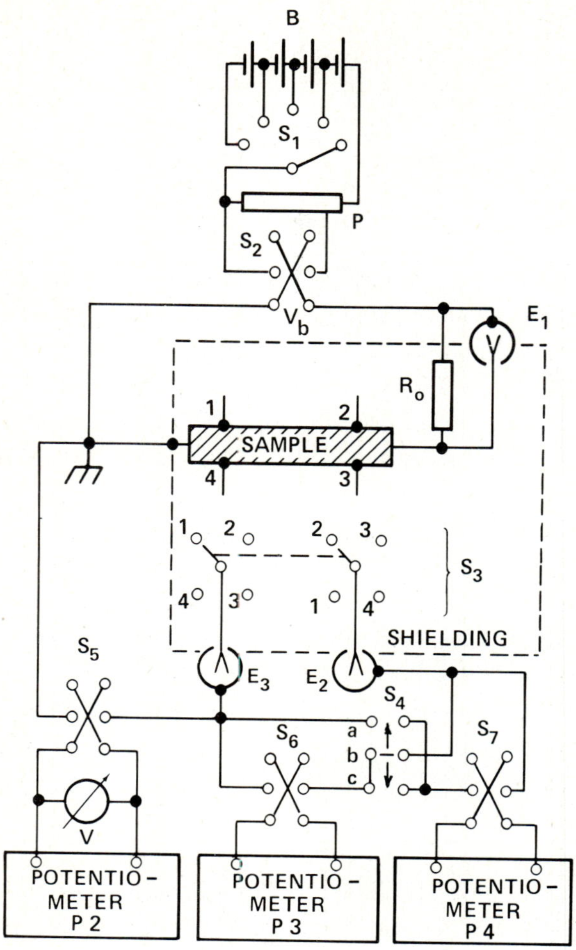

Fig. 8.1. Circuit used for DC resistivity and Hall effect measurement on high resistance rectangular specimens. [after G. Fischer, D. Greig and E. Mooser, Rev. Sci. Instr., 32(1961)842]

the specimen. Then, with S_4 in position b, E_3 is balanced by means of potentiometer P_2 and E_2 is brought to a null by means of potentiometer P_3. The magnetic field is now reversed, E_3 is nulled again by means of P_2 while P_3 is left unchanged; then S_4 is set to position C and E_2 is balanced by means of P_1. The setting of P_1 now corresponds to twice the Hall voltage. The procedures for eliminating error voltages of thermoelectric origin described in Chapter 2 need to be used here as well. Large errors may be introduced in the measurement of high resistance specimens by their contact resistances. To ascertain the magnitude of the contact resistance by means of the apparatus shown in Fig. 8.1 the voltmeter V connected across potentiometer P_2 is used to measure the potential difference between ground and one of the potential probes. By comparing this potential difference with that between the potential probes the fractional contribution of the contact resistance to the total specimen resistance can be determined. For a parallelipiped of length ℓ, width w and thickness δ the resistivity is $\rho=(v_r/i)(w\delta/\ell)$ where v_r is the potential difference measured between the potential probes. The Hall voltage v_h measured between terminals 1 and 4 or 2 and 3 leads to the Hall coefficient $R_h=(v_h/i)(\delta/B)\times10^8(cm^2/V\text{-sec})$. If the charge carriers are assumed to be in a single isotropic spherical conduction band then in accordance with section 2.1 the mobility is

$$\mu = \left(\frac{v_h}{v_r}\right)\cdot\left(\frac{\ell}{w}\right)\left(\frac{10^8}{B}\right), \ cm^2/V\text{-sec.} \qquad (8.1)$$

Suppose that the minimum Hall voltage which can be detected above the noise is $\sim$10 times the thermal noise voltage per unit bandwidth, then

$$\left(v_h\right)_{min} = 20(RkT)^{1/2} \qquad (8.2)$$

where R is the total resistance (including the contact resistance) of the specimen, k is Boltzmann's constant and T the absolute temperature. Suppose that the maximum voltage between the potential probes is

$$\left(v_r\right)_{max} = cV_B \qquad (8.3)$$

The coefficient c is the ratio of the resistance between the potential contacts to the total resistance of the specimen. Its value is usually in the range between 10^{-4} and 1.0. V_B is limited by the Joule heat, $Q=V_B^2/R$ dissipated in the specimen. Therefore the lowest charge carrier mobility, in terms of Eq. (8.3) is

$$\mu_{min} = \frac{2}{c}\left(\frac{kT}{Q}\right)^{1/2}\cdot\left(\frac{\ell}{w}\right)\cdot\left(\frac{10^9}{B}\right), \ cm^2/V\text{-sec.} \qquad (8.4)$$

Fischer *et al.* (ref. 1) found that the minimum measurable Hall mobility is limited primarily by the sensitivity of the electrometer. However, for a heat dissipation of $\sim10^{-3}$W and a specimen resistance $R>10^7$ ohms the maximum potential measured between the potential probes is limited by the supply voltage, $V_B\leq10^2$V and not by heat dissipation. Assuming that $(v_h)_{min}$=2mV and $(v_r)_{max}$=100c, $\ell/w=5$ for a magnetic field of 1 T then $\mu_{min}\geq1$ cm^2/V-sec.

A balanced circuit employing electrometers was described by Heilmeier and Harrison (ref. 2). The Hall voltage is directly coupled to the grids of the electrometer tubes, as shown in Fig. 8.2. Since the capacitively-coupled drift and noise are fed in-phase to the differentially-connected electrometers their

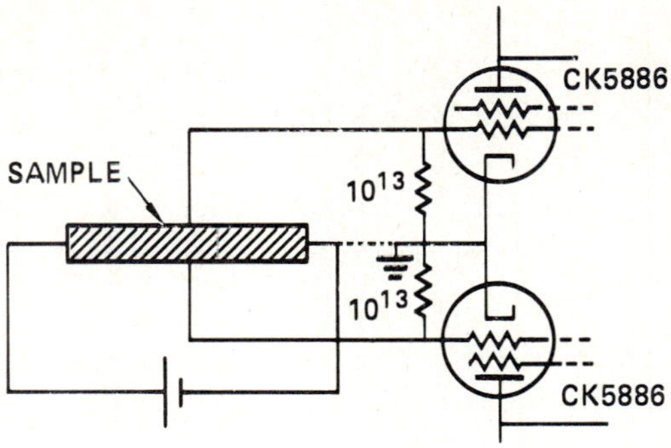

Fig. 8.2. Schematic circuit of differential electrometer circuit for Hall effect measurements. [after G. H. Heilmeier and S. E. Harrison, Phys. Rev., 132(1963) 2010]

contributions are effectively cancelled. A further advantage of this circuit is that the current source to the specimen need not be floated above ground provided that the input resistance of the circuit is much greater than that of the sample resistance. A disadvantage of the circuit shown in Fig. 8.2 is that half of the potential applied to the specimen appears across the grids of the electrometer tubes. This sets a limit on the minimum detectable mobility which, in this instance, is $\sim 9 cm^2/V$-sec.

A servo-controlled potentiometric apparatus considered to be an improvement over that described by Fischer *et al.* (ref. 1) by virtue of greater accuracy, a range extension by one order of magnitude and manipulatory simplicity was described by Fermor and Kjekshus (ref. 3). The current input to the specimen is automatically controlled by a servo loop coupled to an electrometer. An additional servo loop is used to balance the error signal from an electrometer coupled to a potentiometer in the Hall voltage output circuit. Specimen resistances up to 10^{13} ohms can be measured with a current of 10^{-12}A. In order to detect any electrical asymmetry of the specimen all the polarities of the circuit can be reversed by a remotely-operated low thermal-emf switch and another switch which has a low leakage and is adequately shielded allows the electrometer inputs to be transferred between the potential and Hall contacts. A requirement imposed on the circuit is that the electrometers be brought to a null for a maximum initial out-of-balance condition without producing undesirable oscillation in the servo system or overloading the electrometers. This requirement is met by the inclusion of a substantial amount of negative feedback within the electrometer over its operating range. This mitigates variations in input impedance and drift of the electrometer tubes. Furthermore, the output of the electrometer tubes is limited in order to protect the servo system. The electrometers are incorporated in DC-coupled amplifiers and compensation techniques are utilized to minimize zero drift. The main sources of experimental error are:

(a) current injection by the electrometers, this effect is appreciable only for high specimen resistances. A typical error calculation for a specimen resistance of 10^{13} ohms and a contact resistance of 10^{11} ohms leads to error values of 0.5% in specimen current and 0.25% in specimen voltage determinations;

(b) servo loop zero errors, these are usually less than 1 mV and may be reduced or eliminated by conventional screening techniques;

(c) errors produced by the electrometer, this contributes an additional maximum error of $\sim$1 mV for a specimen measured in the voltage range between 1 and 10 V;

(d) calibration errors introduced by the temperature dependence and aging of the standard resistors; the calibration error can be kept to less than 1% by frequent recalibration of the instrument at its operating temperature.

A maximum error of 2% in the determination of resistance is typical; under favorable conditions this can be reduced to 0.5%.

The use of shielded cables connecting the terminals of a high resistance specimen to auxiliary apparatus can lead to serious problems (ref. 4). Two to 3 m of shielded cable connected to each of the terminals of a Hall plate can produce several thousand pF of capacitance and for a specimen with a resistance of 10^8 ohm-cm the time required for the system to reach equilibrium during measurements exceeds 1 min. An improvement in the speed of response can be obtained by grounding one terminal of the Hall specimen and providing unity gain high input impedance amplifiers for each of the other five terminals. Each amplifier drives the shield of the cable connecting the specimen to auxiliary instrumentation; this reduces leakage currents and by eliminating the shunting capacitance of the cables reduces the effective time constant. A voltmeter with a floating ground can be used to measure the potential difference between any two Hall terminals. Since the output impedance of the amplifier is low no special high impedance requirements are made of the voltmeter. Resistivities as high as 10^{10} ohm-cm corresponding to a sample resistance of 10^{12} ohms can be measured with time constants not exceeding 1 min. Figure 8.3 shows such a three-probe system used for routine Hall measurements (ref. 5). Baleshta and Keys (ref. 6) have described van der Pauw-type resistivity and Hall measurements made on high resistivity semiconductors using a single electrometer with a high input impedance and low drift capable of resolving voltages down to 50μV with an accuracy of $\pm$10%. A precision shunt resistance across the electrometer input is used to monitor the current through the specimen. In order to prevent current limiting this shunt should always be selected to be 2 to 3 orders of magnitude smaller than the

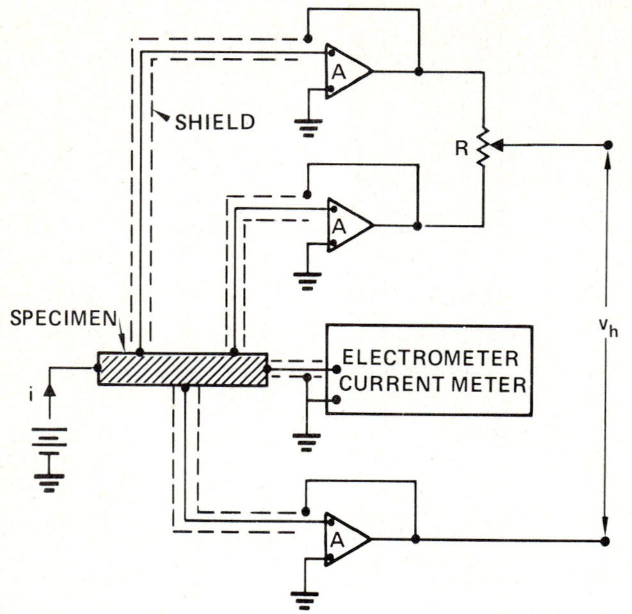

Fig. 8.3. Guarded amplifiers A used to reduce time constant and provide impedance transformation; for Hall effect measurements potentiometer R is balanced in B=0 before v_h is measured.

resistance of the sample. If the sample current is limited due to the resistance of the electrometer shunt, it will not be so limited when used for measuring the potential difference because the current source is now connected directly to the specimen and the electrometer is not part of the current loop. The circuit time constant will affect the length of time to make voltage measurements. For a cable capacity (shunting the electrometer) of 200 pF and a 10^{12} ohm source impedance, steady state will be reached after $\sim$6 time constants; this corresponds to $\sim$20 min. Johnson noise produced by the resistance of the sample does not affect the voltage resolution of the electrometer to a significant degree. In Hall voltage measurements the voltage resolution is limited by electrometer noise of the order of $\pm 10\mu V$; it is larger than the Johnson noise which is estimated to be $\sim 5\mu V$.

The advantages of using high input impedance guarded unity gain amplifiers interposed between a specimen and auxiliary apparatus for van der Pauw-type resistivity and Hall measurements were described by Hemenger (ref. 7). However, commutation of the current and voltage leads is a fundamental requirement and it is this requirement which increases the complexity of the measuring system. Hemenger described the construction of a four unity-gain amplifier system switched to an appropriate current monitor or digital differential voltmeter by means of a well-insulated multi-section rotary switch on the low impedance side of the unity-gain amplifiers. An electrometer is used to monitor the current. Guarded high-impedance reed relays are used to perform the input switching functions. If a dewar enclosure is employed for making measurements on a specimen at cryogenic temperatures then a triaxial cable is used between the dewar and the unity-gain amplifier with the outer shield common to the dewar and circuit ground and the inner or guard-shield run continuously into the dewar with coaxial feedthroughs. Resistance data can be obtained to 10^{13} ohms and Hall measurements can be made on specimens with resistances up to 10^{12} ohms.

In high resistivity materials surface charge carrier transport can represent an appreciable portion of the electrical conductance of a specimen. In some cases surface conductance may be the dominant charge transport mechanism. It is, therefore, desirable to identify the specific surface resistivity contribution to the total resistivity measured on a specimen. A guarded electrode, such as shown in Fig. 8.4.(a) can be used for such purposes (ref. 8). The bulk resistivity is $\rho_v = R_v S/\delta$ where S is the area of the center electrode corrected for edge effects, δ is the thickness and R_v is the volume resistance. The surface resistivity is $\rho_s = R_g p/g$ where R_g is the surface resistance measured between electrodes A and B in Fig. 8.4.(a), p and g are, respectively, the average perimeter and width of the gap between these electrodes. A common method of measurement is to use a single power supply and two sensitive current detectors. However, if the measured resistances differ by several orders of magnitude then the loading effect of the detectors is no longer negligible and guarding is ineffective. If the detector is an electrometer-pico-ammeter then feedback can be applied to the load resistor in order to cancel the voltage drop. An improvement is feasible by placing load resistors successively in series with R_v and R_g. A disadvantage of this method is that two voltage drop measurements with two different circuits are needed to obtain a single value of resistance. A better method is the use of a bridge circuit such as shown in Fig. 8.4.(b). The voltmeter M_2 and the bridge null detector M_1 should, preferably, be electrometers with a high input impedance. At equilibrium, the potential measured by M_1 is zero and

$$R_v = R_1(V_o/V_2 - 1) \qquad R_g = R_2(V_o/V_2 - 1) \tag{8.6}$$

where V_2 is the potential difference measured across R_2. The accuracy of the measurement (ref. 8) is

$$\frac{\Delta R_v}{R_v} \simeq \frac{\Delta R_g}{R_g} \simeq \frac{\Delta R_1}{R_1} + \frac{\Delta V_2}{V_2} + \frac{\Delta E}{V_o}\left(\frac{R_v}{R_1} + \frac{R_v + R_g}{R_B}\right) \tag{8.7}$$

where R_b is an edge resistance in shunt with the series resistance (R_v+R_g) and ΔE is the bridge unbalance voltage. In general $\Delta R_1/R_1$ is less than 1%, $\Delta V_2/V_2$ is of the order of 1 to 2%. If the electrometer is used as a unity-gain amplifier

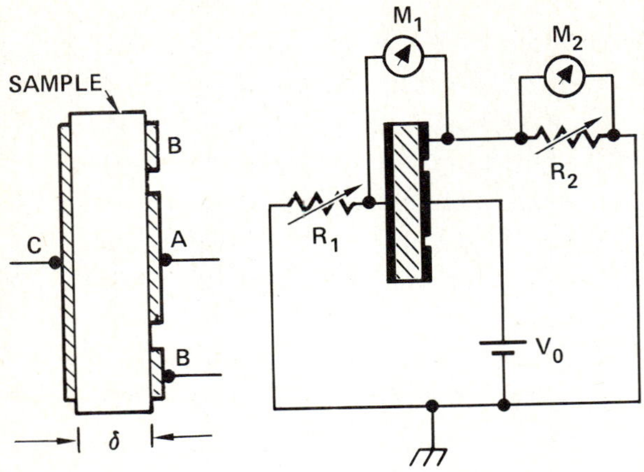

(a) (b)

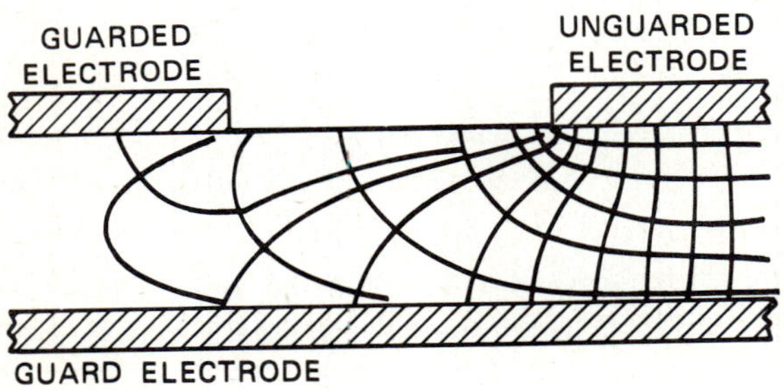

(c)

Fig. 8.4. Guard electrodes used to determine surface and bulk resistivities: (a) center electrode A is guarded, B is guard electrode, and C is common electrode; (b) Wheatstone bridge circuit for measuring surface and bulk resistivities; M_1 and M_2 are high impedance electrometers, R_1 and R_2 are variable resistors; (c) flux lines and equipotential distribution in three-terminal guarded specimen.

and connected to a 0.01% digital voltmeter then the accuracy of measurement is better than 0.025%. V_o must not be too large if carrier injection and sample heating is to be avoided. The factor $\Delta E/V_o$ should be made as small as possible; reduction of ΔE is prevented by internal noise in the resistances, by microphonic mechanical modulation or induction of stray signals. These can be reduced by adequate shielding and by rigid mounting of the components. Practice has shown that an accuracy of the order of 5% can be obtained with this bridge circuit although resistances larger than 1 T-ohm can only be measured with difficulty. If a specimen has a polycrystalline structure made up of large crystallites then the central electrode A can be placed on a single crystal area, so that one can measure bulk resistivities of a single crystal. With a common peripheral electrode B and a common base electrode C, several central electrodes A can be placed upon the same slab to test the homogeneity of the material or to compare single crystal and polycrystalline behavior. For thin layers or films which have a surface conductance much smaller or much greater than that of the bulk material the data obtained from such measurements should be viewed with caution. A high surface conductance may be due to surface accumulation of carriers and this may lower the apparent value of R_v. If the volume conductance is very low then the surface resistance R_g derived from the measurements may contain large errors and the volume resistance may be underestimated.

A four-electrode guarded configuration which provides some advantages over three-electrode guarded structures has been described by Kao (ref. 9). It is particularly useful for the measurement of the surface resistance of dielectric materials. The purpose of the four-electrode guarded structure is:

(a) to divert indirect surface currents flowing across a specimen surface to the area contained between its electrodes;
(b) to divert indirect currents flowing through the specimen bulk; and
(c) to avoid fringing effects caused by the electrode edges and thus to provide a uniform surface electric field.

In a three-electrode guarded structure the resistance of the volume path of the current is a function of the specimen thickness and of the ratio of the electrode separation to specimen thickness (ref. 10). In the four-electrode structure each electrode has a specific function. The first guard electrode in Fig. 8.5.(a) diverts the indirect currents flowing across the remaining surface of the specimen. The second guard electrode not only serves to divert the indirect currents flowing through the bulk, but also to eliminate the fringing effects due to the electrode edges. The guard and the guarded electrodes are at the same potential; therefore, their separation can be made as small as necessary. It is essential to keep the potential at the guard electrodes equal to that at the guarded one in order to avoid any leakage current between them . This can be done by connecting a compensating resistor in series with the guard electrodes, whose value should be equal to that of the metering circuit connected in series with the guarded electrode. The flux lines and equipotentials of such a guarded structure (ref. 11) are shown in Fig. 8.5. Tests performed by Kao on guarded four-electrode and three-electrode structures indicate that the surface resistance of the four-electrode structure is practically independent of the specimen thickness. The surface resistance increases in direct proportion to increasing electrode spacing. For thin specimens the surface resistances of the four-electrode

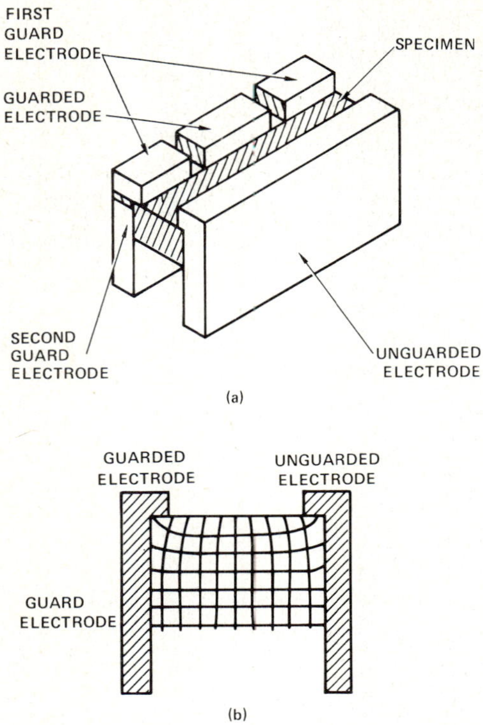

Fig. 8.5. Four-electrode guarded specimen: (a) structure and arrangement of electrodes shown schematically; (b) flux lines and equipotentials. [after K. C. Kao, J. Sci. Instr., 39(1962)208]

system and the three-electrode system are of the same order, but for thick specimens there is marked disagreement. The errors using the three-electrode system may be caused by the shunt resistance through the bulk of the material, and by the fringing effect of the sharp edges of the electrodes.

8.2 AC measurement techniques

A low frequency chopped-DC potentiometric apparatus for measuring resistances up to 10^8 ohms with an error not exceeding 1% has been described by Edwards (ref. 12). In general, the accuracy of low frequency AC Hall effect and resistivity measurements are limited by 1/f noise. A method for circumventing these problems has been described by Pell and Sproul (ref. 13). The specimen is driven with AC and provision is made for automatic polarity reversal of the magnetic field. The Hall voltage is measured by means of a tuned narrow band amplifier and with the magnetic field reversed periodically the Hall voltage is recorded as a symmetric square wave about a quiescent point which corresponds to B=0. Gobrecht *et al.* (ref. 14) have extended the utility of such AC Hall measurements to specimens with resistances up to $2x10^7$ ohms. An AC generator which provides the specimen current drives an amplifier which also contains a compensator used for the elimination of the specimen misalignment potential. The AC Hall output voltage is applied to a preamplifier which also serves as impedance transformer. After further amplification the Hall output voltage is detected synchronously by an amplifier which provides the proper phase compensation with respect to the specimen current. In this manner (ref. 14) they obtained an accuracy of ±12% in their Hall measurements for specimens with resistances of the order of (or less than) 10^6 ohms and an accuracy of ±20% for specimens with resistances $>10^6$ ohms with a reproducibility of ±1% or better.

An AC technique employing frequencies high enough that 1/f noise would appear negligible and a Hall voltage detection bandwidth narrow enough to provide a high signal to noise ratio is considered desirable for performing Hall effect measurements on high resistivity materials. Double AC modulation Hall measurements in which both the magnetic field and specimen current are time variable at different frequencies are limited by the relatively low frequencies and low magnetic induction which can be obtained from an electromagnet without overloading it and without having to allow for magnetic hysteresis of its iron core. An AC apparatus in which the specimen is rotated in a static magnetic field has been described by Ryan (ref. 15). In this manner one can simulate an alternating magnetic field without the penalties imposed by the impedance and time constants of the magnet and its power supply. A similar improved apparatus adapted to the measurement of low mobility and high resistivity semiconductors was described by Hermann and Ham (ref. 16). It is shown schematically in Fig. 8.6. The specimen is rotated at 20 Hz in a static magnetic field by a synchronous motor. The current to the specimen is a 13.33 Hz signal and the magnetic field and specimen current are in phase with the line frequency. A phase-sensitive synchronous amplifier is used to detect the Hall voltage at 33.33 Hz. Its bandwidth is reduced to 0.05 Hz and the misalignment voltage which appears at 13.33 Hz is rejected. Noise inside this bandwidth is a limiting factor in the resolution of the equipment. In practice the observed noise voltage from all sources is a factor of three above the Johnson noise. For semiconductors whose mobility is in the range of 10 cm^3/V-sec. or larger, the sign of the carriers can be determined by nulling the misalignment voltage with a bucking signal and adjusting the position of the sample in the magnetic field. Hall mobilities of the order of 0.2 cm^2/V-sec. can be detected in this manner in specimens whose resistances are smaller than 10^7 ohms.

A double modulation method employing a rotating specimen in a static magnetic field and a modulated current produced by a chopped light beam which modulates the density of photoelectrically-induced carriers in the specimen was also described by Eisele and Kevan (ref. 17). It is intended for Hall measurements at cryogenic temperatures made on high resistivity γ-irradiated KCl specimens with

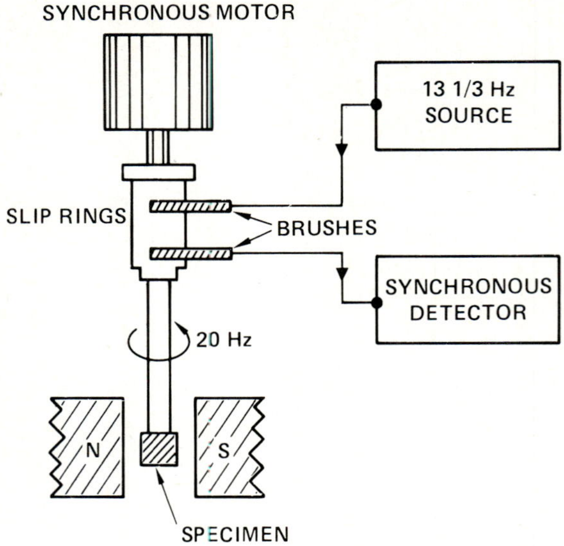

Fig. 8.6. Block diagram of apparatus for AC Hall effect measurements using static magnetic field, rotating specimens and AC current. [after A. M. Hermann and J. S. Ham, Rev. Sci. Instr., 36(1965)1553]

resistances between 10^7 and 10^9 ohms. Figure 8.7 is a block diagram of the apparatus. The charge carrier density in the specimen is modulated by a chopped light beam directed through an aperture in the center of the magnet pole while a DC electric field is applied to it along a direction perpendicular to the magnetic field. The chopper which modulates the light beam is outside the magnet and is phase- and frequency-locked to a stable oscillator driven between 5 Hz and 10^3 Hz. The shaft slip ring assembly and generator are driven by a synchronous motor at a speed which is synchronous with the 60 Hz line frequency. The output voltage which appears at the input of the preamplifier consists of:

(1) Hall voltages at the sum and difference frequencies $f_1 \pm f_2$.

(2) An error signal due to the misalignment voltage at f_1.

(3) A signal due to inductive coupling between the input circuit and magnetic field at f_2.

(4) Intermodulation products of items (2) amd (3) generated at non-ohmic contacts of the specimen at the frequencies $f_1 \pm f_2$ (since these are derived from a voltage proportional to the time derivative of the magnetic field there is a phase difference of $\pi/2$ between them and the Hall voltage).

(5) Thermally-generated noise from the specimen, its contacts and leads.

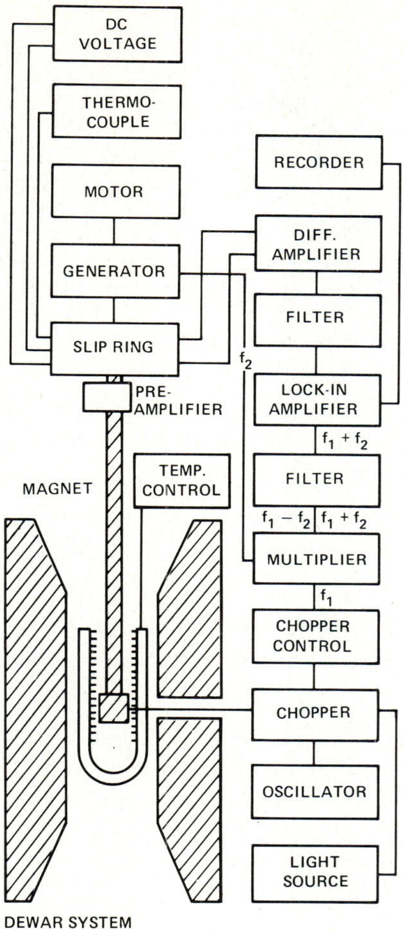

Fig. 8.7. Block diagram of apparatus used to measure Hall mobility of ^{60}Co-irradiated KCl using photoelectric excitation. [after I. Eisele and L. Kevan, Rev. Sci. Instr., 43(1972)189]

(6) Magnetoresistances, Ettingshausen, Nernst, and Righi-Leduc galvano-thermomagnetic effects.

(7) Induced 60 Hz line voltages.

In order to minimize the error voltages described in items (1) to (5) above alignment of the Hall probes should be made with care and the frequency which corresponds to the misalignment voltage should be kept outside the bandwidth chosen for the detector. The Hall voltage leads should be physically positioned to minimize inductive pickup and the frequency f_2 should be as small as possible. Typically f_1 may be between 70 and 110 Hz, chosen to avoid space charges in the specimen and the rotation frequency f_2 may be chosen to be 3.3 Hz.

8.3 Electrical measurements on insulators

The understanding of charge transport in insulators has lagged behind that of metals and semiconductors primarily because the technological importance of the latter has led to more extensive and intensive theoretical as well as experimental investigations. In recent years there has been a surge of interest in the study of insulators as well as in amorphous semiconductors. However, conventional methods of measurement and interpretation of such measurements are not necessarily applicable to insulators. There are several reasons why these insulators need to be considered apart from semiconductors. One of these is the low equilibrium thermal concentration of charge carriers. To make insulators conduct one must generate carriers by external means either by photon injection (photoconductivity), electron injection (such as that produced in an electron microscope) or by electrical injection from suitable electrodes. These methods provide charge carriers near the surface so that their density throughout a specimen is not necessarily uniform. The charge carrier distribution in the presence of an electric field is affected by the space charge of previously introduced carriers and if enough of them are supplied then the current becomes space charge limited (ref. 18). Under these conditions Ohm's law is no longer applicable over the whole specimen since it is valid only if the conductivity is uniform and this requires a uniform carrier distribution.

If a steady state SCL current is passing through a solid then the effect of charge carrier trapping can be represented by some functional relation between the free carrier concentration n_f and the total carrier concentration n. In general, for charge carrier transport along the x-coordinate with

$$n_f = C \exp(E_F/kT) \tag{8.8}$$

where the coefficient C is independent of x and $n_f E_x = J_x (e\mu_x)^{-1}$ where E_x is the electric field and J_x the current density component along the x-axis

$$\exp(E_F/kT) = J_x (Ce\, \mu_x\, E_x)^{-1} \tag{8.9}$$

In the absence of trapping n_f=n and

$$(\partial^2/\partial x^2)E_x{}^2 = 0 \tag{8.10}$$

which has the solution

$$E_x(x) = (\alpha^2 x + \beta)^{1/2} \tag{8.11}$$

where the constants of integration α and β are determined by the boundary condition. If the injecting contact at x=0 can inject an arbitrary number of carriers then $E_x(0)\simeq 0$ and therefore β=0 in Eq. (8.11). The constant α may be determined from the applied voltage along the x-direction between the contacts of the specimen of length ℓ

$$V = \int_0^{\ell} E_x(x)dx \qquad (8.12)$$

with the result that

$$\alpha = 3V/2\ell^{3/2} \qquad (8.13)$$

Therefore, since $\partial E_x/\partial x = en/\kappa_x\varepsilon_0$ where κ_x is the dielectric constant then

$$n_f(x) = (3\kappa_x\varepsilon_0 V/4e\ell^{3/2}).x^{-1/2} \qquad (8.14)$$

which leads to the current voltage relation for SCL currents known as Child's law (ref. 19)

$$J_x(x) = 9\kappa_x\varepsilon_0\mu_x V^2/8\ell^2 \qquad (8.15)$$

It is to be expected that in the absence of trapping Ohmic behavior occurs only at negligible carrier concentrations and insulators change from Ohmic conductance to SCL conductance once the space charge of injected carriers starts to affect the distribution of charges through a specimen. For shallow traps the current density still depends on the square of the applied voltage and the inverse cube of the electrode spacing as shown by Lampert (ref. 20). If an exponential trap distribution is present then most of the space charge is trapped and the current density has a superlinear dependence on V, as shown by Mark and Helfrich (ref. 21). There are several possible variations of space charge limited flow depending on whether the current is due to electrons only or to electrons and holes and whether traps and recombination centers are present. Under double injection, i.e., with electrons injected from the cathode and holes injected from the anode, space-charge limited conditions are at least partially overcome and for intermediate injection levels charge neutrality can be assumed through the insulator. However at high and low injection levels the space charge becomes important again. Other conduction processes in insulators which include internal field emission, Schottky emission and the Poole-Frenkel effect have been described by Lamb (ref. 22).

Hall effect measurements on photoconducting insulators present some difficulties not encountered in metals or semiconductors. The existence of a space charge may distort the applied electric field or reduce it to a negligible value in the region of the Hall electrodes. The effects of the space charge may be minimized by using an alternating electric field, as shown by Olson and Wertz (ref. 23). The input shunt capacitance in the Hall circuit must be reduced as much as possible in order to reduce the loading of the amplifier used to detect the Hall voltage. If, for example, the input shunt capacitance is $\sim$20 pF (a typical value for $\sim$30 cm of shielded cable) then the capacitive reactance at 10^3 Hz is $\sim 10^7$ ohms. This would produce excessive loading of a specimen having an impedance $>10^6$ ohms. A reduction in input capacitance by a factor of 10^3 increases the effective impedance of the Hall detector by the same factor. The requirement of low shunt capacitance and the necessity for shielding the leads are not mutually exclusive if the shield is driven from a low impedance source with a voltage of the same amplitude and phase as that of the input signal (ref. 24). There will then be no current flowing between the input lead and the shield and the capacitance will not degrade the input impedance. The driven shield may be surrounded by a second shield which is grounded; alternatively, it may be driven from an even lower impedance source. Olson and Wertz (ref. 23) employed a high impedance transformer to supply the specimen current thereby isolating the specimen from ground. The Hall voltage detector is similar to that of MacDonald (ref. 24) but with an additional source-follower stage in order to allow a low impedance differential amplifier to convert the Hall voltage to a single-ended signal. The input capacitance measured at the end of a 40 cm-long triaxial cable was found to be reduced to less than 0.2 pF at 10^3 Hz. Hall measurements were made on metal-doped alkaline earth oxides in which charge carriers were induced by strong illumination in order to produce an effective sample resistance of $\sim 10^9$ ohms. Mobilities of the order of 0.05 cm^2/V-sec. were measured in a magnetic field of 1.3T with an AC applied voltage of $\sim$50 V(rms).

An ingenious experimental measurement of the charge carrier mobility of amorphous semiconductors based on Corbino magnetoresistance was described by Carver and Allgaier (ref. 25) and by Carver (ref. 26). A description of the galvanomagnetic properties of a Corbino disc is provided in Chapter 3. It is based on the short-circuit of the Hall voltage by the closed coaxial electrode structure; a radial current i_r flows between the center electrode and the peripheral electrode of the specimen. In the presence of a transverse magnetic field B a circumferential Hall current i_θ appears given by

$$i_\theta = \left(\frac{\mu B i_r}{2\Pi}\right) \ell n(r_2/r_1) \tag{8.16}$$

where r_1 and r_2 are, respectively, the inner and outer radii of the sample contacts. The Corbino disc geometry is analogous to an infinitely wide rectangular sample. In both cases the Hall voltage is short-circuited and the Hall current is rotated through an angle θ with respect to the radial current. If the radial current is AC then the Hall current may be detected inductively by a pickup coil placed near the sample. The Hall current in the sample and the pickup coil become the primary and secondary winding of a transformer circuit. Assuming the pickup coil to be a single turn of radius a at a distance b from a sample of thickness δ and the conditions

$$r_1 << r_2, \qquad b << a, \qquad \delta << r_2 \tag{8.17}$$

the induced voltage (refs. 27, 28) is given by

$$V(a) = i\omega\delta \int_{r_1}^{r_2} M(r,a) \cdot J_\theta(r) dr \tag{8.18}$$

where ω is the frequency of the sample current and $M(r,a)$ is the mutual inductance between the current density element $J_\theta(r)$ and the pickup coil. Evaluation of Eq. (8.18) is feasible for particular frequency ranges (ref. 29). In the low frequency region

$$V(a) = i\omega\mu i_r B \int_{r_1}^{r_2} \frac{M(r,a)dr}{2\pi r} \tag{8.19}$$

While it is possible to evaluate Eq. (8.19) numerically for an accurately known geometry it is more expedient to calibrate the apparatus in terms of a specimen whose mobility is known accurately.

The apparatus used for the Corbino magnetoresistance measurements is shown in Fig. 8.8. It is intended for placement within the ∿3.2 cm air gap of an iron core electromagnet with a static magnetic field. The current electrode is so designed as to distribute the current uniformly over the specimen and to avoid contributions from any Hall currents induced in the electrode itself. Positioning of the coil on the side of the sample opposite to the center current contact serves to reduce capacitive coupling; this is further reduced by wrapping the coil in ∿25μm thick grounded aluminum foil which acts as an electrostatic shield (this is essential at the operating frequency of 10^5 Hz). Further details are shown in the block diagram of Fig. 8.9. The specimen current is supplied by an audio oscillator synchronized to the reference of a phase-sensitive detector. The current source must have good phase stability, low distortion and adequate shielding. The optimum relative phase, between the reference voltage and the and the lock-in amplifier must be determined empirically. The phase shift remains stable and the lock-in phase control is set so as to maximize the signal based on data from mutually orthogonal phase settings. Carver (ref. 26) used a silver foil of known mobility to calibrate the apparatus. He found that the induced signal is linear in frequency, specimen mobility, current and applied magnetic field. A sensitivity of 12 nV was obtained using a 1 mA specimen current in a magnetic field of 2T on a specimen with a mobility of 1 cm^2/V-sec. A range of mobilities from $2x10^4$ to 0.2 cm^2/V-sec. has been measured in a transverse magnetic field of 2T with a reproducibility better than 5% and an accuracy which is determined by calibration errors.

A different method for determining the charge carrier mobility by means of Hall current rather than Hall voltage measurements was described by Dobrovol'skii and Gritsenko (ref. 29). In rectangular specimens whose dimensions are so chosen

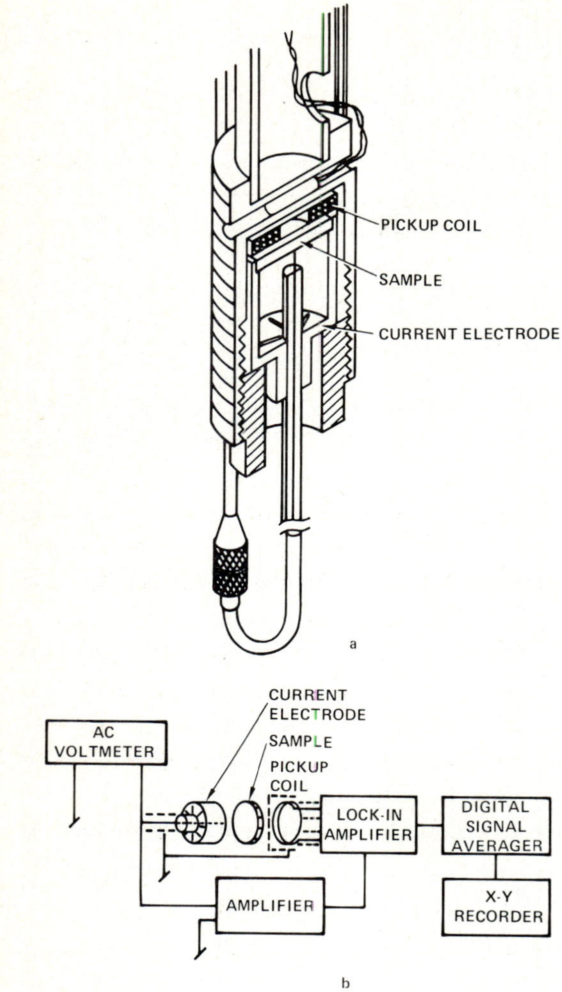

Fig. 8.8. Apparatus for measuring mobility of amorphous semiconductor specimen by means of Corbino magnetoresistance: (a) cutaway view of sample holder, copper current electrode has radial slots cut into it in order to reduce circulating currents; (b) block diagram of sensor including schematic representation of sample and current input. [after G. P. Carver, Rev. Sci. Instr., 43(1972)1257]

that the Hall field is effectively short-circuited, the magnetoresistance dependence on magnetic field is analogous to that of a Corbino disc, as shown in Chapter 3. A measurement of the Hall current can, therefore, provide the data for calculating the charge carrier mobility. The Hall current may consist of a surface component and a bulk component. The circuit shown in Fig. 8.9.(a) is intended to provide an explicit determination of the contribution of each

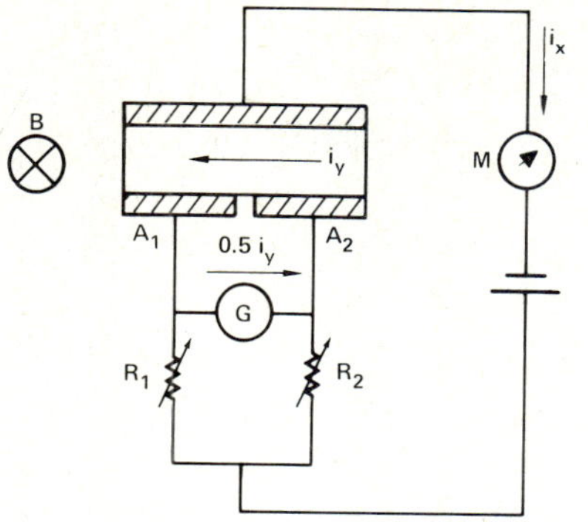

a

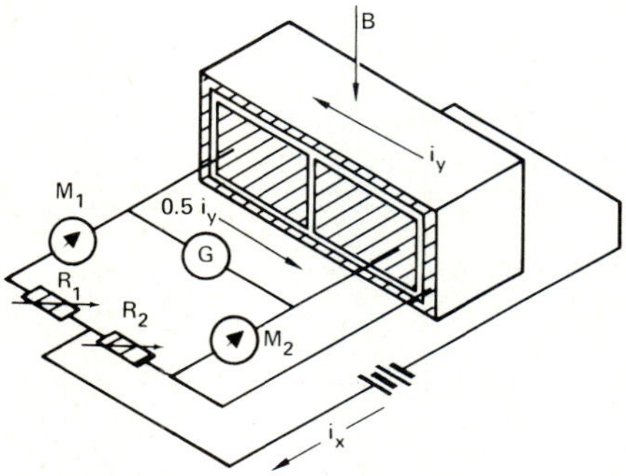

b

Fig. 8.9. Experimental arrangement for measuring surface and bulk Hall currents of high resistivity materials in the presence of a transverse magnetic field B: (a) combination of bulk and surface Hall currents is measured by differential current amplifier or galvanometer G; (b) bulk Hall current measurement by means of guarded electrodes. [after V. N. Dobrovol'skii and Yu. I. Gritsenko, Sov. Phys. Sol. State, 4(1963)2025]

component. Suppose the applied current to the specimen is along the x-coordinate in zero magnetic field; then in a transverse magnetic field a Hall current i_y is produced and the currents to each of the split electrodes is 0.5 i_y. The differential current indicator G in Fig. 8.9 must have an impedance much smaller than that of the specimen measured between electrodes A_1 and A_2 and resistances R_1 and R_2. If these are identical in B=0 then there is no potential difference between them and, therefore, no current flows through G. If this is not the case then R_1 or R_2 should be adjusted for a null. If a magnetic field is now applied then G will indicate a current which corresponds to 0.5 i_y. If a guard electrode is used, as shown in Fig. 8.9.(b) then the current i_y represents the Hall current through the bulk of the specimen. The current i_x is the sum of the currents measured by M_1 and M_2. For a correct interpretation A_1, A_2 and the guard electrode must be at the same potential and the physical separation between A_1 and A_2 and between these and the guard electrode must be much smaller than either the length ℓ or the width w of the specimen. A comparison of mobilities derived from Hall current and Hall voltage measurements made on specimens of cuprous oxide yielded good agreement (ref. 29) if surface charge carrier transport was negligible. However, if surface conduction was significant then a discrepancy appeared between the mobilities determined from Hall current and Hall voltage measurements. The latter yield values smaller by a factor of 3 compared to the former and this was attributed to surface charge carrier transport.

A comparison of Hall current vs. Hall voltage methods used to determine the charge carrier mobility of insulators was given by Mortensen *et al.*(ref. 30). He indicates that maximum sensitivity is more easily obtained by means of the Hall current technique for the following reasons:

(a) a slight asymmetry between the two portions of the split electrode (ref. 29) used to measure the Hall current does not affect the accuracy of the measurement if the gap between the electrode is small and the resistivity of the specimen is large. In the voltage method a slight asymmetry in the placing of the Hall probes produces large errors because of the high electric fields necessary for Hall voltage measurements on insulators;

(b) the current measurements require a differential current amplifier or a low impedance high sensitivity galvanometer; in contrast with this the voltage method requires a detection system with a very high input impedance because insulators have a high internal resistance;

(c) the Hall current method can be used on very thin specimens whereas the Hall voltage method may lead to problems in making or placing electrodes which do not perturb the current streamlines in their vicinity;

(d) following the application of a magnetic field to an insulator several periods comparable to the relaxation time must pass before the full Hall voltage is established, whereas the Hall current is established at once.

The dielectric relaxation time in insulators may be quite long so that the AC Hall voltage measurements can be used only over a limited frequency range; no such limitation applies to the Hall current method. Furthermore, such measurements can be made without having to consider the limitations imposed by charge carrier trapping; detrapping of trapped carriers can be induced by means of modulated photon injection or by the application of a pulsed voltage to the input electrodes. Such techniques are used (ref. 31) for determining drift mobilities in materials in which space charge limited currents are dominant. Current transients can be used for Hall measurements as well and the magnetic field may also be applied in the form of a pulse. Such a method is, however, not suitable for Hall voltage measurements because of the time lag between the time the current is applied and the Hall voltage reaches its steady-state value.

In order to measure the sheet resistance of insulating layers an electron beam may be used to advantage (refs. 32-35) in place of a top contact of the usual metal-insulator-metal structure or a metal-insulator-semiconductor structure. Chester and Kosicki (ref. 35) have developed a method in which a low velocity electron beam is used to inject charge into an insulating layer. The injected charge drifts under the influence of its self-induced field to metal electrodes deposited on the layer where it is collected and measured as a time-dependent current. If this current is ohmic and if the injected carrier density is much smaller than the density of the charge carriers present in the insulating layer before irradiation then a current decay curve may be calculated whose time constant τ is the only adjustable parameter. Chester and Kosicki (ref. 35) have made such sheet resistance measurements on 5 to 40 nm thick GaAs layers. Sheet resistance values between 10^{11} and 10^{14} ohms have been determined. These correspond to $\sim 10^{12}$ electrons/cm^2 and an electron density of $\sim 10^{18}$/cm^3 in layers $\sim$10 nm in thickness. The mobility was found to be low, of the order of 10^{-4} to 10^{-7} cm^2/V-sec.; a plausible model for such charge transport is hopping or tunneling between neighboring trapping centers (ref. 36) either at the surface or within the insulating layers.

REFERENCES

1 G. Fischer, D. Grey and E. Mooser, Rev. Sci. Instr., 32(1961)842.
2 G. H. Heilmeier and S. E. Harrison, Phys. Rev., 132(1963)2010.
3 J. H. Fermor and A. Kjekshus, Rev. Sci. Instr., 36(1965)763.
4 D. Colman, Rev. Sci. Instr., 39(1968)1946.
5 Y. S. Park, P. M. Hemenger and C. H. Hung, Appl. Phys. Lett., 18(1977)45.
6 T. M. Baleshta and J. D. Keys, Am. J. Phys., 36(1968)23.
7 P. M. Hemenger, Rev. Sci. Instr., 44(1973)698.
8 J. P. Zielinger, M. Tapiero and C. Noguet, J. Phys. E:Sci. Instr., 6(1973)579.
9 K. C. Kao, J. Sci. Instr., 39(1962)208.
10 W. G. Amey and F. Hamburger, Proc. Amer. Soc. Test. Mat., 49(1949)1079.
11 E. C. Salthouse, J. Sci. Instr., 40(1963)49.
12 W. D. Edwards, J. Sci. Instr., 42(1965)432.
13 E. M. Pell and R. L. Sproul, Rev. Sci. Instr., 23(1952)548.
14 H. Gobrecht, K. H. Franke, F. Niemeck and K. E. Boeters, Z. Angew. Phys., 13(1961)261.
15 F. Ryan, Rev. Sci. Instr., 33(1962)76.
16 A. M. Hermann and J. S. Ham, Rev. Sci. Instr., 36(1965)1553.
17 I. Eisele and L. Kevan, Rev. Sci. Instr., 43(1972)189.
18 R. H. Tredgold, Space charge conduction in solids, Elsevier Publishing Co., Amsterdam (1966).
19 A. Rose, Phys. Rev., 97(1955)1538.
20 M. A. Lampert, Phys. Rev., 103(1956)1648.
21 P. Mark and W. Helfrich, J. Appl. Phys., 33(1962)205.
22 D. R. Lamb, Electrical Conduction Mechanisms in Thin Insulating Films, Methuen and Co., London (1967).
23 E. E. Olson and J. E. Wertz, Rev. Sci. Instr., 41(1970)419.
24 J. R. MacDonald, Rev. Sci. Instr., 25(1954)144.
25 G. P. Carver and R. S. Allgaier, J. Noncryst. Sol., 8/9(1972)347.
26 G. P. Carver, Rev. Sci. Instr., 43(1972)1257.

27 P. W. Shackle, Phil. Mag., 21(1970)987.
28 A. Fortini, A Le Bourgeois, J. Phys.(Paris), 25(1964)175A.
29 V. N. Dobrovol'skii and Yu. I. Gritsenko, Sov. Phys.-Sol. State, 4(1963)2025.
30 O. Sonnich Mortensen, R. W. Munn and D. F. Williams, J. Appl. Phys., 42(1971) 1192.
31 G. T. Pott and D. F. Williams, J. Chem. Phys., 51(1969)1901.
32 W. Tantraporn, J. Appl. Phys., 39(1968)2012.
33 K. A. Pickar, Solid-State Electron., 13(1970)303.
34 J. Lampert, J. Vac. Sci. Technol., 6(1969)753.
35 A. N. Chester and B. E. Kosicki, Rev. Sci. Instr., 41(1970)1817.
36 P. M. Hill, Thin Solid Films, 1(1967)39.

CHAPTER 9

THERMOELECTRIC AND THERMOMAGNETIC MEASUREMENTS

9.0 Introduction

In the previous Chapters charge carrier transport in metals and semiconductors in electric and magnetic fields has been considered primarily under isothermal conditions. This Chapter is concerned with the effects of thermal gradients on transport phenomena and with the interaction between thermal, electrical and magnetic effects in electronic materials.

In a homogeneous isotropic electronic material with a single conduction band, a spherical Fermi surface, density n and effective mass m_n^* characterized by an average relaxation time $<\tau>$ the current density vector

$$J_x = \frac{ne}{m_n^*}\left[eE_x + \frac{\partial}{\partial x}(E_F/T)\right]<\tau E> -\left(\frac{n}{m_n^* T}\right)\frac{dT}{dx}<\tau E^2> \tag{9.1}$$

for the electric field E_x, the Fermi level E_F, the absolute temperature T, the electron energy E and the temperature gradient dT/dx. The Thomson coefficient ψ represents the proportionality between the electric field and the temperature gradient

$$E_x = -\psi(dT/dx) \tag{9.2}$$

For $J_x=0$ Eq. (9.1) becomes

$$E_x = T \cdot \frac{d}{dT}\left[\frac{<E\tau> - E_F<\tau>}{eT<\tau>}\right] \cdot \frac{dT}{dx} \tag{9.3}$$

and from Eq. (9.2)

$$\psi = - T \cdot \frac{d}{dT}\left[\frac{<E\tau> - E_F<\tau>}{eT<\tau>}\right] \tag{9.4}$$

If two materials are joined to form two junctions A and B, as shown in Fig. 9.1, then the open-circuit voltage v_o measured between terminals C is given by (ref. 1)

$$v_o = \int_{T_1}^{T_2} S(T)dT \tag{9.5}$$

where S is the thermoelectric power or the Seebeck coefficient.

In general, the Seebeck coefficient is a second order tensor $\hat{S}$. In cubic crystals, in the absence of a magnetic field, the symmetry of the crystal requires that $\hat{S}$ be diagonal and, furthermore, that all its components be equal. In terms of Fig. 9.1, if $T_2=T_1+\Delta T$, for small ΔT

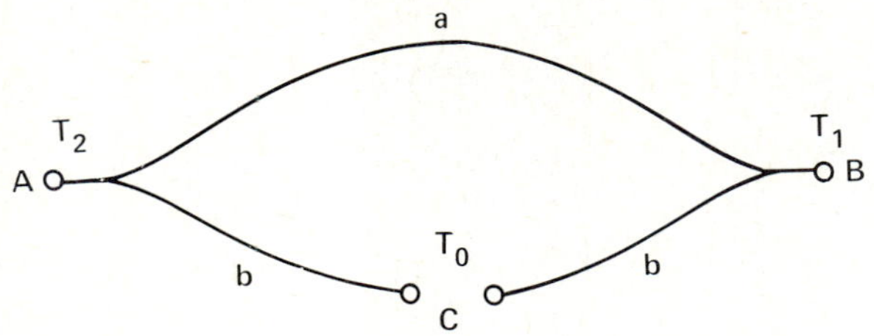

Fig. 9.1. Circuit containing two junctions A and B at temperatures T_1 and T_2 of two materials a and b; the open circuit voltage is measured at C.

$$V_0 = S_{ab}\Delta T \tag{9.6}$$

and if Ψ_a and Ψ_b are, respectively, the Thomson coefficients of the two materials a and b then

$$\Psi_a - \Psi_b = T(dS_{ab}/dT) \tag{9.7}$$

If a material b is so chosen that its Thomson coefficient is zero, (lead, for example), then S_{ab}=S is sometimes defined as the absolute thermoelectric power or Seebeck coefficient. Therefore, from Eqs. (9.4) and (9.7)

$$S = \frac{E_F\langle\tau\rangle - \langle\tau E\rangle}{e\langle\tau\rangle T} \tag{9.8}$$

For a non-degenerate n-type semiconductor with a scalar effective mass and an energy-dependent relaxation time of the form E^{-s} the Seebeck coefficient (ref. 2) is

$$S_n = -\frac{k}{e}\cdot\left[\frac{E_F}{kT} + \frac{5}{2} - s\right] \tag{9.9}$$

If the two materials a and b in Fig. 9.1 are of the same n-type semiconductor with different carrier concentrations, respectively, n_a and n_b then, since

$$(E_F)_a - kT\ell n(N_c/n_a) \qquad\qquad (E_F)_b = -kT\ell n(N_c/n_b) \tag{9.10}$$

where N_c is the density of states function of the semiconductor, it follows that $S_{ab}=-(k/e)\ell n(n_b/n_a)$. The density of states function, $N_c=2(2\pi m_n^* kT/h^2)^{3/2}$ where h is Planck's constant and Eq. (9.9) can be expressed in the form

$$S_n = -\frac{k}{e}\left[\ell n(N_c/n) + \frac{5}{2} - s\right] \tag{9.11}$$

The first term is dependent uniquely on the bandstructure and the free carrier concentration, the second depends on the electron scattering mechanisms. For a non-degenerate electron distribution and in the absence of a magnetic field the second term has a value between 2 and 4. The lower value holds for pure acoustic scattering and increases for increasing ionized impurity scattering. Now suppose that the junctions in Fig. 9.1 are brought to the same temperature and by inserting a battery at C a current i is made to flow around the circuit. If there is a rate of heating q=dQ/dt at junction A then junction B must be cooled at the same rate and the differential Peltier coefficient $\Pi_{ab}=q/i$. In defining the Thomson coefficient ψ for one of the conductors, it is supposed that in addition to the flow of current there is a temperature gradient dT/dx which leads to a heating or cooling rate per unit length dq/dx such that $\psi=(dq/dx)[i(dT/dx)]^{-1}$. The Peltier coefficient is related to the Seebeck coefficient through the Kelvin relation $\Pi_{ab}=S_{ab}T$. Thus the Peltier effect can be expressed in terms of the Seebeck effect and Seebeck measurements are much easier to make than Peltier effect measurements.

Fredericks (ref. 3) and Geballe and Hull (ref. 4) encountered unusually large Seebeck coefficients measured on Ge at low temperatures; these exceeded the values expected from the classical theory by more than one order of magnitude. The interpretation provided independently by Frederickse (ref. 5) and by Herring (ref. 6) is that phonons exert pressure on the motion of electrons and holes in the presence of a thermal gradient; such a phonon drag had been first predicted by Gurevich (refs. 7,8) with reference to metals. Any disturbance in the electron distribution of a material is imparted to its phonon system and vice versa. For example, an electric field, although it does not affect the phonon system directly, does disturb the phonon spectrum through its interaction with the electron distribution. The thermoelectric power or Seebeck coefficient may be considered to have an electronic component S, such as given by Eqs. (9.1) to (9.11) and a phonon drag component S_p. For low charge carrier concentrations at low temperatures the phonon drag component has to be taken into account in high purity specimens free of crystalline imperfections. The net departure of the phonon distribution from equilibrium is a function of two opposing factors: (a) the rate at which the electron-phonon momentum exchange occurs and (b) the rate at which the phonons can lose excess momentum and return to equilibrium. Steady state is obtained when a balance is established between these two rates. The first factor is connected with the electrical resistivity due to electron-phonon collisions, the second factor can be described in terms of an effective relaxation time for phonons τ_p. For the theoretical foundations of the phonon drag effect the reader is referred to the paper by Herring (ref. 9).

For a p-type non-degenerate semiconductor with an energy-dependent relaxation time of the form $E^{-s'}$ the Seebeck coefficient is (ref. 2),

$$S_p = \frac{k}{e}\cdot\left[\left(\frac{5}{2} - s'\right) + \ln\left(\frac{N_v}{p}\right)\right] \tag{9.12}$$

where p is the hole density and $N_v = 2(\pi m_p^* kT/h^2)^{3/2}$ is the density of states function of the valence band. The sign of the Seebeck coefficient, comparing Eqs. (9.11) and (9.12) thus provides a simple, direct test of the dominant charge carrier concentration of a semiconductor.

For two-band conduction of electrons and holes the Seebeck coefficient is

$$S = \frac{S_n\sigma_n + S_p\sigma_p}{\sigma_n + \sigma_p} \tag{9.13}$$

The contributions of each band combine as if the thermoelectric voltages were in parallel each being produced by a generator whose admittance is proportional to the conductivity $\sigma_{n,p}$ of a single band.

The presence of a magnetic field affects thermoelectric phenomena in two ways: it changes the magnitudes of the thermoelectric voltage along a temperature gradient and it produces a component of the thermoelectric voltage at right angles to the temperature gradient and to the magnetic field B. The former effect is usually an even function of B, the latter an odd function of B. The transverse effect [see Fig. 2.6, Chapter 2] is called the Nernst-Ettingshausen or Nernst effect. The Nernst coefficient P_N is defined by

$$\overline{E}_N = - P_N \overline{B} x \overline{\nabla T} \tag{9.14}$$

where E_N is a transverse electric field developed by the temperature gradient $\overline{\nabla T}$ under zero current conditions. Eq. (9.14) is, in fact, only appropriate in the weak magnetic field limit since for large B the contribution to E_N does not necessarily have the direction $\overline{B}x\overline{\nabla T}$. In the presence of a magnetic field oriented along the z-axis of a cartesian system with an electric field E_x applied to a homogeneous isotropic n-type semiconductor with a spherical Fermi surface, the current density vectors in the x-y plane are

$$\begin{aligned} J_x &= \sigma_{11}E_x + \sigma_{12}E_y + \alpha \frac{\partial T}{\partial x} + \beta \frac{\partial T}{\partial y} \\ J_y &= - \sigma_{12}E_x + \sigma_{11}E_y - \beta \frac{\partial T}{\partial x} + \alpha \frac{\partial T}{\partial y} \end{aligned} \tag{9.15}$$

The isothermal Nernst effect is obtained for $J_x=0$, $\partial T/\partial y=0$, and $J_y=0$ from Eqs. (9.15) for the weak magnetic field approximation from

$$E_y = \frac{\beta\sigma_{11} - \alpha\sigma_{12}}{(\sigma_{12})^2 + (\sigma_{11})^2} \cdot \frac{\partial T}{\partial x} \tag{9.16}$$

Substituting for σ_{11}, σ_{12}, α and β from the constitutive Ohm Hall equations [Chapter 2, Eq. (2.6)] and Eq. (9.14) the Nernst coefficient is

$$(P_N)_n = \frac{\mu_n}{eT}\left[\frac{\langle E\tau^2\rangle}{\langle\tau^2\rangle} - \frac{\langle\tau^2\rangle\langle E\tau\rangle}{\langle\tau\rangle^3}\right] \tag{9.17}$$

If τ is an energy-dependent relaxation time of the form E^{-s} then

$$(P_N) = -(k/e)\mu_n r_B s \tag{9.18}$$

where μ_n is the electron mobility and r_B is the Hall factor of Table 4.1 in Chapter 4.

For a p-type semiconductor with an energy-dependent relaxation time τ proportional to E^{-m},

$$(P_N)_p = -(k/e)\mu_p r'_B m \tag{9.19}$$

where r'_B is the appropriate Hall factor and μ_p is the hole mobility. The Ettingshausen effect leads to the appearance of a transverse temperature gradient $\partial T/\partial y$ produced by a longitudinal current density J_x and a transverse magnetic-field B_z for $J_y=0$

$$\frac{\partial T}{\partial y} = -PB_zJ_x \tag{9.20}$$

For an n-type semiconductor with a spherical Fermi surface the Ettingshausen coefficient is

$$(P)_n = -(k/e)\mu_n r_B\, s(T/\kappa_T) \tag{9.21}$$

where κ_T is the thermal conductivity and s is the exponent of the energy dependence of the relaxation time. For a p-type material,

$$(P)_p = -(k/e)\mu_p r'_B m(T/\kappa_T) \tag{9.22}$$

For a detailed description of the foundations of thermoelectric and thermomagnetic phenomena and their applications the reader is referred to the review by Jan (ref. 10), and the monographs of Tauc (ref. 1), Goldsmid (ref. 11), and Harman and Honig (ref. 12).

9.1 Seebeck measurements

Experimental measurements made of the temperature dependence of the Seebeck coefficient of metals and semiconductors can provide the data for determining the density of states at the Fermi level and the energy dependence of the charge carrier relaxation times (ref. 13). It is usually easier to perform measurements in large thermal gradients than in large magnetic fields and the Seebeck coefficient is sensitive to two-band or multi-band conduction processes. Therefore Seebeck measurements might be made in conjunction with and, in some cases, in preference to Hall effect measurements.

The apparatus shown in Fig. 9.2 is intended for making Seebeck measurements (ref. 4) in the temperature range between 20°K and 375°K. The device shown in this figure is an ultrasonically-processed 8-arm bridge bulk Ge specimen also used for resistivity and Hall effect measurements. The two side-arms F are used as Hall probes and the two outside arms E as electrical resistivity probes. The

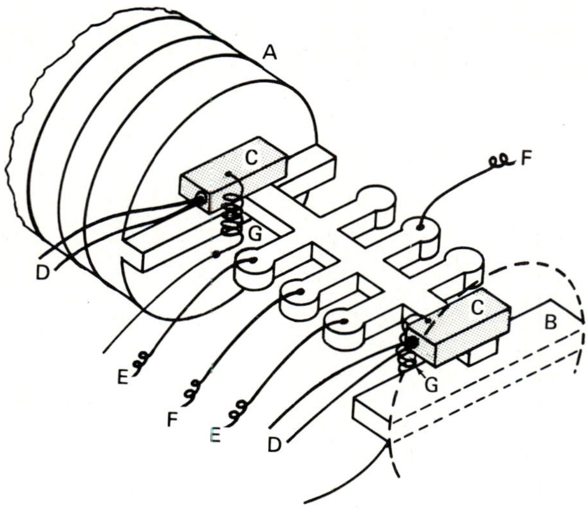

Fig. 9.2. Apparatus for making Seebeck and Hall measurements, A is a heater, B is a thermal sink, C are thermocouple blocks, D are thermocouples, E and F are, respectively, electrical resistivity and Hall probes. [after T. H. Geballe and G. W. Hull, Phys. Rev., 94(1954)1134]

lightweight hollow heater A is supported by the specimen to be measured. The thermal sink B is made of copper-filled lead in order to provide the requisite thermal stability at cryogenic temperatures. The copper thermocouple blocks C are soldered to the upper surfaces of the specimen. To facilitate assembly, they are held in place by electrically and thermally insulated phosphor-bronze springs. The thermocouples are made of copper-constantan and soldered into 0.06 cm holes drilled close to the lower inside corner of each block. The differential thermocouples G are included in order to measure the thermal impedance across each junction. The entire system is surrounded by a radiation shield and placed in an enclosure which is connected to a high vacuum pump maintained at 5×10^{-6}Torr. The Seebeck effect is measured in the steady state with a few hundredths of a watt dissipated at the heater A. For a period of about 10 minutes, the power dissipated at the heater, the thermocouple readings, and the Seebeck voltage across the two thermocouple copper leads are recorded as a function of time. The system inside the container drifts upward about 0.02° per minute at room temperature with a gradient of 2° between the thermocouples when in a vacuum. The decreasing thermal conductivity of the specimen causes the drift to increase as the temperature is lowered. To maintain the 2° gradient between the junctions, the heat introduced causes a 0.05° drift at nitrogen temperatures and 0.3° drift per minute at 30°K. However, it is possible to stabilize the system between 20 and 30°K, and also in the liquid nitrogen range, by introducing a small pressure of helium into the container. The temperature gradient can be changed several fold, the container temperature can be changed from room temperature to liquid nitrogen or liquid helium temperatures (this shows that radiation and heat-conduction losses are not affecting measurements) and the specimen can be removed and then remounted without changing the Seebeck coefficient by more than 2%. Above 250°K good agreement was found between Hall and thermoelectric measurements made on Ge specimens correlated with classical theory (ref. 4). Below 250°K the measurements are consistent with a strong phonon drag effect (ref. 6).

Gee and Green (ref. 14) have described the construction of thermoelectric probes for measuring the Seebeck coefficient of bulk and thin film semiconductors. The apparatus employs chromel-alumel thermocouples held in a quartz jacket made by drawing a 6 mm diameter twin-bore silica tube down to 0.5 mm with this end ground concave. Side holes are blown about 75 mm from the tip with the overall length of the probe of ∿10 cm. The thermocouple wires are electrically-welded in an argon atmosphere and bent into a V-shape with the junction at the apex of the V; a gold wire is melted in the form of a ball on the apex of the junction. This is intended primarily for improving the thermal contact between the probe and the specimen to be measured. The silica tube intended for use as the hot probe is coated on the outside with an opaque layer of paint which prevents direct illumination of the specimen surface by the incandescent projection lamp used to heat the junction. The silica tube transmits and guides a substantial fraction of the spectral output of the lamp down to the probe tip where the light is absorbed by heating the gold ball. In the apparatus shown in Fig. 9.3 the hot probe is tensioned against the lamp in the lamp housing 3. Section AA shows the means by which the hot probe can be adjusted so as to receive the optimum light flux. The ambient probe 2 is tensioned against the platform. The support rod 5 is used for minimizing heat conduction from the lamp housing to the sample platform. The temperature difference between the probes is adjusted by varying the voltage applied to the lamp. The circuit used to perform the Seebeck measurements is shown in Fig. 9.4. Four thermoelectric potentials can be measured. The terminal pair 1,1 gives the hot-probe temperature T_H; pair 2,2 gives the ambient temperature T_e; pair 3,3 gives the thermoelectric potential V_A, developed across the semiconductor plus the alumel wire; pair 4,4 gives the thermoelectric potential V_c, developed across the semiconductor plus the chromel wire.

The Seebeck coefficient of the semiconductor for the mean temperature $(T_H+T_e)/2$ is given by

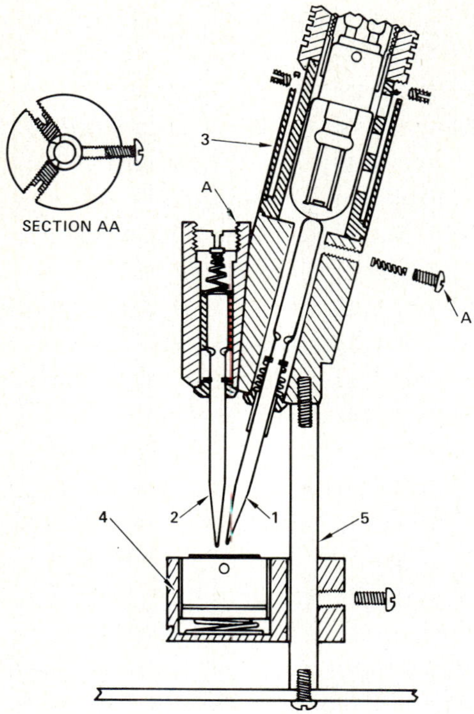

Fig. 9.3. Apparatus for Seebeck measurements in which the hot probe is heated by light conducted down a silica pipe. [after W. Gee and M. Green, J. Phys. E, Sci. Instr., 3(1970)135]

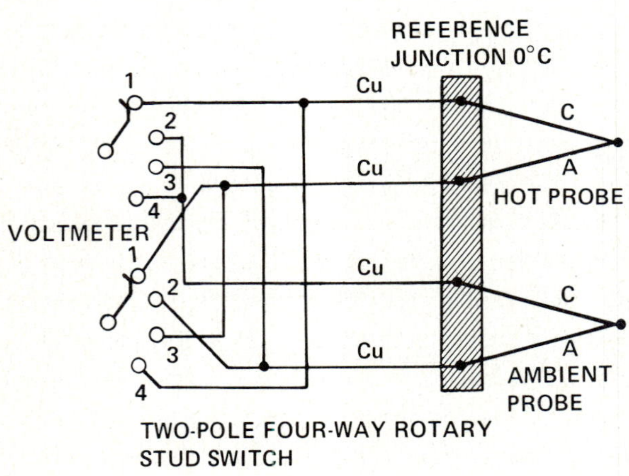

Fig. 9.4. Circuit and wiring arrangement for measuring Seebeck coefficients.

$$S_{cs} = (\pm)\,\frac{40\ V_c}{V_A - V_c} + 21.5\ (\mu V\ K^{-1})$$
$$S_{sc} = (\pm)\,\frac{40\ V_A}{V_A - V_c} - 18.5\ (\mu V\ K^{-1}) \qquad (9.23)$$

where the sign is read from the voltmeter: 40 is the thermoelectric potential per degree for the chromel-alumel thermocouple at room temperature, and +21.5 is the absolute Seebeck coefficient of chromel at room temperature.

Techniques used for measuring the Seebeck coefficient may be classified as integral or differential procedures (ref. 15). In the integral method, one end of a rod-like specimen is held at some fixed temperature and the other end is heated to a temperature T and made to vary through some temperature range of interest. The thermoelectric potential v(T) between the specimen-lead thermocouple is measured and the temperature dependence of the Seebeck coefficient is determined from $S(T)=dv(T)/dT$. In the differential method, a small temperature differential ΔT is maintained across a specimen producing the thermoelectric potential Δv such that $S(T)=\Delta v(T)/\Delta T$. For this relation to be valid ΔT must be small enough, the specimen must be homogeneous, the temperatures in the vicinity of junctions between dissimilar metals in the measuring circuit must be uniform, and spurious voltages of thermal origin must be eliminated. The latter can be accomplished by a sequence of measurements of Δv_i for incremental changes in the temperature gradient ΔT_i; thereafter, the thermoelectric voltage vs. the temperature difference values are plotted and the slope of the linear region provides the Seebeck coefficient. Such a procedure is time consuming since a steady state must be established for each ΔT and several increments are required for an evaluation of the Seebeck coefficient at any one temperature. A rapid procedure for the differential evaluation of S(T) was developed by Caskey *et al.* (ref. 16). It is based on the simultaneous measurement of ΔT and Δv as the temperature gradient is increased slowly. The procedure was tested (ref. 17) by measuring the Seebeck coefficient of nickel in the temperature range between 4.2°K and 300°K. The apparatus (ref. 16) is shown in Fig. 9.5.(a). Rod-shaped 3 to 4 cm long specimens were placed in an evacuated stainless steel enclosure immersed in liquid helium or liquid hydrogen. The heaters were also enclosed within the same chamber; the specimen temperature was controlled by the current applied to the main heater and the thermal gradient was produced by the current applied to a small heater attached to the free end of the specimen.

The thermoelectric voltage generated by the specimen and the voltage output of the differential thermocouple were amplified by Keithly model electrometer null detectors. The signals from the temperature sensor and the attenuated outputs from the two electrometers were recorded sequentially at 4 sec. intervals with a

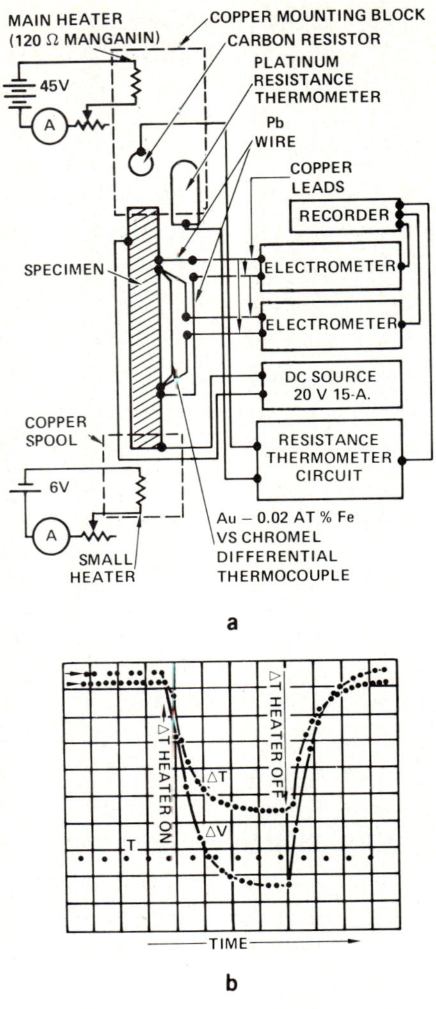

Fig. 9.5. (a) Schematic circuit for measuring the Seebeck coefficient of rod-shaped specimens; (b) example of recorded data from specimen, differential thermocouple and platinum or carbon resistance thermometer. [after G. R. Caskey, D. J. Sellmyer and L. G. Rubin, Rev. Sci. Instr., 40(1969)1280]

six-point recorder operating at a chart speed of 5 cm/min. The sequence of operations performed with the instrument was to measure the specimen temperature, then the voltage of the differential thermocouple followed by the measurement of the thermoelectric voltage of the specimen and the remeasurement of the voltage of the differential thermocouple. Spurious thermal voltages from the specimen and differential thermocouple appeared. They varied as the specimen temperature was changed but remained constant after an equilibrium temperature was achieved. They were suppressed with the zero suppression circuits in the electrometer amplifier null detectors. A temperature gradient of slowly increasing magnitude was then generated by passing current of between 5 to 30 mA through the small heater on the free end of the specimen. The deviation of the signals as a function of time from the arbitrary points for the specimen and differential thermocouple is shown as recorded in Fig. 9.5.(b). The Seebeck coefficient was obtained from the data by drawing a smooth curve through the points representing the temperature gradient interpolated to obtain values which correspond in time to the specimen voltage points. Thereafter a linear plot was made of the specimen voltage vs. the voltage output of the differential thermocouple and its slope was determined. The Seebeck coefficient is

$$S(Pb/X) = (\text{slope}) \cdot \frac{R(X)}{R(\Delta T)} S(AuFe) \tag{9.24}$$

where S(Pb/X) is the Seebeck coefficient of the specimen X with respect to lead, R(X) is the range setting on the Keithley amplifier for the specimen voltage and R(ΔT) is the range setting on the other Keithley amplifier used for measuring the output of the differential thermocouple; S(AuFe) is the Seebeck coefficient of the Au-Fe/chromel differential thermocouple. The estimated error in the measurement of the slope used in Eq. (9.24) is ±2%. The uncertainty in the calibration of the gold-iron/chromel thermocouple is approximately ±2% and the Seebeck coefficient of lead is known only to ±1%. A reduction in time and effort of making such measurements and analyzing the data might be effected by converting the analog signals to digital data by means of an analog-to-digital converter for real-time data processing by a digital computer and using a coupled microprocessor for controlling the sequence of the measurements.

An AC method for the measurement of the Seebeck coefficient of metals was described by Freeman and Bass (ref. 18). Its inherent characteristics eliminate the effects of spurious thermally-induced DC potentials and its sensitivity is high enough so that it might be used to detect small emf's produced by a couple made of a given metal and a slight compositional perturbation introduced into the same metal such as that produced by changes in the Seebeck coefficient by surface scattering (size effect), cold work or various dilute impurity concentrations. Alternatively, it might be used for measurements requiring the use of small temperature differences which require the determination of the Seebeck coefficient temperature dependence near a phase transition or a critical point. If the respective Seebeck coefficients of these two metals a and b are S_a and S_b at some average temperature T, then a thermocouple between them with a temperature difference ΔT produces the potential

$$\Delta V_{ab} = (S_a - S_b)\Delta T \tag{9.25}$$

If the same temperature difference is produced across a reference thermocouple having a known Seebeck coefficient S_{cd} and the voltage output of this couple is

$$\Delta V_{cd} = S_{cd}\Delta T \tag{9.26}$$

then the ratio of their voltages is equal to the ratio of their Seebeck coefficients

$$\Delta V_{ab}/\Delta V_{cd} = (S_a - S_b)/S_{cd} \tag{9.27}$$

The AC apparatus for measuring Seebeck coefficients (ref. 18) is based on the fact that the same equation holds if the differential temperature has a sinusoidal time dependence of the form $\Delta T \cdot \sin\omega t$ provided that ΔV_{ab} and ΔV_{cd} are considered to be the amplitudes (or the rms amplitudes) of the AC voltages produced by the thermocouples. The apparatus used for the measurements is shown in Fig. 9.6. The reference thermocouple was made of 0.076 mm chromel and alumel wires, the specimen thermocouple was made of a 0.025 mm Al foil to which two gold wires were spot-welded at a distance of ~4 cm. The two thermocouples were connected in parallel and supported by an electrically-insulated support within a furnace. The cold junctions of the two couples were attached to the opposite sides of a thin sheet of mica and the hot junctions were spot-welded together. Another chromel-alumel thermocouple next to the cold junction was used to measure its temperature. Chopped radiation at 21 Hz obtained from a tungsten lamp was focused on the Al foil; this produced the periodic variation in ΔT. The output ΔV_{ab} from the lock-in amplifier coupled to the sample thermocouple was fed to the Y input of the X-Y recorder. The output ΔV_{cd} from its lock-in amplifier coupled to the parallel chromel-alumel thermocouple was fed into the reference input of the recorder. When modified as described by Aspness (ref. 19), the Y channel of the recorder displayed the ratio $\Delta V_{ab}/\Delta V_{cd}$. The output of the third, temperature measuring thermocouple was amplified and fed to the X channel of the recorder. The data obtained by Freeman and Bass indicate that accuracies of a few percent using a temperature difference of 0.03°K can be obtained with this method. With better lock-in amplifiers which might respond to frequencies of ~2 Hz it is to be expected that temperature differences of 0.01°K might be used thus providing Seebeck coefficient measurements accurate to ~1%.

Equipment for measuring the Seebeck coefficient of materials either as a function of temperature or as a function of pressure was developed by Shubha and Ramesh (ref. 20). Figure 9.7 shows a block diagram of the apparatus used to perform such measurements with an accuracy of ~1% with temperature gradients as small as 0.05°C. If S_a and S_b are the Seebeck coefficients of the thermocouples in Fig. 9.7 then the absolute Seebeck coefficient of the specimen,

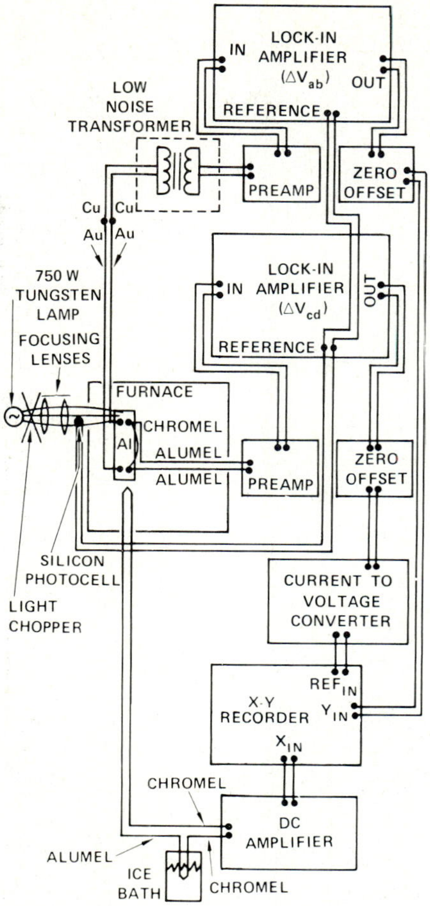

Fig. 9.6. Schematic of AC apparatus for Seebeck measurements. [after R. H. Freeman and J. Bass, Rev. Sci. Instr., 41(1970)1171]

$$S_x(T) = S_a(T) + (S_a - S_b)_T \cdot \left\{ \frac{\Delta V_{13}}{\Delta V_{13} - \Delta V_{24}} \right\} \tag{9.28}$$

The circuits in Fig. 9.7 provide the means for plotting directly the ratio $v_\alpha = \Delta V_{13} \cdot (\Delta V_{13} - \Delta V_{24})^{-1}$ as a function of the temperature T for a differential temperature $\Delta T = T_2 - T_1$. The output of the analog divider yields the ratio v_α applied to the Y-input of the X-Y recorder and the amplified output V_{34} added to $(1/2)(\Delta V_{13} - \Delta V_{24})$, v_β, is applied to the X-input of the recorder. Alternatively if the Seebeck coefficient is measured as a function of pressure (with the mean temperature held constant) the output of a pressure transducer is applied to the X-input of the recorder.

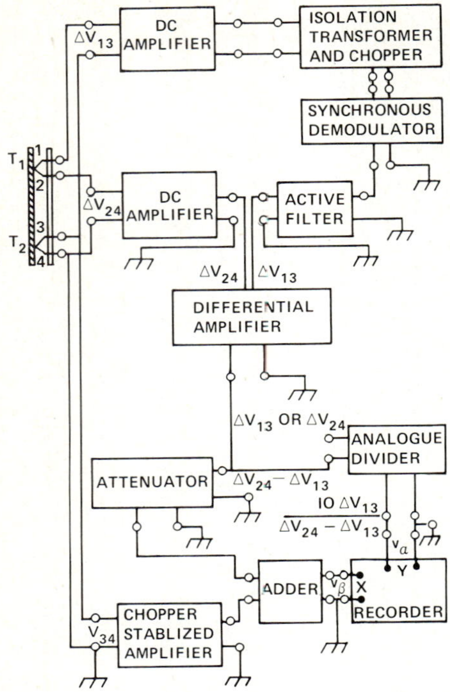

Fig. 9.7. Block diagram of apparatus used for automatic recording of Seebeck coefficient as a function of temperature or pressure. [after V. Shubha and T. G. Ramesh, J. Phys. E, Sci. Instr., 9(1976)435]

A high resolution apparatus for Seebeck-effect measurements in the temperature range between 1.5°K and 300°K was described by Zdrusky and Showalter (ref. 21). It employs two continuously measuring DC nanovolt amplifiers, one used for measuring the thermoelectric voltage and the other for measuring the thermocouple output voltage which is proportional to the temperature difference across the specimen. By using the heat generated by a low-mass electrostatically-guarded heater to produce a temperature difference across the specimen it is possible to determine the absolute thermoelectric power (vs. Pb) or the differential thermoelectric power of two nearly similar metal or alloy specimens. Using an additional independent heater provides a reduction in the thermal time constant and thermoelectric power measurements can then be made in a few seconds with a resolution of $\pm[10^{-3}(\Delta T)^{-1}$, $\mu V/°K+1\%]$ for thermal gradients in the range between 0.01 and 1.0°K.

9.2 Thermomagnetic measurements

A transverse magnetic field B_z acting on a longitudinal current density J_x generates a temperature gradient dT/dy in a perpendicular direction to both J_x

and B_z. The Ettingshausen coefficient is

$$P = \frac{(dT/dy)}{J_x \cdot B_z} \tag{9.29}$$

The Nernst effect (also called the Nernst-Ettingshausen effect) is a thermogalvanomagnetic phenomenon in which a transverse electric field is produced across a conductor which is subjected to a longitudinal thermal gradient dT/dx and an orthogonal magnetic field B_z,

$$P_N = \frac{E_y}{B_z \cdot dT/dx} \tag{9.30}$$

The Nernst and Ettingshausen effects are related to each other by the Bridgman equation $P=P_N T$ which is analogous to the Kelvin equation, $\Pi=ST$ between the Peltier effect Π and the Seebeck coefficient S. The longitudinal temperature gradient dT/dx and the orthogonal magnetic field B_z also produce a transverse thermal gradient dT/dy which corresponds to the Righi-Leduc coefficient

$$R_L = \frac{dT/dy}{B_z \cdot dT/dx} \tag{9.31}$$

A magnetic field and temperature gradient may produce two longitudinal effects: the magneto-Seebeck effect (also known as the longitudinal Nernse effect) and the magnetothermal resistance effect (also known as the Maggi-Righi-Leduc effect). These two effects appear as changes of the Seebeck coefficient and of the thermal conductivity in magnetic fields.

It is usually desirable to investigate thermomagnetic phenomena over a wide range of temperatures and magnetic fields. In general, it is not possible to determine the coefficients under isothermal conditions (i.e., with no transverse temperature gradient when a longitudinal temperature gradient is applied to a specimen). It is desirable to approach the adiabatic condition for most experimental measurements (i.e., no transverse heat flow and no transverse current flow). The specimen should be maintained in an evacuated chamber in order to minimize heat losses. Thermocouples should be made of relatively poor thermal conductors such as chromel and constantan rather than copper. The use of magnetic alloys such as alumel should be avoided because such materials tend to move in a strong magnetic field. There appears to be no good reason why only the adiabatic Nernst coefficient should be measured without also measuring the Righi-Leduc coefficient. If both are measured then the isothermal Nernst coefficient can be calculated.

Two useful geometrical arrangements have been proposed for measuring the Nernst effect (refs. 22,23). Both of these involve cylindrical geometries. The arrangement proposed by O'Brien and Wallace (ref. 22) employs a radial magnetic field with heat flow in the z-direction. The current path is in the form of a spiral.

The configuration proposed by Norwood (ref. 23) uses a radial heat flow and an axially-directed magnetic field. This confines the current to a path which resembles a screw thread. Unfortunately, these structures are not well adapted for Seebeck or Peltier effect measurements.

Nernst effect measurements on principally polycrystalline bulk or non-degenerate or weakly degenerate InSb specimens were made by Zhuze and Tsidil'kovski (ref. 24) and on single crystal degenerate InSb specimens by Emel'yanenko *et al.* (ref. 25). They found that between 100°K and 600°K in magnetic fields between 0.06T and 0.4T the Nernst coefficient changes sign from negative to positive at some specific temperature which shifts to lower temperatures with increasing degeneracy (ref. 26). Other measurements have been made by Ed-Saden and Thomas (ref. 27) however their validity has been questioned by Mette and Loscoe (ref. 28).

The Righi-Leduc effect is the thermal analog of the Hall effect and may be regarded as due to the Hall angle rotation of electrons involved in heat conduction; a transverse temperature gradient dT/dy is established in the presence of a magnetic field B when heat flows in the longitudinal direction, dT/dx. The experimental difficulty consists principally in the measurement of temperature differences of millidegrees across specimens which have longitudinal temperature gradients of ~10°K/cm. The Righi-Leduc effect has been described by Jan (ref. 10) and Putley (ref. 29). The coefficient R_L should be equal to the Hall mobility; in materials which have an appreciable lattice thermal conductivity R_L is diminished by the ratio of the electronic thermal conductivity to the total thermal conductivity. An elementary analysis is based on the assumption (ref. 30) that the magnetic field induces a transverse electron heat flux q_{RL} of the form

$$q_{RL} = \mu B_z K_{TE} \cdot \partial T/\partial x \tag{9.32}$$

where K_{TE} is the electronic thermal conductivity. In order that no heat flow take place at equilibrium q_{RL} must be balanced by the opposing heat flow

$$q_y = K_T(\partial T/\partial y) \tag{9.33}$$

where K_T is the total thermal conductivity. From Eqs. (9.32 and 9.33)

$$\mu B_z K_{TE}(\partial T/\partial x) = K_T(\partial T/\partial y) \tag{9.34}$$

This leads to a solution for the Righi-Leduc coefficient,

$$R_L = (K_{TE}\mu)/K_T \tag{9.35}$$

Putley (ref. 29) solved the Boltzmann transport equation under the assumption that $\mu B \ll 1$, that classical statistics apply and that electron scattering is by acoustic lattice vibrations;

$$R_L = -\left(\frac{21\pi}{32}\right)\left(\frac{k^2T}{eK_T}\right)\cdot n\mu^2 \qquad (9.36)$$

using the classical expression, $K_{TE}=2(k/e)^2\sigma T$,

$$R_L = (7/8)(K_{ET}/K_T)\mu \qquad (9.37)$$

He also devised general expressions which may be used to compute R_L for any degree of degeneracy and for any values of B for which the Boltzmann equation is valid. Mette (ref. 31) investigated the Righi-Leduc effect in InSb between 25°C and 300°C. He found that $R_L=2.2\times10^{-2}T^{-1}$ at room temperature and for $\mu=7.72\times10^4 cm^2/V$-sec; from Eq. (9.35), $K_{TE}/K_T=2.85\times10^{-3}$.

9.3 Heat transport parameters

The inhomogeneous distribution of the current density in a rectangular Hall plate subjected to a transverse magnetic field leads to an inhomogeneous temperature distribution which is a function of the geometrical aspect of a Hall plate, its thermal parameters and the magnetic field. Gruetzmann (ref. 32) investigated analytically the temperature distribution in such Hall plates in which a Peltier effect appears at the contact between a metal electrode and a semiconductor. The heat generated per unit volume per unit time is $|J|^2\sigma^{-1}$ thus the distribution of the current density J in the plate must be determined in conjunction with the thermal conductivity K_T (in W/cm-°C) and a heat transfer coefficient k' (in W/cm^2-°C) to an appropriate substrate-heat sink. He found that for a Hall angle $\theta\simeq45°$ a uniform temperature distribution along an electrode appears only for plate widths <3 mm if $(k'/K_T)\cdot d\simeq10/cm^2$ where d is the thickness of the InSb Hall plate. Subsequent experimental measurements (ref. 33) confirmed, to a large extent, these theoretical calculations. For Hall plate length-to-width ratios, $\ell/w=0.1$ and a Hall angle $\theta\simeq60°$ the temperature differences are due primarily to the Peltier effect. For Hall plates with $\ell/w=2$ and large Hall angles, $\theta\simeq80°$, the relative temperature difference along the electrodes increased by a factor of 19 over that present in the Hall plate for $\theta=0$. Mimizuka and Ito (ref. 34) have calculated the two-dimensional temperature distribution in a rectangular Hall plate using a relaxation procedure based on the finite difference method. They found that the Hall electrodes have a significant effect on the temperature distribution in their immediate vicinity and the correlation between the electric field and temperature distribution is a weak one.

The thermal resistance of thin specimens with a high heat conductivity can be measured by means of a procedure such as described by Savvides and Goldsmid (ref. 35). Figure 9.8.(a) shows schematically the experimental set-up which is an adaption of that first described by Harman (ref. 36) and developed by Penn (ref. 37). The specimen to be measured, B, is placed between two outer wafers, A,

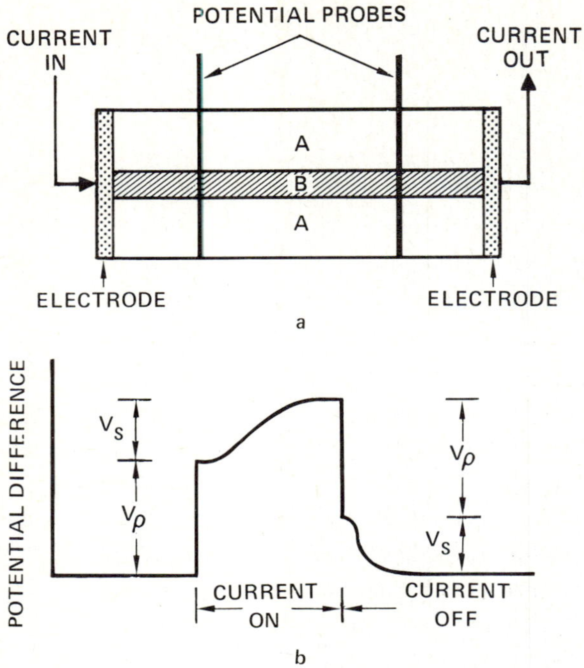

Fig. 9.8. (a) Three-layer sandwich for thermal conductivity measurements; (b) time dependence of potential difference showing resistance components V_ρ and thermoelectric component V_s. [after N. Savvides and H. J. Goldsmid, J. Phys. E, Sci. Instr., 5(1972)553]

made of a large Seebeck coefficient material such as bismuth telluride. The electrical resistance of the wafer A is sufficiently small that the specimen may be regarded as an electrical insulator. If an electric current is applied between the terminals of the three-layer sandwich shown in Fig. 9.8.(a) then the potential difference between the potential probes is made up of a resistive component V_ρ and a thermoelectric voltage V_s due to the Peltier effect. The thermal conductivity $(K_T)_B$ of the thin specimen can be determined from the relation

$$(K_T)_B = \frac{S_A{}^2 T}{R_A} \cdot \left[\left(\frac{V_\rho}{V_s}\right)_b - \left(\frac{V_\rho}{V_s}\right)_o \right] \tag{9.38}$$

where S_A is the Seebeck coefficient of the bismuth telluride layers and R_A is their combined resistance. The subscripts b and o correspond, respectively, to a measurement made with the specimen in place, as shown in Fig. 9.8.(a), and with the specimen removed and the same measurement made with the bismuth telluride plates joined across their interface. V_ρ and V_s can be determined by observing the potential difference developed across such a structure as a function of time after an electric current is applied to it. Immediately following the current "turn-on" the potential difference is V_ρ whereas after equilibrium is established it is equal to $(V_\rho+V_S)$ as shown in Fig. 9.8.(b). The results obtained with 50μm and 20μm-thick Si wafers placed between 5 mm thick bismuth telluride plates indicate that K_B=1.45/cm-°K with a reproducibility of ±3%.

If heat flows through a slab of thickness δ and thermal conductivity K_T then a temperature difference ΔT appears along the thickness dimension for a constant heat flux input per unit area q=Q/A,

$$q = K_T(\Delta T/\delta) \tag{9.39}$$

Thus the heat flux can be determined from a measurement of the temperature difference between the two surfaces and the known thermal conductivity, surface area and thickness of the slab. The application of thin film resistance thermometry for the construction of such a heat flux gauge was used by Brown *et al.*(ref. 38). Heat flux meters made of a bar of silver telluride in which a voltage is developed across the bar in consequence of a potential gradient between its terminals has been described by Hatfield and Wilkins (ref. 39) and de Jong and Marquenie (ref. 40) have described a thermopile for the measurement of low heat fluxes. Weickhardt *et al.*(ref. 41) have investigated the feasibility of using a semiconducting oxide film on a metal surface to measure the heat flux through that surface. If the heat is made to flow through a small thermal impedance which covers the surface uniformly then the heat flux at any point can be determined from the temperature drop across the thermal impedance at that point. The use of an oxide is advantageous since it can be deposited on a metal or semiconductor surface by a variety of methods and since many oxides have large energy gaps they have large Seebeck coefficients. A thermoelectric potential v is measured between the specimen and a probe applied to the oxide for a transverse temperature gradient with the heat flux q (in W/cm^2),

$$q = vK_T\cdot(S\delta)^{-1} \tag{9.40}$$

where K_T is expressed in W/cm-°K, δ is the thickness of the oxide in cm and S in V/°K.

Goldsmid *et al.*(ref. 42) have compared heat flux meters employing the Nernst effect with those employing the Seebeck coefficient and indicate the former to be superior to the latter because unlike the Seebeck effect in which the output voltage is proportional to the temperature difference across the specimen the Nernst output voltage is proportional to the temperature gradient and is therefore independent of the specimen thickness. The figure-of-merit of a Nernst heat flux meter may be expressed in terms of its sensitivity σ_N (in V/W) divided by the thermal contact resistance R_C; thus σ_N/R_C is expressed in V/°K. Goldsmid *et al.*(ref. 42) found that a two-phase specimen of InSb-NiSb (InSb with needle-like NiSb inclusions) in a magnetic field of 0.5T, at room temperature, had a response of 1.5 mV/°K for a device cross-sectional area in the heat flow direction of ~3.5 mm^2 and a thickness of 1.6 mm with the semiconductor bonded to anodized aluminum heat-exchange plates.

9.4 Thermogalvanomagnetic properties of inhomogeneous materials

Inhomogeneous materials such as those which consist of a metallic phase embedded in a semiconductor matrix have electrical, galvanomagnetic and thermomagnetic properties which may deviate to a considerable extent from those of the same homogeneous semiconducting material. Wagini and Weiss (ref. 43) have shown that the two-phase material InSb-NiSb which consists of an anisotropic-ordered array of NiSb needles embedded in a matrix of InSb has a Nernst coefficient which is ~18 times larger than that of homogeneous InSb. InSb-NiSb has a large magnetoresistance coefficient induced by the short-circuit of the Hall field in the InSb matrix by the metallic NiSb needles. Furthermore, the absorption coefficient of this two-phase material between 9 and 15μm is about 30 times larger than that of homogeneous InSb (ref. 44) and the frequency response of a Nernst radiation detector made of this material in this spectral range with the heat flux parallel to the needle axis is nearly constant to 10^3 Hz if the specimen thickness is no greater than 100μm. Kaila and Goldsmid (ref. 45) have made Seebeck and Nernst effect measurements on two-phase InSb-In films in which filamentary indium inclusions are present in an InSb matrix. Such films are grown (ref. 46) by the recrystallization from the liquid phase of vacuum-deposited InSb films. With the InSb-In films oriented in the x-y plane the Seebeck coefficient was measured for a temperature difference along the x-coordinate and of the Nernst coefficient for heat flow along the y-coordinate with the magnetic field oriented in either case along the z-coordinate. Kaila and Goldsmid interpreted their measurements of the magnetic field-dependence of the 300°K Seebeck coefficient and Nernst coefficient in terms of a model which takes into account the fractional change in the components of the conductivity tensor of InSb introduced by the metallic indium inclusions. The relevant current density equations are

$$
\begin{aligned}
J_1(x) &= \sigma_1[E(x) - S_1\, dT/dx + R_{h_1} BJ_1(y)] \\
J_2(x) &= \sigma_1 k_x E(x) \\
J_1(y) &= \sigma_1[E(y) - S_1 dT/dy - R_{h_1} BJ_1(x)] \\
J_2(y) &= \sigma_1 k_y E(y)
\end{aligned}
\tag{9.41}
$$

Here the subscripts 1 and 2 refer to the parameters of the semiconductor and metallic phases respectively. With the mobility μ of the InSb matrix expressed in terms of its Hall coefficient R_{h_1} and conductivity σ_1, $\mu=|R_{h_1}\sigma_1|$, the Seebeck coefficient for $J_x=J_y=0$ and $dT/dy=0$ is

$$S = \frac{S_1}{1 + k_x(1 + k_y\mu^2B^2)\cdot(1 - k_x)^{-1}} \tag{9.42}$$

and the Nernst coefficient for $J_x=J_y=0$ and $dT/dx=0$ is

$$P_N = \frac{\mu|S_1|k_y}{1 + k_x(1 + k_y\mu^2B^2)\ (1 - k_x)^{-1}} \tag{9.43}$$

the magnetoresistance is

$$\frac{\Delta\rho}{\rho_0} = \frac{\mu^2B^2k_y(1 - k_x)}{1 + \mu^2B^2k_xk_y} \tag{9.44}$$

Kaila and Goldsmid fitted their experimental measurements of $(\Delta\rho/\rho_0)$ as a function of B with μ_1 = 6.1x10^{4}cm^2/V-sec., k_x=0.38 and k_y=0.43. The same parameters in conjunction with S_1=-261μV/°K were used to fit the dependence of the Seebeck coefficient and Nernst coefficient on B up to 0.6T in good agreement with the experimental measurements.

A similar model with k_x=0 was used by Goldsmid and Sidney (ref. 47) to interpret the experimentally-measured dependence of the thermomagnetic properties of Cd_3As_2-NiAs. This material consists of a Cd_3As_2 matrix containing an ordered array of NiAs needles. The Nernst effect for heat flow in a direction parallel to the axis of the needles in a transverse magnetic field at room temperature can be used as a thermal radiation detector. The thermomagnetic sensitivity of such a Cd_3As_2-NiAs detector is greater than that of an InSb-NiSb radiation detector (with the same resistance and response time) by a factor of ∿3. Goldsmid *et al.* (ref. 48) have measured at 300°K the Nernst coefficient of Cd_3As_2-NiAs with the heat flux oriented either parallel or perpendicular to the needle inclusions and with magnetic fields up to ∿5T. At low fields P_N is negative for both orientations and becomes positive for B$\gtrsim$1.6T when the heat flow is parallel to the metallic inclusions. The Nernst effect is strongly anisotropic and is a maximum for heat flow perpendicular to the axis of the needles.

Not only ordered two-phase structures but randomly-distributed metallic inclusions in semiconducting matrices can have a significant effect on their thermoelectric and thermomagnetic properties. Kudinov and Moizhes (ref. 49) have investigated analytically the effects of random inhomogeneities in cubic crystals on their Seebeck and Nernst coefficients in strong magnetic fields. They found that the relative effect of random inhomogeneities on the Nernst effect increases with increasing transverse magnetic field; their effect on the transverse and longitudinal Seebeck coefficient does not increase with B. The large sensitivity of the Nernst effect to the concentration of the randomly-distributed inhomogeneities was ascribed to the fluctuations of the Hall fields in adjacent regions. For materials with large Seebeck coefficients due account must be taken of the eddy currents induced by the inhomogeneities. Drichko and Mochan (ref. 50) have performed thermoelectric and thermomagnetic measurements on InSb containing a random distribution of microinhomogeneities. Their data provides a substantial confirmation of the theoretical calculations of Kudinov and Moizhes. The Seebeck coefficient is only slightly affected by microinhomogeneities, however, the Nernst effect in n-type InSb is entirely determined by microinhomogeneities such as those produced by random fluctuations in the impurity concentration. Dislocations can also produce anisotropic Nernst coefficients in semiconductors as shown by Kaila *et al.*(ref. 51). They used a static couple technique for the plastic deformation of InSb specimens at 300°K producing dislocations with a preferential alignment along the [100]-direction with densities estimated to be of the order of 10^6 to $10^7/cm^2$. Measurements of the Seebeck coefficient as a function of temperature between 100°K and 300°K indicate that a virgin n-type InSb has a negative Seebeck coefficient over this temperature range. Plastically-deformed specimens undergo a transition from a negative to a positive Seebeck coefficient at low temperatures. The transition through zero of the Seebeck coefficient occurs at a temperature ∿15°K higher for heat flow perpendicular in comparison with that of heat flow parallel to the dislocations. The Nernst coefficient in B=0.6T is negative for the virgin (unstrained) InSb in the intrinsic temperature region and it is positive for the plastically-deformed material. It is presumed that the change in sign is due primarily to the generation of acceptors centers by the dislocations and dislocation scattering which reduces the effective mobility of the electrons.

REFERENCES

1 J. Tauc, Photo and Thermoelectric Effects in Semiconductors, Pergamon Press, Oxford (1962)145.
2 R. A. Smith, Semiconductors, Cambridge Univ. Press, Cambridge (1968)172.
3 H. P. R. Frederickse, Phys. Rev., 92(1953)248.
4 T. H. Geballe and G. W. Hull, Phys. Rev., 94(1954)1134.
5 H. P. R. Frederickse, Phys. Rev., 91(1953)491.
6 C. Herring, Phys. Rev., 92(1953)857.
7 L. Gurevich, J. Phys. USSR, 9(1945)499.
8 L. Gurevich, J. Phys. USSR, 10(1946)67.
9 C. Herring, Phys. Rev., 96(1954)1163.
10 J. P. Jan, Solid-State Phys., 5(1957)3.
11 H. J. Goldsmid, Thermoelectric Refrigeration, Plenum Press, New York (1964).
12 T. C. Harman and J. M. Honig, Thermoelectric and Thermomagnetic Effects and Applications, McGraw-Hill Book Co., New York (1967).
13 D. K. C. MacDonald, Thermoelectricity, Wiley and Sons, New York (1963) pp. 1-36.

14 W. Gee and M. Green, J. Phys. E, Sci. Instr., 3(1970)135.
15 G. K. White, Experimental Techniques in Low Temperature Physics, Oxford Univ. Press, London (1959).
16 G. R. Caskey, D. J. Sellmyer and L. G. Rubin, Rev. Sci. Instr., 40(1969)1280.
17 G. R. Caskey and D. J. Sellmyer, J. Appl. Phys., 40(1969)1476.
18 R. H. Freeman and J. Bass, Rev. Sci. Instr., 41(1970)1171.
19 D. E. Aspness, Rev. Sci. Instr., 38(1967)1663.
20 V. Shubha and T. G. Ramesh, J. Phys. E, Sci. Instr., 9(1976)435.
21 D. R. Zrudsky and A. B. Showalter, Rev. Sci. Instr., 44(1973)497.
22 B. J. O'Brien and C. S. Wallace, J. Appl. Phys., 29(1958)1010.
23 M. H. Norwood, J. Appl. Phys., 34(1963)594.
24 V. P. Zhuze and I. M. Tsidil'kovski, Sov. Phys. Tech. Phys., 3(1958)2177.
25 O. V. Emel'yanenko, F. P. Kesamanly and D. N. Nasledov, Sov. Phys. Solid State, 4(1962)397.
26 O. V. Emel'yanenko, N. V. Zotova and D. N. Nasledov, Sov. Phys. Solid State, 1(1959)1711.
27 M. R. El-Saden and F. W. Thomas, J. Appl. Phys., 36(1965)181.
28 H. Mette and C. Loscoe, J. Appl. Phys., 36(1965)2970.
29 E. H. Putley, The Hall Effect and Related Phenomena, Butterworths and Co., London (1960).
30 C. R. Whitset, J. Appl. Phys. Suppl., 32(1961)2257.
31 H. Mette, Z. Physik, 176(1963)329.
32 S. Gruetzmann, Frequenz, 18(1964)42.
33 S. Gruetzmann, Frequenz, 18(1964)265.
34 T. Mimizuka and S. Ito, Solid-State Electron., 15(1972)1197.
35 N. Savvides and H. J. Goldsmid, J. Phys. E, Sci. Instr., 5(1972)554.
36 T. C. Harman, J. Appl. Phys., 29(1958)1373.
37 A. W. Penn, J. Sci. Instr., 41(1964)626.
38 E. A. Brown, R. J. Charlson and D. L. Johnson, Rev. Sci. Instr., 32(1961)984.
39 H. S. Hatfield and F. J. Wilkins, J. Sci. Instr., 27(1950)1.
40 J. de Jong and L. Marquenie, Instr. Practice, 16(1962)45.
41 P. L. Weickhardt, I. M. Ritchie, R. J. Batterham and D. G. Wood, J. Sci. Instr., J. Phys. E, 2(1969)1010.
42 H. J. Goldsmid, T. Knittel, N. Savvides and C. Uher, J. Phys. E, Sci. Instr., 5(1972)313.
43 H. Wagini and H. Weiss, Solid-State Electron., 8(1965)241.
44 B. Paul and H. Weiss, Solid-State Electron., 11(1968)979.
45 M. M. Kaila and H. J. Goldsmid, Phys. Stat. Sol. (a), 28(1975)K167.
46 H. H. Wieder and A. R. Clawson, Solid-State Electron., 8(1965)467.
47 H. J. Goldsmid and K. R. Sydney, J. Phys. D, Appl. Phys., 4(1971)869.
48 H. J. Goldsmid, N. Savvides and C. Uher, J. Phys D, Appl. Phys., 5(1972)1352.
49 V. A. Kudinov and B. Ya. Moizhes, Sov. Phys.-Sol. State, 7(1966)1868.
50 I. L. Drichko and I. V. Mochan, Sov. Phys.-Sol. State, 7(1966)2634.
51 M. M. Kaila, L. B. Harris and H. J. Goldsmid, Phys. Stat. Sol.(a), 37(1976)K43.

CHAPTER 10

AUTOMATIC AND COMPUTER-AIDED MEASUREMENTS

10.0 Introduction

Automatic instrumentation systems are used to measure and record parametric variations in voltage, current, temperature, pressure, etc., as a function of experimentally-imposed constraints or as a function of environmental perturbations. An important application of computers and computer-aided instrumentation is to monitor, control and process data from experiments. In these applications the computer is programmed to sense parametric variations or to produce them by means of appropriate control signals.

In general, analog data systems are used if a low accuracy can be tolerated and a wide dynamic range is required for data processing. Digital data systems are more complex, they provide higher accuracy, a narrower dynamic range and a lower cost per information channel in comparison with analog data acquisition systems. Digital data acquisition and control systems may consist of a single channel such as used for measuring and recording the variation of one parameter as a function of another or of a multi-channel system which measures a large number of input parameters, compares them against preset limits and performs computation on the data. A digital signal may consist of a pulse train composed of discrete and discontinuous equal amplitude electrical pulses whose temporal relationship contains information about the functional dependence of related variables. Such signals are used to perform logical operations in a computer "language;" this is used to write instructions and commands and various programs which constitute the "software" of a computer. The computer is connected to experimental apparatus by means of interfaces which consist of digital electronics "hardware" programmed by appropriate software. The user communicates with the computer by means of various input and output (I/O) components such as those shown in Fig. 10.1. The computer interrupt is a process whereby an external device can get the immediate attention of a computer. Figure 10.2 represents a signal flow graph of real-time signals. Digital signals can be handled directly by a computer. Analog signals require analog-to-digital (A/D) conversion. The output of a digital computer may have to be converted into analog form by means of a digital-to-analog (D/A) converter if the computer is used in a feedback loop to control the parameters of an experiment. Non-electrical input parameters such as temperature, pressure, velocity, acceleration, etc., require transducers to convert them into equivalent electrical quantities. In any case,

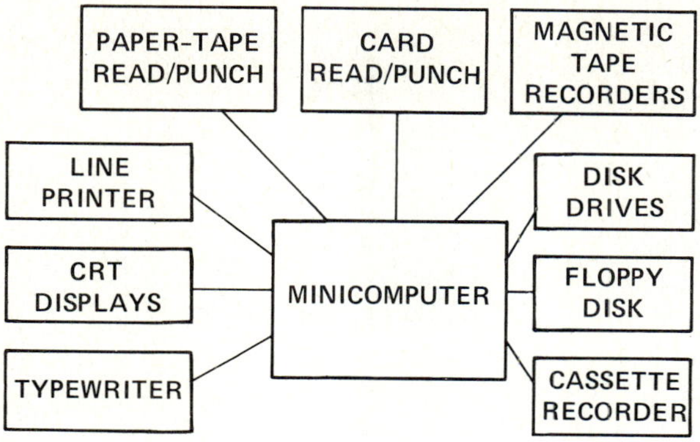

Figure 10.1. Minicomputer and peripheral input/output components.

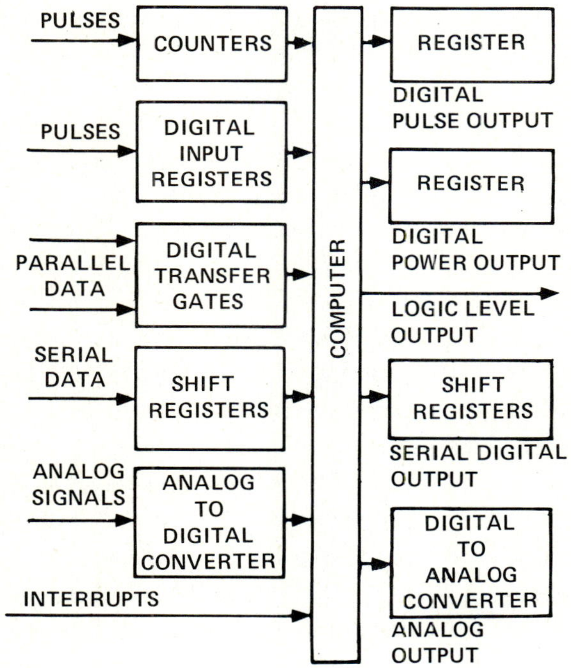

Figure 10.2. Signal flow in computer systems.

it is desirable to convert all of the analog signals into a common form and to group them into a suitable range by means of signal conditioners. It is usually expedient to share a single A/D converter between a number of analog signal sources using a multiplexing switch to select the appropriate address of the input signal, as shown in Fig. 10.3. An ideal multiplexer should transfer a

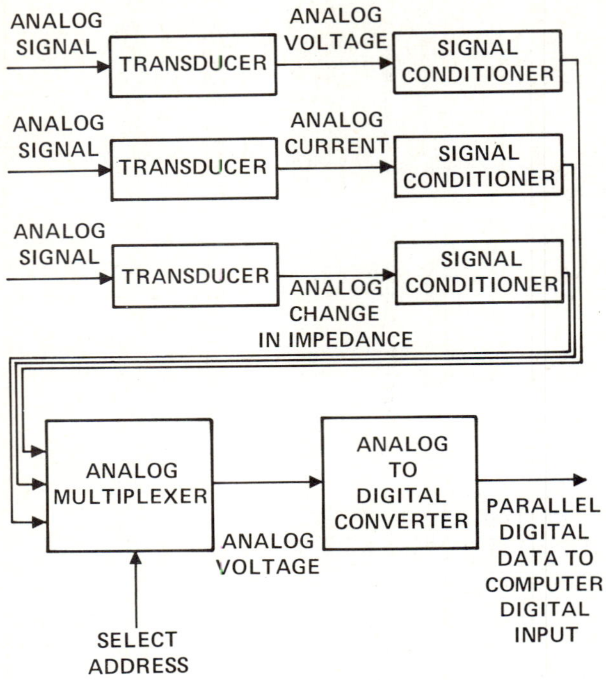

Fig. 10.3. Analog input signals converted to digital output signals.

selected analog input signal, with complete fidelity, from the input to the output without loading the signal source. A relay multiplexer is essentially a multipole single-throw switch. When the relay is energized by passing current through its armature the input signal is connected through its closed contacts to the output of the multiplexer. The relays have, as a rule, mercury-wetted contacts or are reed relays and can be operated in periods of $\sim$2 to 10 msec. Solid-state multiplexers usually employ transistors or integrated circuits so arranged as to act as a switch between two resistors. They are used for switching and commutating high frequency high level signals usually at speeds of 2.5×10^4 to 10^6 operations/sec. In a sequential scanning system the computer provides a starting address to the analog multiplexer and a sequential scanning-mode signal starts a continuous scan through the input channels. After each channel is scanned the multiplexer address selection is automatically incremented by one and the next channel in the sequence is connected to the A/D converter. A stop control can be implemented at any time in the scan sequence or after a preset number of scans. Random scanning implies an address for each selected channel and these channels can be selected in any sequence. A sampling sequence schedule can be constructed as a table of addresses in the memory of the computer. With the advent of low cost A/D integrated circuit converters it is sometimes more economical to provide a separate converter for each input signal. Separate A/D converters may also be desirable if the analog signals are generated at positions remote from the computer. In that case it may be desirable to transmit the signals in digital form. In either case the digital outputs of all the A/D converters are multiplexed on a common digital data bus.

Most A/D converters contain an analog signal comparator, a D/A converter and logic circuitry. An analog input signal is compared with the output of the D/A converter. The difference signal and its polarity are used to control the D/A converter and to make its output equal to the analog signal. When the A/D converter completes its operation a termination pulse is generated. This indicates that the data can be read by the computer and the multiplexer can be reset for the next cycle. A/D converters can be obtained in bipolar form with a conversion resolution >15 bits (1 part in 32768). For high precision conversion a sample and hold amplifier is used at the input of an A/D converter. The sample and hold amplifier acts as an analog storage device; under the control of the A/D converter it accepts a command to store an input signal at some particular instant in time. Sample and hold times of the order of 50 msec or less may be used. Several advantages are obtained by using sample and hold techniques; one of the most significant of these is the extremely high precision in the timing of the A/D converter and the resultant accuracy of the conversion process.

The same basic techniques are used for both A/D and D/A conversion. The basic component of a D/A converter is a ladder network made up of precision resistors connected as a switched parallel network. The resistors are selected by either mechanical or electronic switches. When a switch is actuated current flows through that resistor branch and the sum of the current is the analog equivalent of the digital input.

Figure 10.4 is a block diagram of the principal components of a digital computer. The memory stores binary-coded "words." Each word consists of n binary "bits" and each word has an address associated with it and the information contained within can be recalled or stored in a random manner by specifying its address. The information stored in the memory can be in the form of data, instructions or addresses. The speed with which information may be stored or retrieved is the basic memory cycle time. The central processing unit (CPU) controls the overall operation of the computer. It contains arithmetic registers and logic circuits which execute logical and arithmetic operations. The CPU retrieves each instruction from the memory, interprets and executes it and then moves on to the next instruction. The arithmetic registers are high-speed electronic two-state devices (like flip-flops), which can be used to accumulate

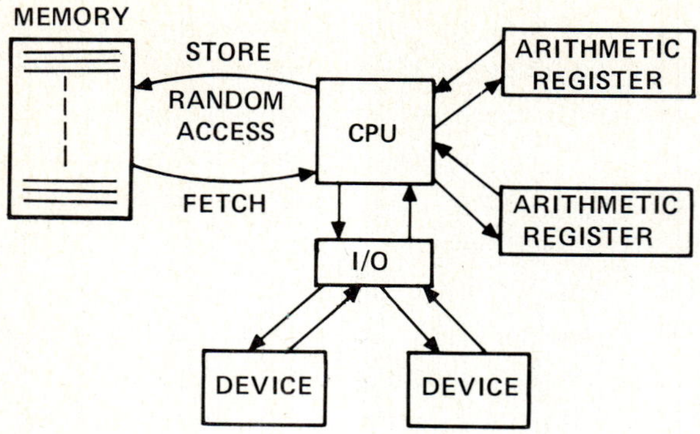

Figure 10.4. Block diagram of a digital computer configuration.

intermediate results of binary arithmetic involving n-bit data. Nearly all the arithmetic and logical operations of the CPU are carried out on data contained in the arithmetic registers. Binary information can be transferred to or from memory and the arithmetic registers by the execution of appropriate machine instructions. The console switch registers and displays are used for communication between the operator and computer. The switch register consists of an array of switches used to select a memory address, to store binary instructions or to communicate directly with the central processor.

Computer-aided instrumentation involves control of some or all of the aspects of an experimental set-up including data acquisition, processing and presentation of the final results in printed form or as graphs or charts. A computer-aided system may be considered desirable because:

(1) a computer can provide rapid response;

(2) complex experimental control operations are involved, and

(3) a desire to eliminate tedious, inefficient manual operations.

The computer to be chosen may be a minicomputer for dedicated operation of a single instrument or a large time-sharing computer servicing several smaller satellite computers and/or several different experimental set-ups while simultaneously engaged in the execution of off-line programs. Many applications fit somewhere between these extremes.

It should be recognized *a priori* that digital data processing of experimental data does not necessarily improve the results. Regardless of the dynamic range and fidelity of the digitizing devices the digital data is neither more accurate nor is it more precise than the analog signal. Another serious problem which must be considered by the user is the manner in which the experimental data is to be transmitted to the computer. This question is generally resolved on the basis of the frequency and level of the information to be transmitted. For data in which the dominant frequency is somewhat less than 1 Hz, it is possible, with filtering, to transmit millivolt-level analog signals over long distances through well-shielded cable. If the signal is first amplified to a level in the range of a few volts to minimize problems of electronic pickup, then analog signals up to about 1 kHz may be transmitted to the computer.

In the following sections some specific examples are presented of computer-aided measurements of electrical and galvanomagnetic properties of solids. For a comprehensive description of laboratory applications of computers the reader is referred to the monographs of Finkel (ref. 1), Perrone and Jones (ref. 2) and Wilkins *et al.*(ref. 3).

10.1 Automatic resistivity and Hall measurements

A semi-automatic digital recording system for measuring the temperature dependence of the conductivity and Hall coefficient of semiconductors has been described by Carter *et al.*(ref. 4). It is based on the use of a self-balancing potentiometer (ref. 5) for measuring small DC potentials such as those encountered in Hall effect measurements. The semiconductor specimen is mounted either in a cryostat or a furnace depending on the temperature range of the measurements. Three input parameters, derived from the specimen and its enclosure, corresponding to the temperature, conductivity and Hall coefficient are applied to the selector switch which connects these sequentially to the electronic potentiometer. As the potentiometer is nulled the digitizer follows the slidewire setting and a readout signal generated by the selector switch is fed to the tape encoder which then accepts the output of the digitizer translating it into a format suitable for punched tape recording. The tape thus produced is applied to a computer for processing. The computer then provides the functional dependence of the variables either in printed form or graphically as the logarithm of the conductivity, Hall coefficient and Hall mobility as a function of the reciprocal temperature.

An automatic programmable data acquisition system intended for the measurement of the Hall coefficient and resistivity of semiconductors was developed by Duga *et al.*(ref. 6). The system can be used to measure particular charge carrier transport properties of more than one specimen as shown in Fig. 10.5. The leads and thermocouples for measuring the resistivity, Hall coefficient and temperature are attached to each of the four specimens A, B, C, and D and mounted within a thermostated furnace or cryostat S. The lead connections to specimens B, C, and D are the same as those shown for specimen A. The auxiliary thermocouple T is intended to provide only an approximate temperature of the specimen environment. Its output voltage, after amplification by a factor of 10^2, is compared against a reference chosen, for example, to provide $\sim 10^2$ equal increments on a $\ell n(1/T)$ scale over the range of measurements. When the difference voltage is zero then an instruction is sent to an input selector which connects the appropriate leads

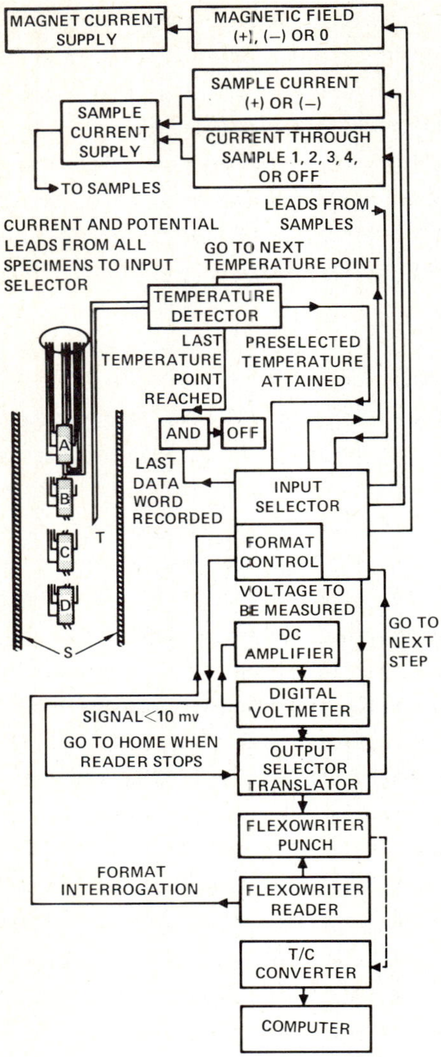

Fig. 10.5. Automatic data acquisition system for Hall effect and resistivity measurements. [after J. J. Duga, J. J. Freundlich and R. T. Bate, Rev. Sci. Instr., 33(1962)365]

sequentially to the voltage measurement apparatus. Instructions are relayed to the magnet power supply to raise, lower, reverse, or turn off the magnetic field at certain steps of the cycle. A voltage chosen by the input selector is applied directly to the digital voltmeter if it is larger than 10 mV; if it is less than 10 mV it is first amplified by a factor of 10^2 and then redirected to the voltmeter. The final voltage to be recorded is relayed to the output selector which translates the binary code of the voltmeter to the binary code of the typewriter-punch. The typewriter records the measured voltage in a specific format and the same data is punched into a five-channel paper tape. The input selector then steps to the next point in the cycle and the operation is repeated. The information on the tape can be converted to punched cards fed to a digital computer whose output is the tabulated or plotted desired information.

The temperature sensor employs two 50-position mechanical switches to provide up to 100 desired temperature measurements and the input selector uses two 50-position stepping switches to provide a maximum of 100 measured quantities at any given temperature. The voltages to be measured and recorded do not pass through the stepping switches; they are directed instead through hermetically-sealed relays with gold contacts energized by the stepping switches. The voltages are measured by means of a 4-digit ranging voltmeter used in conjunction with a chopper-stabilized DC amplifier with a gain of 100. It is capable of 50 readings/sec. with a resolution of $1 \mu V$, an accuracy of $\pm 0.01\%$ with an input impedance of 10^9 ohms (at balance). The DC amplifier which is automatically switched-in for input voltages <10 mV has a gain accuracy of $\pm 0.2\%$ and an input impedance of 10^8 ohms.

A relatively simple re-programming of the input selector allows measurements to be made of the magnetoresistance (at a fixed magnetic field), thermoelectric power, magneto-thermoelectric power and other transport parameters. It may also be used to determine rapidly the temperature distribution in a medium where the independent parameter may be temperature, time, pressure, displacement, etc.

An automatic data acquisition controller (ADAC) system has been described by Loebner *et al.*(ref. 7). The data acquisition system consists of an input scanner, a guarded data amplifier with an input impedance $>10^{10}$ ohms, an integrating digital voltmeter whose output may be applied to a teletype punch through an appropriate coupler, a punched card coupler or magnetic tape recorder which provide the input for the data-processing computer shown in Fig. 10.6. The block labeled environment represents the magnetic field, electric field, temperature, etc., to which the experimental sample is exposed. These parameters are controlled by ADAC in a predetermined sequence which is set by its program control matrix and programming switches. An example of the sequence of events in the measurement is as follows:

(1) a scan is initiated by the interval timer;
(2) ADAC sets the environment (turns on the magnetic field, for example);
(3) ADAC provides a scan-command signal to the input scanner;
(4) the data acquisition measures and records up to 25 separate voltages between preselected pairs of terminals and these voltages are generated both by the experimental sample and by the environment sensors;

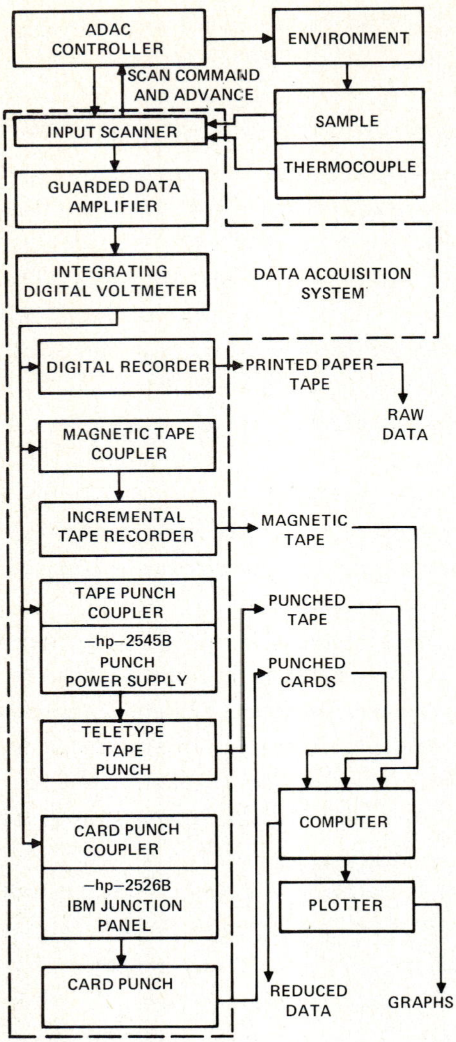

Fig. 10.6. Automatic data acquisition controller system for Hall effect and conductivity measurements. [after E. Loebner, T. J. Diesel and C. M. Schade, Hewlett-Packard J., 18(1966)9]

(5) after completion of the measurement and recording of the data the data acquisition system gives a scan-advance signal to ADAC which proceeds to change the environment;

(6) events 2-5 are repeated until the specimen has been subjected to the appropriate changes of its environment;

(7) when ADAC has completed its environment-controlling cycle it resets the interval timer (which can be set from 1 to 30 min.) and the apparatus is at rest until this time expires when another cycle is initiated; and

(8) after the expiration of the time set by another (total acquisition time) timer ADAC turns the entire system off.

Because of this automatic mode of operation, which may last to 30 hours, it is possible to leave the apparatus unattended overnight or on weekends allowing the specimen temperature to drift up slowly from liquid nitrogen (for example) to room temperature.

An automated data-collection system applied to Hall effect and resistivity measurements has been described by Thomas (ref. 8) for measuring the electrical and galvanomagnetic properties of silicon and cadmium sulfide bulk and thin film specimens between 4.2°K and 400°K. A block diagram of the system is shown in Fig. 10.7. The primary components consist of a 50-step shielded 2-wire input scanner, a timer panel, a 50-step function scanner, a relay-operated function controller with manual or automatic switching modes, a 50-step function programmer, a DC preamplifier, an oscilloscope signal monitor, a self-balancing 4-digit autoranging digital voltmeter, a data translator and a recorder that provides both typewritten copy and computer-compatible punched paper tape outputs. The lead wires of the specimen holder were connected to a terminal board that provides up to fifty 2-wire shielded inputs to the input scanner. The latter collects the appropriate signal pairs for measurement and connects them to the preamplifier input. At the same time the function programmer switches all the functions that are programmed for that particular step. The function scanner tracks the input scanner and distributes the 28V DC source to the appropriate relays in the function control circuit energizing them. The selection of the relays to be energized for a particular input is determined by the function-programming board. This is of the pin-board type and allows the choice of either an "on" or "off" mode of 10 functions for each of the 50 steps or inputs selected by the input scanners. The preamplifier is controlled by the function programmer when the system is operating in the automatic mode. The input signal is routed through the preamplifier if required and its appropriate gain of 10, 10^2 or 10^3 can be thus chosen. The amplified signal is then routed to the autoranging digital voltmeter where it is measured and applied to the output data translator. The latter takes the voltmeter reading along with other 10-line parallel information which includes the input scanner channel identification, voltage exponent polarity, decimal point location and voltage polarity, serializes it to conform to the data word format, converts it to the 1-2-4-8 binary-coded decimal code and applies this data to the recorder. For each temperature the sequence of operations continues until all the voltages are measured and recorded. The input scanner then resets to start the cycle again; a timer is programmed-in on the

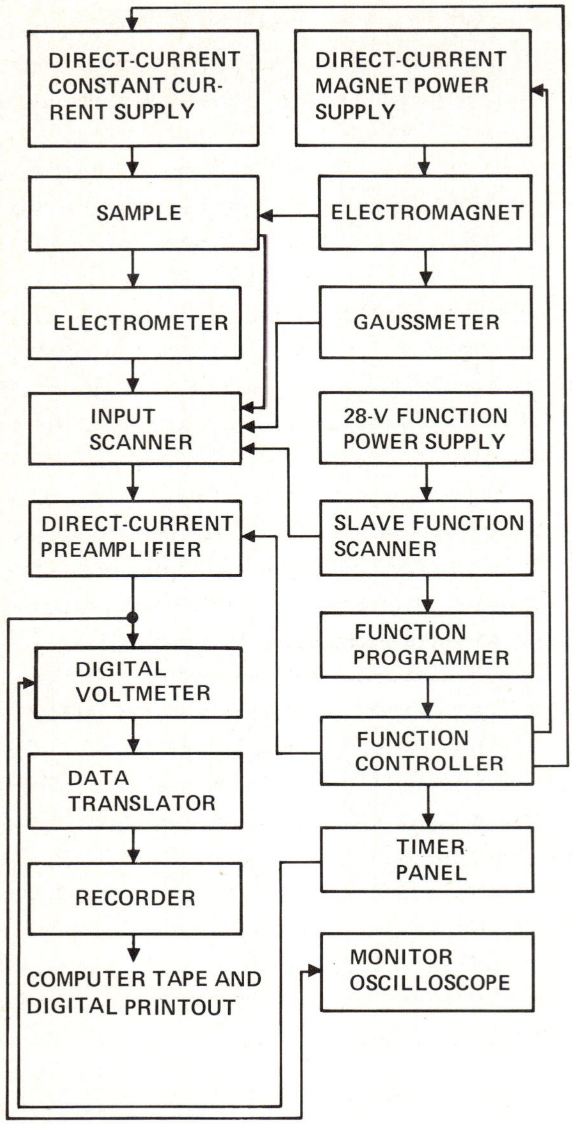

Fig. 10.7. Block diagram of function-programmable automated Hall effect and resistivity measuring system. [after R. D. Thomas, NASA Tech. Memo. TMX-1464 (1967)]

first step of the cycle to control the frequency of the cycles. Extension of the system to include the measurement of high resistivity materials was achieved by introducing a high input impedance electrometer between the sample leads and the input scanner.

Elliott and Wilson (ref. 9) have described a data acquisition system for making Hall coefficient and conductivity measurements on semiconductors between liquid helium and 400°K in a single 6-hour run. All the voltage measurements were made with an automatic-ranging digital voltmeter which accepts potentials up to 10^3V and a preamplifier with an input impedance $>10^{10}$ ohms. The sensitivity of the voltmeter-preamplifier combination was $\sim 1\mu V$. A bank of reed relays with diffused gold contacts was used to switch the input channels. The relays were found to perform satisfactorily for low level DC signals and have adequate leakage resistance. To initiate an operation sequence the input channel is selected and the appropriate environment (direction of current and magnetic field) is selected by the sequence generator. A delay of 10 sec. after the magnetic field has been reversed or 2 sec. for other operations is established before a read-instruction is given to the voltmeter. The binary-coded decimal information from the voltmeter, together with a channel-identifier is fed via a serializer to a tape punch for entry; following this operation a signal is generated which operates the sequence generator. The sequence generator employs digital-integrated circuits in the form of a "ripple-through" binary counter consisting of five J-K flip-flops which is reset to zero after a count of 17.

The temperature of the cryostat was servo-controlled using an appropriate time-varying reference signal. A suitable sensor for the temperature controller is a forward-biased GaAs diode whose sensitivity from 400°K to 70°K is nearly constant at 2 mV/°K falling to 0.7 mV/°K at 40°K if biased with 1 mA. A temperature stability of ±0.2°K can be obtained at 10°K and ±0.1°K at or above 100°K.

A remotely programmable electromagnet and associated data acquisition system for galvanomagnetic measurements was developed by Clawson (refs. 10-12). The power supply which provides the regulated current for the electromagnet is an adaptation of a design described by Garwin (refs. 13, 14); it can maintain the current stable to 1 part in 10^5. The magnetic field intensity can be programmed externally by means of a reference voltage. This voltage provides a reference for operating the magnet in the current controlled mode by means of a meter shunt or in a field-controlled mode using a Hall generator sensor. The block diagram shown in Fig. 10.8 shows the basic building blocks of the controller and data acquisition system (ref. 11). The commutation of four electrical contacts on a specimen is provided for by the system and the polarity of the current applied to the specimen is reversible for both the van der Pauw resistivity and Hall voltage measurement configurations. Four computer controlled relay switches provide the command logic for the contact commutation. Internal switching logic maintains the same output voltage polarity for every choice of current contacts and commutates the voltage contacts for either van der Pauw resistivity measurements or Hall voltage measurements depending on whether adjacent or alternate contacts are chosen for the input current I. Manual operation is provided by

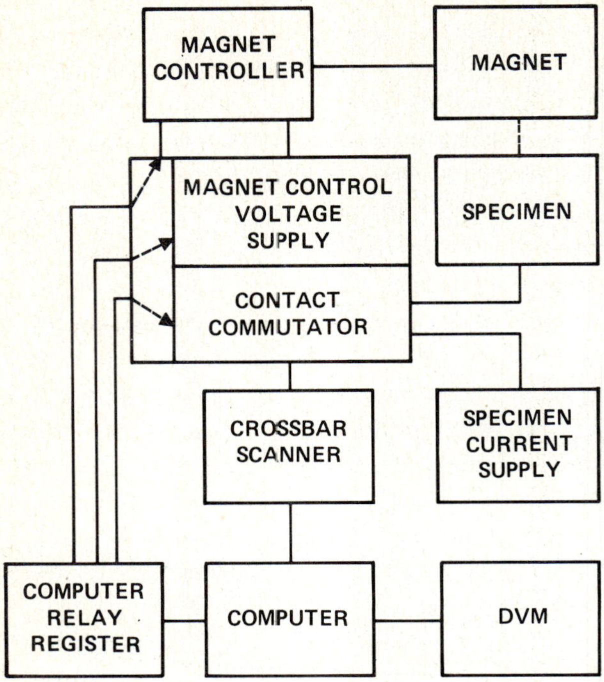

Fig. 10.8. Block diagram of apparatus for conventional and van der Pauw-type resistivity and Hall measurements. [after A. R. Clawson, NOLC Tech. Memo 613-35 (1969)]

two five-position rotary switches, one for I+ and one for I-. There is an "off" position and four contact positions. A remote-operation-override disables the switches when the mode is automatic (computer controlled).

The specimen current is obtained from an external constant current supply. A precision resistor is provided to allow monitoring of the current by the DVM. The voltage drop across the resistor is fed to one channel of the crossbar scanner. Two bits of binary information for each lead allow it to be connected to any one of the four specimen contacts. This allows twelve different current configurations in the specimen (the other four possibilities of the $N^2=16$ commutations are for I+ and I- applied to the same terminal).

A semi-automatic system intended for van der Pauw-type Hall measurements has been described by Eden and Zakrewski (ref. 15). It consists of a programming switch with 38 single-pole double-throw contacts and 60 program positions. The switch can be programmed to take measurements on one or more samples. It is driven by a stepping motor which is energized by two electronic timers. Three relays are used to turn the magnetic field on and off and to reverse its polarity. The relays are actuated by contacts located on the programming switch in a sequence chosen to minimize the effects of inductive transients generated by switching of the magnetic field.

Shewchun *et al.*(ref. 16) have described a computer-controlled automatic conductivity and Hall measurement apparatus, shown in Fig. 10.9, which employs a small process control computer (PCC) to operate and control a temperature-programmable cryostat and a digital data acquisition system. The control functions include setting the sample current, the temperature and magnetic field,

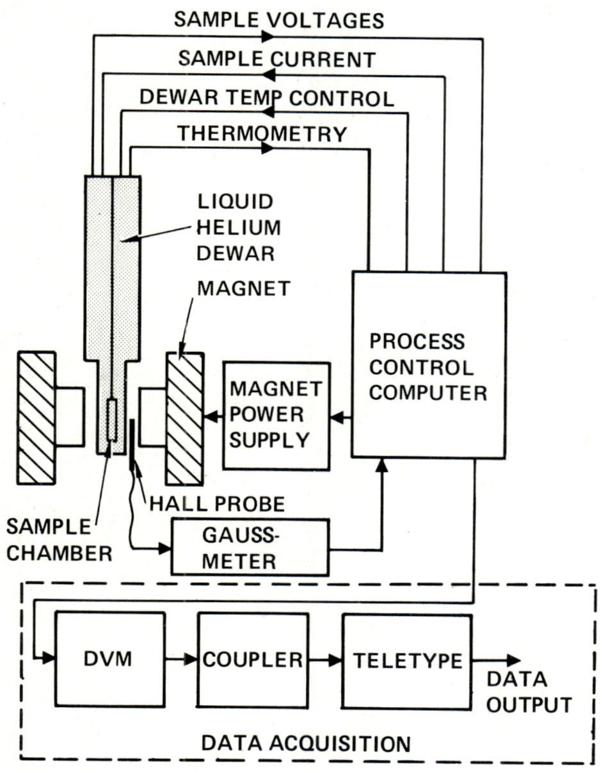

Fig. 10.9. Apparatus for automatic conductivity and Hall measurements between 4.2°K and 400°K showing liquid helium dewar, process control computer, and data acquisition system. [after J. Shewchun, K. M. Ghanekar, R. Yager, H. D. Barber and T. Thomson, Rev. Sci. Instr., 42(1971)1797]

reading and recording these parameters and following the completion of a sequence proceeding to the next temperature in the program and repeating the cycle. The PCC has six modes of operation: automatic temperature control, semi-automatic temperature control, manual temperature control, automatic temperature drift, semi-automatic temperature drift and manual temperature drift. It performs the following measurements, monitoring and control functions:

(1) adjustment of sample temperature with a sequence delay to provide for temperature equilibrium;
(2) adjustment of sample current so that voltage drop across the sample is maintained between preprogrammed limits;
(3) monitors for system malfunction and operator errors;
(4) provides time sequence control with proper time intervals;
(5) provides magnet control including turn on, turn off, and polarity reversal; and
(6) adjusts the electrical configuration of the sample for measurement and read commands (sync. or activate signals) to the data acquisition system.

In the automatic temperature-controlled mode all six functions are executed by the PCC and the system shuts down when all preprogrammed temperatures have been scanned. In the semi-automatic mode the system holds at the end of a sequence until the next temperature is selected manually. In the temperature drift modes the system operation is the same as in the temperature-controlled modes except that the cryostat is allowed to warm up slowly without any temperature control.

A fully automatic Hall and resistivity measurement system designed for performing measurements between 5°K and 300°K has been described by Blood and Headon (ref. 17). It employs hard-wired logic circuits for performing the measurement and control functions and can stabilize the specimen temperature to within ±0.01°K at 50 preprogrammed temperatures in the above described temperature range. This excellent temperature control feature is achieved by careful design of the thermal impedances between the sample, sample holders, and temperature sensors.

10.2 Time-sharing computer-controlled measurements

A time-shared computer has been used by Bullis *et al.*(ref. 18) to perform resistivity and Hall effect measurements and to process the data derived from them. In order to use a time-shared computer for experimental measurements the process must not place a heavy work load on the computer. A time-shared computer provides services for a number of independent on-line users by multiplexing various functional blocks in the computer. If any one block is employed for a period which is relatively short with respect to the response time expected by an individual user, then each user is given the impression that he alone is using the system.

Interaction between a user and a time-shared computer usually consists of a sequence of message pairs: an instruction to and a response from the computer. Response times of seconds or minutes are usually adequate while the computer cycle time may be of the order of μsec. Experimental apparatus is connected to the computer by means of an interface which consists of a buffer and hard-wired logic circuitry. If only a few simple tasks require fast response times and the rest can be accommodated by a considerably longer response time, then a simple interface can provide the interconnection to the computer.

A time-shared computer system contains one or more central processor units (CPU) for exercising control and for performing arithmetic and other operations determined by its cycle time, number of internal registers, instruction repertoire and organization. Main-frame random access memory is usually expensive compared to other computer components. Such a memory is usually shared by a number of users and only rarely is there enough storage space available to accommodate all users. It is, therefore, necessary to swap programs or positions of a program in and out of the main-frame memory after they have used their share of CPU time. A buffer or intermediate memory such as a drum or disc file can be used for such purposes.

A variety of interconnections between experimental apparatus and a time-shared computer system are feasible:

(1) apparatus may be connected by means of a specially-designed hard-wired interface to a special purpose computer which services other specially interfaced apparatus;

(2) apparatus may be connected via a special purpose hard-wired interface to a large general purpose time-shared computer system. The interface should be designed so that the burden of the experiment on the computer is of the same order as that placed on the system by the conventional I/O components; and

(3) an experiment may be connected to a small computer which is itself connected as a remote terminal to a larger time-shared computer.

An experiment which requires servicing a number of critical real time needs yet still requires extensive computation "on-line" but without a critical response time is a candidate for this type of configuration.

Bullis *et al.*(ref. 18) have described the configuration of the apparatus for Hall effect measurements, the associated interface and its connection to the computer with on-line remote devices (CORD) time-sharing system using a teletypewriter (TTY) communication line as shown in Fig. 10.10.(a). Interconnections between the apparatus and the communication controller are made with a two-wire shielded cable which provides duplex communication. The switch box provides local or computer-controlled operation. In the local mode everything typed on the TTY is transmitted to the interface and the output of the interface is printed on the TTY.

In the computer-controlled mode everything typed on the TTY is transmitted to the experimental apparatus and to the computer; the computer output is transmitted to the apparatus and printed on the TTY and experimental data generated by the specimen is transmitted to the computer and printed on the TTY.

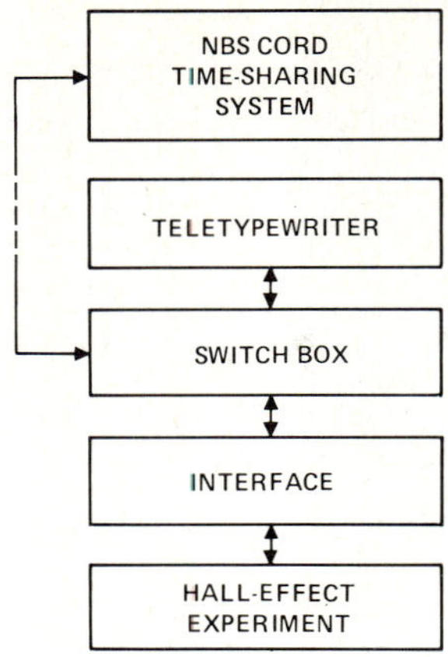

(a)

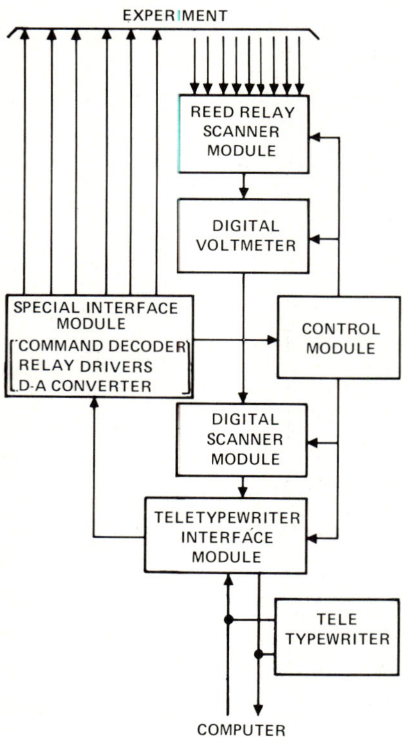

(b)

Fig. 10.10. (a) Block diagram of computer configuration used for time sharing; (b) block diagram of interface. [after W. M. Bullis, W. R. Thurber, T. N. Pyke, Jr., F. H. Ulmer and A. L. Koenig, Nat'l Bureau of Std's Tech. Note 510(1969)]

The interface consists of four general purpose modules, one special purpose module and an autoranging digital voltmeter. Figure 10.10.(b) shows the interconnection of these modules and their relative positions in the system. The voltmeter is connected to the apparatus through the reed-relay scanner module. The output from the voltmeter is converted to a bit-serial signal transmitted to the computer in TTY-compatible form. If a scan command is received then the control module assumes its supervisory role checking status signals and issuing commands to the other modules until all the voltage readings have been transmitted to the computer. Control of the environment (magnetic field and temperature) is performed by the interface module which acts as a command decoder and relay driver and controls the specimen temperature in either of two modes:

(1) a single character code indicates a current pulse to a heater with the duration and amplitude of the pulse preset manually;

(2) a continuously supplied heater current with the heater current controlled by the computer; this requires two code characters, the first indicates that the next is to be processed as a heater current, the second is switched to a D/A converter which, in turn, drives a programmable power supply source for the heater.

The reed-relay scanner is used as a sequential multiplexer. It can scan up to 20 analog data channels and connect them to a common output line. The multiplexing is done by means of double-pole single-throw normally-open reed relays. The digital voltmeter with a rated accuracy of ±0.01% uses a dual slope integration technique which minimizes the effects of long-term drift in component value in the measuring circuit. Voltage ranges of ±0.1, 1.0, 10, 100, and 1000 V are selected automatically by remote control or manually. The digitizing is done with four full decades plus a fifth digit to allow a maximum reading of 1600 V on all ranges and parallel readout is provided by the voltmeter in binary-coded decimal output. The digital scanner module is a multiplexing device for connecting parallel digital data sources to serial-by-character output devices. It can switch sequentially up to 16 inputs of 8 lines each to a common set of 8 output lines.

The primary function of the TTY interface module is the conversion of the data from the parallel lines to serial form, its secondary function is that of code conversion. The module is also used for remote control of punched paper tape reading and for the generation of busy signals which inform the system controller of its status.

The control module exercises supervisory control over the other interface modules. Depending on the status of the system (as indicated by the busy signals on the condition lines, time signals, switch positions) the control module generates "do" commands to the other modules in the system. The actions of the control module may be determined by an internally stored program and can perform an instruction such as energize, test, go to the next step every 50 μsec. It has a capacity of 16 instructions, 8 commands and 14 condition lines.

The major function of the special interface module is impedance matching and voltage level shifting required for the digital voltmeter. Another function is that of command decoding of characters received on the TTY from the computer. Decoding is done by a diode matrix whose inputs are connected to the outputs of the buffer registers in the TTY interface module. The outputs of the decoders are used to drive relays, signal the control module or to initiate other control sequences in the interface module. Figure 10.11 shows the general program flow for a Hall measurement run. The parameters include the specimen geometry and a set of stop points for thermocouple voltages representing temperature values. The specimen is brought to the first temperature and a full set of data is taken from the experiment. During the data input sequence the heater current, which controls the specimen temperature, is adjusted to minimize temperature drift. The magnetic field, resistivity, Hall coefficient and mobility are then computed and printed on the TTY. When the last temperature stop point has been reached the program is halted. Reproducibility of the computer-controlled measurements at 77°K was found to within four significant figures. Normally-made measurement and measurements made with the time-shared system on the same specimens were found to agree within 1% for resistivity and 2% for Hall measurements. The uncertainty in the temperature reading is ±1°K at 20°K decreasing to ±0.5°K at 100°K and ±0.2°K above this temperature.

Crossley and Ham (ref. 19) have described a data collection and analysis system used to obtain parametric data for characterizing integrated circuit manufacturing processes. The system is designed to collect and analyze results of electrical tests on specialized test structures fabricated on epitaxial silicon-on-sapphire or bulk silicon epilayers. The tester can sense currents as small as 1 pA, it can measure capacitance to 0.1 pF at frequencies between 10 kHz and 1 MHz and can generate currents as small as 1 nA. The tester is controlled by a special purpose computer (70-25) which can be connected to a much larger computer (70-45) operated in the time-sharing mode. Figure 10.12 shows a block diagram of the system. The larger multi-purpose computer can control the tester but a better solution is the use of the magnetic tape recorder which eliminates the dependence of the system on the time-sharing link; when the data gathering is complete then the 70-45 can be used to analyze it by a number of methods. Large volumes of data may be reduced in the form of histograms. Correlation between independently measured parameters can also be made in the form of two dimensional correlation plots. Alternatively, a perspective three-dimensional plot of parameter values or their combination can be represented as a function of wafer position. This type of analysis requires a minicomputer, as shown in Fig. 10.12.

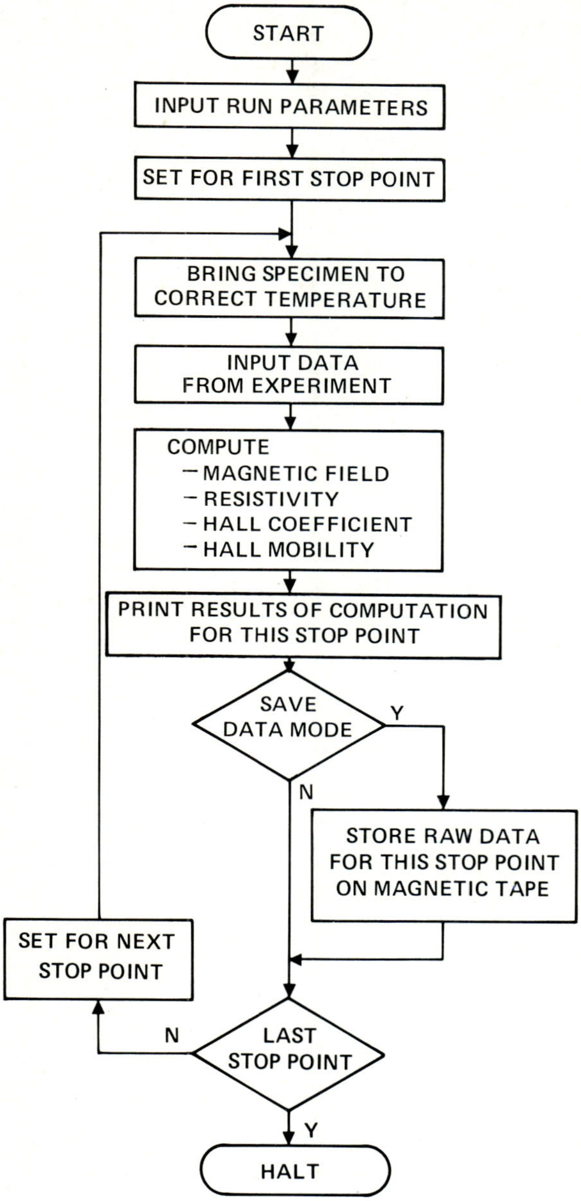

Fig. 10.11. General program flow graph. [after W. M. Bullis *et al.*, Nat'l Bureau of Std's Tech. Note 510(1969)]

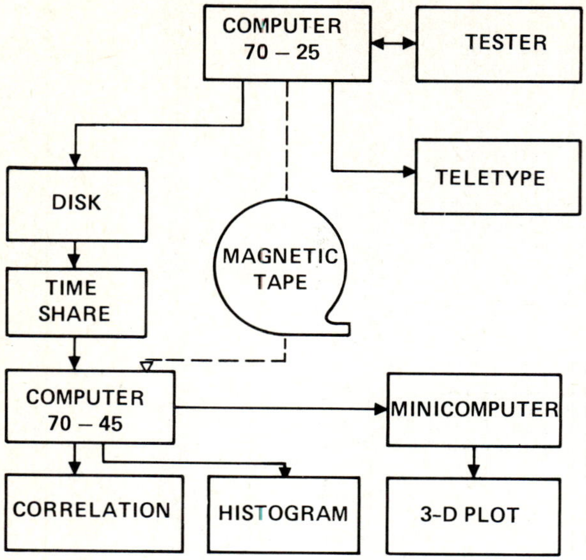

Fig. 10.12. Layout of data collection and analysis system for the investigation of integrated circuit processes. [after P. A. Crossley and W. E. Ham, J. Electron. Mat., 2(1973)465]

When operated in the diagnostic mode, programs written for the special purpose computer can:

(1) provide comprehensive rejection of invalid data;

(2) can reduce large numbers of instrumental readings to a single parameter value;

(3) can provide tests weighted and conditioned by the results of previous measurements;

(4) can alter the sequence of tests based either on external controls or test results;

(5) can provide a variety of error messages to specify which tests are invalid and why; and

(6) can accommodate a wide range of test structure sizes, each of which has the same general configuration, without requiring reprogramming.

The data generated in this manner can provide a characterization of the capabilities and limitations of many aspects of integrated circuit fabrication processes.

10.3 Ovens, cryostats and temperature controllers

The measurement of the electrical and galvanomagnetic coefficients usually require a temperature-controlled environment, a furnace for high temperature measurements and a cryostat for low temperature measurements. Furthermore, the dimensions of these enclosures must usually be such as to fit between the pole-pieces of a magnet. The construction of a furnace suitable for investigating Ge specimens at temperatures up to 925°C has been described by Dunlap *et al.*(ref. 20). The design and construction of cryostats suitable for electrical and galvanomagnetic measurements down to 1.2°K have been described by Putley (ref. 21). The properties of various cryostats, techniques of handling liquid helium, and temperature measurement procedures are described in the excellent monograph of Rose-Innes (ref. 22). An automatic temperature controller and cryostat providing constant temperatures to within ±0.01°K in the range between 4°K and 400°K was described by Arajs and Conroy (ref. 23). The cryostat employs a controlled heat leak between two concentric chambers and a motor-driven proportional temperature controller. A more elaborate double-chamber cryostat with electronic temperature control was described by Hennephof *et al.*(ref. 24). The double-chamber cryostat shown in Fig. 10.13.(a) is immersed in liquid nitrogen or liquid helium contained in a dewar vessel. Each of the two chambers is closed by soldering its respective cap with Cerroseal solder. The caps are provided with a ring to which the electrical leads are anchored and the chambers are separated by a length of 11.8mm x 12.6mm stainless steel tubing whose lower section fits into a bush on the copper cap. Bifilar-wound constantan wire is attached to this bush and constitutes the heater which controls the temperature of the inner chamber. Between both caps an Au+0.03 at.% Fe-Chromel differential thermocouple measures the temperature difference between the cap of the chamber containing the specimen and the cryogenic liquid. Figure 10.13.(b) shows a block diagram of the temperature control system. The differential thermocouple output voltage and a reference voltage are applied to the photochopper which converts their DC sum into a 50 Hz AC signal whose phase is controlled by the sign of the DC signal. The temperature of the inner chamber can be maintained to within 0.05°K for more than 0.25 hours.

A low cost temperature-controlled oven of long term (more than 20 hours) stability of ±0.6μ°K, a regulation of 3.5×10^{-5} and a reproducibility of $\Delta T/T=10^{-3}$ was described by Beigang and Telle (ref. 25). The circuit is based on the use of a filamentary heater which is also the temperature sensor. Its resistance is compared against a high precision reference resistance in a Wheatstone bridge network. The voltage output of the bridge is amplified by a low-drift operational amplifier and thereafter by a power amplifier whose output is then applied in a feedback loop to the input of the bridge circuit. A similar temperature controller has been described by Weidman and Brown (ref. 26). A stable low cost proportional temperature controller using a thermocouple sensor to control temperatures to 1000°C with a stability of ±1.5°C has been described by Benkiki and Penchina (ref. 27). It employs an integrated circuit differential

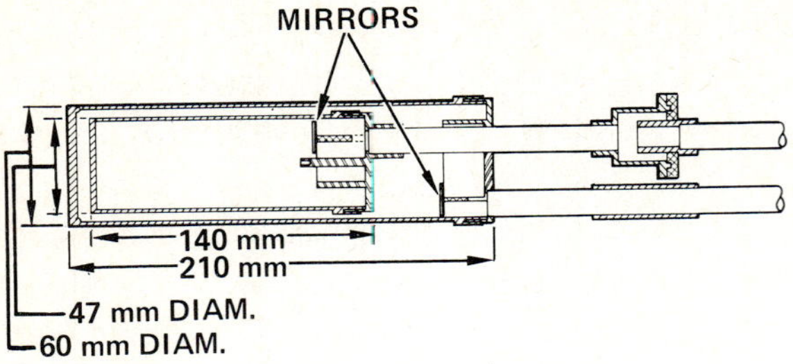

(a)

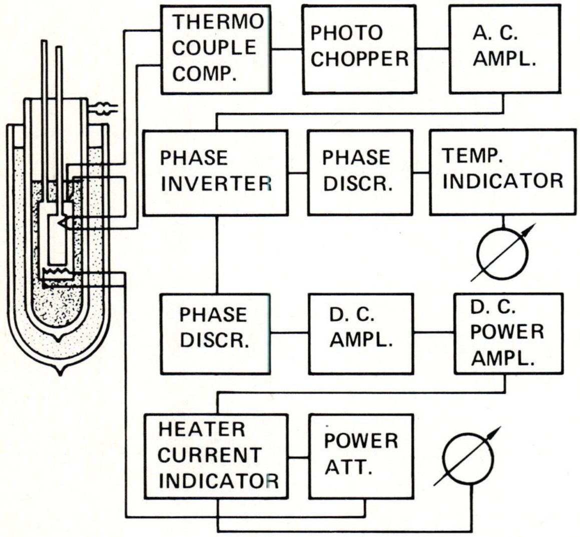

(b)

Fig. 10.13. (a) Double-chamber cryostat systems; (b) temperature control using a thermocouple sensor. [after J. Hennephof, G. Hoekstra and J. van Overbeeke, J. Sci. Instr., (J. Phys. E, Ser. 2)2(1969)]

amplifier with a zero crossing switch and a triac power control circuit. Chromel-alumel thermocouples have been used to control 1.8kW furnaces with such controllers. Ultra-high precision temperature controllers which can control temperatures to within $\pm 15\mu°K$ have been described by Williams (ref. 28) and by Sarid and Cannell (ref. 29).

10.4 Digitally-programmable components

Many experimental measurements require a voltage or current which can be changed in predetermined steps and DC power supplies, which can be programmed remotely with binary-coded decimal (BCD) signals, are available commercially. Hummer and Burns (ref. 30) have described the use of a low cost hand-held calculator for programming such a power supply. They designed an interface which converts the light-emitting diode (LED) display of the calculator to a multiplexed BCD signal which is then used to program the power supply. It is possible to choose an initial voltage and an appropriate step size to produce a stair-step voltage ramp with the dwell time at each voltage step determined by an external clock or other control signal. The interface can generate an increasing or decreasing ramp voltage. Using such an interface and a precision voltage source Hummer and Burns were able to sweep the power supply from 0 to 10^3V with any step size in excess of 10^{-3}V or 0.01%.

A programmable low-cost calculator for analytical processing of data derived from digital instrumentation was described by Olofsson and Edwall (ref. 31). Five-digit BCD-coded data obtained from a digital voltmeter (DVM) was applied to a hand-held programmable calculator in order to perform a variety of mathematical operations upon it and then displaying the result. The design took into account the fact that such calculators accept input data sequentially while DVMs present their output data with all digits in parallel. Using five shift registers, up to five digits from a DVM can be entered into the calculator in a specific sequence; numbers and instructions from the calculator keyboard and four internal keys can be entered into the program by means of a diode matrix encoder.

Multichannel analyzers when used in the multichannel scaling mode (MSC) provide a signal scaled with some parameter such as voltage, magnetic field, temperature, etc. A voltage which is proportional to the address at which the scaled signal is stored can be derived from a multichannel analyzer (MCA) or from the address lines of a D/A converter. Risley (ref. 32) has described the construction of such apparatus using the LED display of a programmable calculator. A block diagram of its components is shown in Fig. 10.14. Such digitally-controlled voltage sources can be used to drive an auxiliary power supply which, in turn, can control the temperature, magnetic field, pressure, linear or angular displacement, etc., of an experiment and with the addition of a voltage-to-frequency converter it can be used as a digitally programmable oscillator. The calculator establishes the required stepping routine and displays a number proportional to the voltage. The number is decoded into BCD-format and a D/A converter provides the output voltage through an operational amplifier. Signals from five digits of the display on the calculator are used and the voltage levels from the display are converted into TTL levels in the display interface. Each segment of each digit in the display is turned on sequentially, the data is stored in the segment latches and then decoded in the seven-segment-to-BCD converter. When a different voltage is required the inputs to the D/A converter

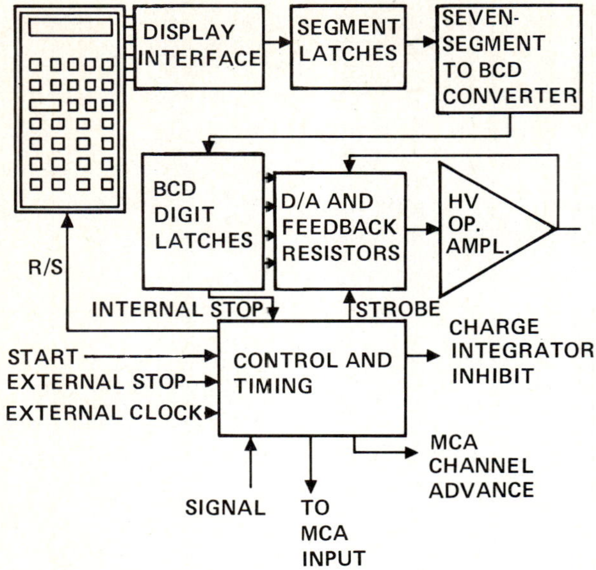

Fig. 10.14. Programmable power supply for use with a multichannel analyzer. [after J. S. Risley, J. Phys. E., Sci. Instr., 10(1977)464]

are strobed. The output from the D/A converter controls a high voltage operational amplifier. The control and timing circuitry determines when to strobe the D/A converter, gates the input pulses to the MCA, gates the charge integrator accepts an external clock pulse to cycle the system and activates the run/stop (R/S) key on the calculator.

Computer-aided and computer-controlled laboratory instruments are still in their infancy. Microprocessors are generating a minor revolution in the concepts and operation of instruments used for electrical measurements. The reader is referred for pertinent information to the paper by Randle and Keith (ref. 33).

REFERENCES

1 J. Finkel, Computer-Aided Experimentation: Interfacing to Minicomputers, John Wiley and Sons, New York (1975).
2 S. P. Perone and D. O. Jones, Digital Computers in Scientific Instrumentation, McGraw Hill Book Co., New York (1973).
3 C. L. Wilkins, S. P. Perone, C. E. Klopfenstein, R. C. Williams and D. E. Jones, Digital Electronics and Laboratory Computer Experiments, Plenum Press, New York (1975).
4 R. H. A. Carter, D. J. Howarth and E. H. Putley, J. Sci. Instr., 35(1958)116.
5 E. H. Putley, J. Sci. Instr., 33(1956)164.
6 J. J. Duga, J. J. Freundlich and R. T. Bate, Rev. Sci. Instr., 33(1962)365.
7 E. Loebner, T. J. Diesel and C. M. Schade, Hewlett-Packard J., 18(1966)9.
8 R. E. Thomas, "Automated Data Collection Systems Applied to Hall Effect and Resistivity Measurements," NASA TMX-1464(1967), available from Clearinghouse for U.S. Fed. Sci. and Tech. Info.

9 C. T. Elliott and D. J. Wilson, J. Phys. E, Sci. Instr., 2(1969)956.
10 A. R. Clawson, Remotely Programmable Laboratory Magnet, NWC Tech. Memo 613-34(1969), available by request from author, NOSC, San Diego, CA.
11 A. R. Clawson, Remote Commutation of the Contacts on a Galvanomagnetic Specimen and Computer-Controlled Reference Voltage for Programming the Magnetic Field, NWC Tech. Memo 613-35(1969), available by request from author, NOSC, San Diego, CA.
12 A. R. Clawson, Van der Pauw Resistivity and Hall Measurements - A Basic Language Data Acquisition System, NOSC(1977), available by request from author, NOSC, San Diego, CA.
13 R. L. Garwin, Rev. Sci. Instr., 29(1958)223.
14 R. L. Garwin, D. Hutchinson, S. Penman and G. Shapiro, Rev. Sci. Instr., 30(1959)105.
15 R. C. Eden and W. H. Zakrewski, Rev. Sci. Instr., 41(1970)1030.
16 J. Shewchun, K. M. Ghanekar, R. Yager, H. D. Barber and D. Thompson, Rev. Sci. Instr., 42(1971)1797.
17 P. Blood and R. F. Headon, J. Phys. E, Sci. Instr., 8(1975)958.
18 W. M. Bullis, W. R. Thurber, T. N. Pyke, Jr., F. H. Ulmer and A. L. Koenig, Use of a Time-Shared Computer System to Control a Hall Effect Experiment, Nat'l Bureau Std's, U.S., Tech. Note 510(1969) available from Superintendent of Documents, U.S. Gov't Printing Office, Washington, D.C.
19 P. A. Crossley and W. A. Ham, J. Electron. Mat., 2(1973)465.
20 W. C. Dunlap, Jr., D. A. McMillan and R. A. Brooks, Phys. Rev., 93(1953)911.
21 E. H. Putley, The Hall Effect and Related Phenomena, Butterworths, London (1960)34.
22 A. C. Rose-Innes, Low Temperature Techniques, D. Van Nostrand Co. Inc., Princeton (1964).
23 S. Arajs and J. W. Conroy, Japan. J. Appl. Phys., 8(1969)614.
24 J. Hennephof, G. Hoekstra and J. van Overbeeke, J. Sci. Instr., (J. Phys. E, Ser. 2)2(1969)96.
25 R. Beigang and H. Telle, J. Phys. E., Sci. Instr., 10(1977)881.
26 P. D. Weidman and F. K. Brown, J. Phys. E., Sci. Instr., 8(1975)553.
27 M. Benkiki and C. M. Penchina, J. Phys. E., Sci. Instr., 9(1976)95.
28 J. Williams, Electron. Design, 10(1974)102.
29 D. Sarid and D. S. Cannell, Rev. Sci. Instr., 45(1974)1082.
30 C. R. Hummer and D. J. Burns, Rev. Sci. Instr., 47(1976)921.
31 A. Olofsson and G. Edwall, Rev. Sci. Instr., 47(1976)779.
32 J. S. Risley, J. Phys. E., Sci. Instr., 10(1977)464.
33 W. C. Randle and N. Keith, Proc. IEEE, 66(1978)172.